Current and Future Developments in Nanomaterials and Carbon Nanotubes

(Volume 2)

(Nanocomposite Materials for Sensors)

Edited by

Manorama Singh
Department of Chemistry,
Guru Ghasidas Vishwavidyalaya,
Bilaspur, CG-495009
India

Vijai K Rai
Department of Chemistry,
Guru Ghasidas Vishwavidyalaya,
Bilaspur, CG-495009
India

&

Ankita Rai
School of Physical Sciences,
Jawaharlal Nehru University,
New Delhi-110067
India

Current and Future Developments in Nanomaterials and Carbon Nanotubes

Nanocomposite Materials for Sensors

(Volume 2)

Editors: Manorama Singh, Vijai K Rai & Ankita Rai

ISSN (Online): 2589-2207

ISSN (Print): 2589-2193

ISBN (Online): 978-981-5050-98-1

ISBN (Print): 978-981-5050-99-8

ISBN (Paperback): 978-981-5049-00-8

Published by Bentham Science Publishers – Pte. Ltd, Singapore. All Rights Reserved.

need for a court order if at any point you breach any terms of this License Agreement. In no event will any delay or failure by Bentham Science Publishers in enforcing your compliance with this License Agreement constitute a waiver of any of its rights.

3. You acknowledge that you have read this License Agreement, and agree to be bound by its terms and conditions. To the extent that any other terms and conditions presented on any website of Bentham Science Publishers conflict with, or are inconsistent with, the terms and conditions set out in this License Agreement, you acknowledge that the terms and conditions set out in this License Agreement shall prevail.

Bentham Science Publishers Pte. Ltd.
80 Robinson Road #02-00
Singapore 068898
Singapore
Email: subscriptions@benthamscience.net

BENTHAM SCIENCE

CONTENTS

Preface

Nanocomposites are rapidly emerging as novel materials for sensor technology; therefore, the scientific community has recently focused on the advancement in the development of innovative methods and materials relied upon efficient composites. The implementation of nanocomposite materials for the development of specific and sensitive sensing platforms receives good attention. This book focuses on the reviews of important reported literature for new approaches of nanocomposite material preparation and their applications in the development of physical, chemical, electrochemical, biological, and optical sensors, *etc.* These nanomaterials have been extensively used widely (to amplify the signal) in the detection of heavy metal ions, vital signs (*i.e.*, glucose, *etc.*), explosives, hydrazine, humidity, *etc.*

This book focuses on representing some state-of-the-art review chapters based on reported works in the last few decades, outlining the synthesis, role, and progress of nanocomposite materials in fabricating flexible and multifunctional sensing platforms in sensor technologies. The book is intended to prepare a highly compiled knowledge for designing novel nanocomposite materials to be used as sensing platforms in sensor technologies. A broad range of readers such as graduates and post-graduates, Ph.D. scholars, faculty members, professionals working in the area of material science, the healthcare industry, biological sciences, medical sciences, environmental science will be benefitted from the topics preferred in the proposed book.

Manorama Singh
Department of Chemistry
Guru Ghasidas Vishwavidyalaya
Bilaspur, CG-495009
India

Vijai K Rai
Department of Chemistry
Guru Ghasidas Vishwavidyalaya
Bilaspur, CG-495009
India

&

Ankita Rai
School of Physical Sciences
Jawaharlal Nehru University
New Delhi-110067
India

List of Contributors

A Rajapriya	Department of Nanoscience and Technology, Bharathiar University, Coimbatore, Tamilnadu, India
A Rebekah	Department of Nanoscience and Technology, Bharathiar University, Coimbatore, Tamilnadu, India
Anu Rose Chacko	Mahatma Gandhi University, Kottayam, Kerala 686560, India
Archana Aravind	NSS Hindu College, Changanashery, Kerala, India
Ashoka S	Dayananda Sagar University, Banglore, India
Beena Mathew	Mahatma Gandhi University, Kottayam, Kerala, India
Chandra Shekhar Kushwaha	Department of Polymer Science, Bhaskaryacharya College of Applied Sciences, University of Delhi, Delhi, India
D Amilan Jose	Department of Chemistry, National Institute of Technology, Kurukshetra, India
D Navadeepthy	Department of Nanoscience and Technology, Bharathiar University, CoimbatoreBharathiar University, Coimbatore, Tamilnadu, India
G Srividhya	Department of Nanoscience and Technology, Bharathiar University, Coimbatore, Tamilnadu, India
J Debbarma	Department of Chemistry, National Institute of Technology, Agartala, Tripura (West), India
Juhi Srivastava	Department of Chemistry, MMV, Banaras Hindu University, Varanasi, Uttar Pradesh, India
M Swapna Sai	Centre for Applied Research, Chennai Institute of Technology, Chennai, India
M Varsha Shree	Centre for Applied Research, Chennai Institute of Technology, Chennai, India
Manju Srivastava	Dayalbagh Educational Institute, Agra, Uttar Pradesh, India
Meenakshi Singh	Department of Chemistry, MMV, Banaras Hindu University, VaranasiUttar Pradesh, Uttar Pradesh, India
Mitali Saha	Department of Chemistry, National Institute of Technology, Agartala, Tripura (West), India
N Ponpandian	Department of Nanoscience and Technology, Bharathiar University, Coimbatore, Tamilnadu, India
Nancy Sharma	Department of Chemistry, National Institute of Technology, Kurukshetra, India
Nazia Siddiqui	Dayalbagh Educational Institute, Agra, Uttar Pradesh, India
Prashanth S. Adarakatti	SVM Arts, Science and Commerce College, Karnataka, India
Pratibha Singh	Department of Polymer Science, Bhaskaryacharya College of Applied Sciences, University of Delhi, Delhi, India
S Keerthana	Department of Nanoscience and Technology, Bharathiar University, Coimbatore, Tamilnadu, India

Sachin Saxena   Dayalbagh Educational Institute, Agra, Uttar Pradesh, India

Sajini T   St Berchmans Autonomous College (Affiliated to Mahatma Gandhi University), Kottayam, India

Saroj Kr Shukla   Department of Polymer Science, Bhaskaryacharya College of Applied Sciences, University of Delhi, Delhi, India

Srushti Gadiyaram   Department of Chemistry, National Institute of Technology, Kurukshetra, India

Suma B. Patri   Department of Chemistry, Bangalore University, Central College Campus, Bengaluru, India

V Dhinakaran   Centre for Applied Research, Chennai Institute of Technology, Chennai, India

CHAPTER 1

Nanocomposites: Introduction, Structure, Properties and Preparation Methods

V Dhinakaran[1,*], M Swapna Sai[1] and M Varsha Shree[1]

[1] Centre for Applied Research, Chennai Institute of Technology, Chennai-600069, India

Abstract: The production of composites and materials based on nanocellulose has attracted considerable attention in the last few decades since their abundance, renewability, high strength and rigidity, environmental friendliness, and low weight are all unmissable and potentially useful. This analysis deals with crucial factors in the manufacture of nanocellulose composites and presents and explores different composite processing techniques. Rare combinations of features and new design opportunities are seen in high-performance nanocomposites. Their potential is so high that their utility in different fields, ranging from packaging to biomedicine, with an annual growth rate projected at around 25% and a standardized summary emphasizes the need for such products, their methods of fabrication, and several recent studies on structure, properties and potential applications. There is a focus on the possible use of naturally occurring materials like clay-based minerals, chrysotile and lignocellulose fibers. In this chapter, an overview of nanocomposites is deliberated in detail and the nanocomposite applications provide new technology and business options for different industries in the aerospace, vehicle, electronics, electrical and biomedical engineering sector as they are naturally friendly.

Keywords: Carbon nanotubes, Nanocomposites, Nanometers, Polymer matrix, Scanning tunnel microscope, Sensors.

INTRODUCTION

At the atomic or molecular level, nanotechnology engineering is the collective term for a wide variety of processing technologies and measurements involving the smallest scale handling of matter from 1 to 100 nanometers. The processing of particles and materials at nanoscale dimensions is concerned with nanotechnology [1]. Nanocomposites are composites in which the nanometer range of at least one of the phases is 1 nm = 10^{-9} m. Because of their outstanding Properties, nanocomposites are potential alternatives to micro composites and monolithic and consist of two or more distinct constituents or phases of physical and chemical

* **Corresponding author V. Dhinakaran**: Centre for Applied Research, Chennai, Institute of Technology, Chennai-600069, India; Email: dhinakaranv@citchennai.net

Manorama Singh, Vijai K Rai and Ankita Rai (Eds.)

properties, which are separated by a separate interface [2]. However, nanocomposite preparation techniques face challenges due to the regulation in the nanophase of elemental composition and stoichiometry. The constituent, which is normally more current, is known as the matrix. In order to improve the mechanical characteristics of nanocomposites, the component is called reinforcement in the matrix material or nanomaterials [3]. Strengthening is normally made of nanosized fillers. In general, anisotropic nanocomposites occur because of the distinct properties of constituents depending on the direction and because the reinforcement is inhomogeneous [4]. In addition, as dimensions reach the level of nanometers, interactions at interfaces are much better and appropriate for improving the material's properties. In these cases, the surface area or volume ratio of the materials used for the preparation of nanocomposites is essential for understanding the structure-property links [5]. Furthermore, the discovery and subsequent use of Carbon Nanotubes (CNTs) for the manufacturing of composites showing some of the special mechanical, thermal, and electric characteristics of CNT introduced a new and fascinating dimension [6]. Further advances in the production and application of CNT-containing nanomaterials were rendered by the possibility of spinning CNTs into composite products and textiles [7]. In addition to being environmentally sustainable, nanocomposites now deliver new technologies and market opportunities for all industries. A large range of materials, where one of the dimensions belongs to a nano range, is defined by nanocomposite. In certain cases, nanocomposites are stronger than typical composites [8]. Because of their excellent properties, nanocomposites are extremely good alternatives to traditional composite materials and are used in many areas [9].

STRUCTURE OF NANOCOMPOSITES

Nanocomposite architecture usually comprises a matrix of particle, whisker, fiber and nanotube nanosized reinforcement components. Several researchers have used various equipment and techniques for characterizing nanocomposites, including microscopy Atomic Force Microscopy (AFM), Scanning Tunnel Microscopy (STM), Fourier Transformed Infrared Spectroscopy (FTIR), X-ray photoelectron, Nuclear Magnet Resonance (NMR), Differential Calorimetry Scan (DSC), and scanning and transmission of microelectrons [10]. The AFM is a powerful method to research the surface up to the nanometer level. Simultaneous experiments have been used on quantitative characterizations of nano-structuring and crystallite structures of some nanocomposites at Small Angles of X-Ray Dispersion (SAXS) and X-Ray Diffractometry (XRD). Furthermore, theoretical calculations and simulations were developed to predict the force properties, including stress and strain curves [11]. A brief description of the CNTs will be given here due to their distinctive properties, compared with other refurbishments, before the structure and properties of nanostructures such as CNTs are discussed [12]. In short, SWCNTs have a metal density of less than one-sixth, while MWCNT is roughly half the metal density. Tensile strengths of SWCNT and MWCNT have been stated to be considerably higher than steel with high resistance, while the values of Young's diamond module are comparable [13]. They show exceptional resilience because the deformations of plastic metal and carbon fiber fractures are different from bowing and rejuvenating without damage. Also, thermal and electrical theoretical conduciveness with an almost zero coefficient of thermal expansion is equal to diamond [14]. In contrast to lower metal wires in microchips and high magnetic parallel perpendicular susceptibility, they ensure good thermal stability in both air and vacuum. Theoretically, these materials have surface values of 3000 m2/g, although the calculated gas value is different [15].

PROPERTIES OF NANOCOMPOSITES

The properties of nanocomposites depend not on the characteristics of each component but on the parameters (processes used in the development of nanocomposites) (type and orientation of filling materials, improvement of mechanical efficiency of the parental material, improved clarity due to small scale, small, high looks and therefore large area particles) [16]. Particles should adequately be dispersed and spread into matrix material in order to attain the improved nanocomposite characteristics, as otherwise the particulate matter will agglomerate and the nanocomposite characteristics will deteriorate. Particles must be properly distributed and distributed to the matrix [17]. The nanocomposite's most common feature is the layering of the interface between the matrix and the filler material. The interface properties, composition and microstructure of the filler vary from the interface matrix [18]. The interface between the nanofillers and the polymer matrix optimizes the interactors and can therefore be adapted to fit the superficial bonding surface, so the overall properties of the nanofillers are quite profound [19]. The interface region is highly interconnected with the matrix and filler. In relation to the relationships, the surface energy filling and matrix ratio are calculated [20]. The properties of nanocomposite depend on their microstructure. The relation between the structural characteristics of nanocomposite's polymer nanoplatelets defines the morphological nature of the composite system [21]. A good nanomaterial dispersion is hard to achieve, especially for non-polar polymers, but a consistent distribution of nanoplatelets guarantees good nanomaterial quality [22].

CLASSIFICATION OF NANOCOMPOSITES

Nanocomposites are graded in accordance with the forms of material reinforcement and matrix used in their construction [23]. Nanocomposites are usually categorized into three different groups according to the form of the matrix material, as shown in Table **1**.

Table 1. Different types of nanocomposites [34].

Class	Typical Composites
Polymer	Thermoplastic or thermoset polymer material, layered silicates, polyester or TiO_2, polymer and CNT, polymer or layered double hydroxides
Ceramic	Al_2O_3/SiO_2, SiO_2/Ni, Al_2O_3/TiO_2, Al_2O_3/SiC, Al_2O_3/CNT
Metal	$Fe-Cr/Al_2O_3$, Ni/Al_2O_3, Co/Cr, Fe/MgO, Al/CNT, Mg/CNT

Polymer Matrix Nanocomposites

Polymer nanocomposites are materials used as polymer materials and nano-additives in reinforcement form. Additives can have single sizes (such as fibre and nanotubing), 2D (such as clays and layers) or 3D particles of the spheres. In the academy and industry, polymer nanocomposites were considered significant because of their exceptional mechanical qualities, such as high elastic stiffness and strength with small nanoadditive concentrations [24]. Additional good properties of nanocomposite polymers include barrier strength, flame retardance, resistance to wear, magnetism, electrical and optical applications. The polymer (matrix) and a filler mixture are typical of a composite (reinforcement) [25]. Polyamide is a polymer of thermoplastic and is typically used as carbon and glass fibre reinforcement material. Carbon fibers are used in the aerospace industry for refurbishment purposes [26]. Polymers have excellent characteristics, including lightness, high resistance, fast processing, resistance to corrosion, ductility and low costs. Polymers have relatively poor mechanical, thermal and electrical characteristics compared to pottery and metals. Low gas barrier, thermal strength and fire efficiency characteristics are also present in polymers [27]. Polymers are not as dense as ceramics and metals, but have low coordinated carbon and hydrogen atoms and are as light as the backbones, allowing them to be used as construction components or building materials or in lightweight automotive, defence, air transportation, and electronic construction [28].

Ceramic Matrix Nanocomposites

Ceramic nanocomposites with nano-dimensional measurement represent a new age with a wide variety of industrial applications. The nanoceramic composite microstructures produce exceptional electrical and mechanical properties [29]. Various methods of producing ceramic matrix nanocomposites have been published in the literature. Generally speaking, the development of composite materials like solgel, colloidal, or precipitation approaches and template synthesis is common with traditional powder methods, polymer pathways, pyrolysis and chemical methods. After the discovery of carbon nanotubes, they have been widely used in nanocomposite growth; the typical ceramic matrix of nano-composites is Al_2O_3/SiO_2, SiO_2/Ni, Al_2O_3/TiO_2 and Al_2O_3/Sic (CNTs). Al_2O_3/CNT, $MgAl_2O_4/CNT$ and MgO/CNT are ceramic matrix basic nanocomposites [30].

Metal Matrix Nanocomposites

Nanoparticles consist of a metal matrix composed of ductile metal or a matrix of alloys in which nanoparticles enhancement is introduced are nanocomposites reinforced. These compounds consist of the nanoparticular filled metal/alloy matrix, which exhibits completely different physical, chemical and mechanical properties than the matrix [31]. Generally, the nanoparticles boost wear strength, mechanical properties and features of damping. Metal matrix nanocomposite researchers are recently exploring a wide variety of applications in structural components due to their superior features due to nanoparticles embedded [32]. At the nano stage, the particle interaction with dislocations is significant and the mechanical properties are greatly enhanced. Nanoparticles serve as a barrier to dislocation and thus improve mechanical characteristics [33]. The techniques used to process nanocomposites include spray pyrolysis, fluid metal penetration, steam technology, rapid consolidation, electronic positioning, and chemical methods such as colloids and sol-gel methods. The standard metal nanocomposites are Fe-Cr/Al_2O_3, Ni/Al_2O_3, Fe/MgO, Al/CNT and Mg/CNT.

PREPARATION METHODS

Sol-Gel Method

The sol-gel method is a technique for the wet chemical processing of glassy as well as ceramic materials. The sol (or solution) is progressively evolving in this process into a gel-like network that contains a liquid and a solid phase. The common precursors are metal alcohols and metal chlorides that are subjected to colloids by hydrolysis and polycondensation [35]. The sol-gel chemical process is shown in Fig. (**1**). The basic structure or morphology of the solid phase can be employed for anything from discrete colloidal particles to continuous polymer strands. The solid nanoparticles in the solution form the colloidal interconnection of solid nanoparticulate (the soles), together with hydrolysis reactions, which form interconnected networks between the phases (gels). The 3D polymer network encompasses the entire fluid [36]. The polymer acts as a central agent and makes layered crystals easier to produce. As the crystals expand and form a nanocomposite, the polymer is filtered among layers.

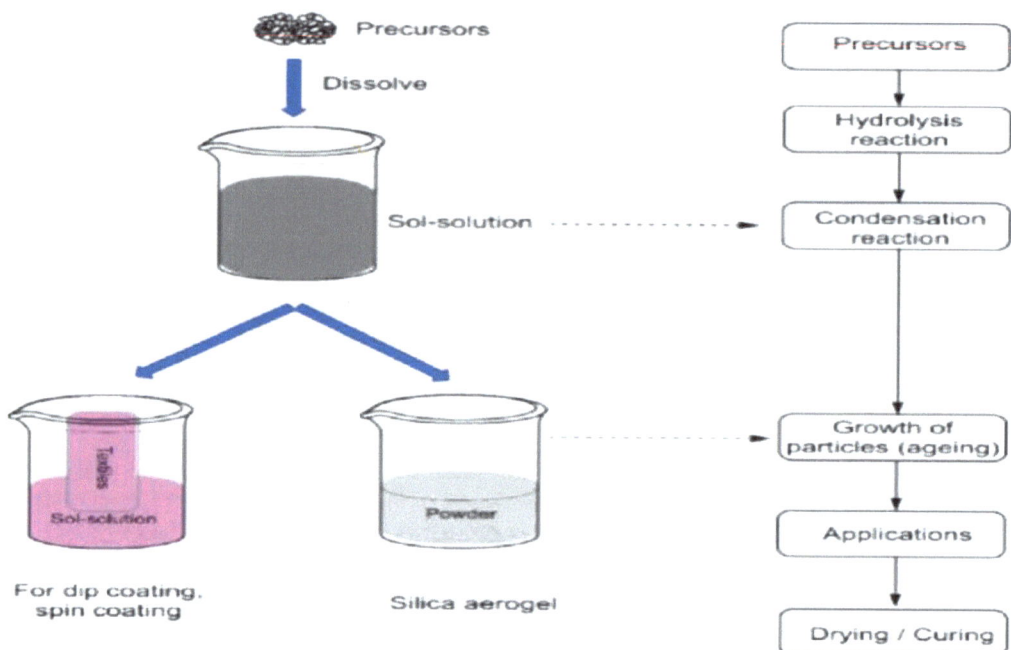

Fig. (1). Sol-gel chemical process.

Electrospinning Method

Fiber spinning is a method of production using electro power to draw polymer solutions or melts of polymer threads to the order of a hundred nanometers of fiber diameter. Electrospinning Schubert provides an overview of models related to fiber diameter, method, and solution parameters, while a recent hypothesis suggests that the fiber diameter is predicted and distributed [37]. Electrospinning has electrospinning characteristics and traditional fiber dry spinning solution. In order to generate solid danger from a solution, it is not appropriate to use coagulation chemistry or high temperatures. This makes the method especially appropriate for the processing of large and complex molecules of fibers [38]. Molten precursor electrospinning is often used to ensure that no solvent is passed into the finished product. As a liquid droplet receives appropriate voltage, the liquid body will be charged, and electrostatic repulsions reverse the surface voltage, and the droplet will be stretched; the surface is exposed to a flood of liquid at a critical point [39]. The Taylor cone is known as this point of eruption. If the molecular cohesion of the liquid is high enough, there will be no breakdown of the stream (if it is, electro-sprayed droplets) and a charged liquid jet will create [40]. The current flow mode switches from ohms to convective as the aircraft dries in the flight, as the charges move to the fibre. The jet is extended to the grounded collector by whipping, which induces an electrostatic repulsion that is initiated by a slight bend in the fiber, as shown in Fig. (**2**). The extension and dilution of the fiber induced by this bending instability contribute to the development of nanometer-size uniform fibers.

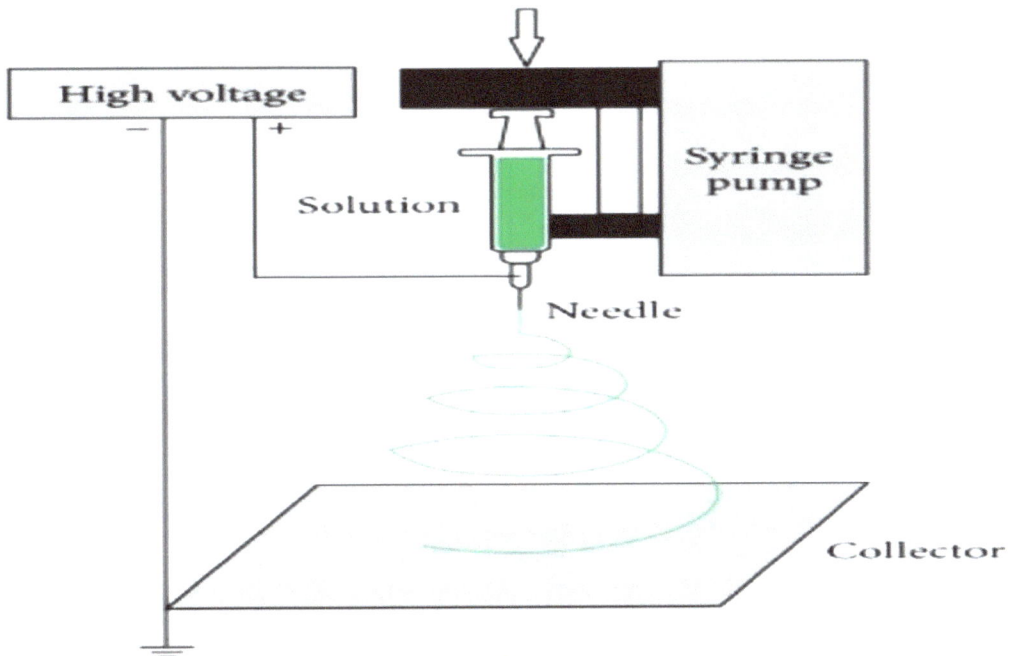

Fig. (2). The electrospinning processes.

In Situ Polymerization Method

Polymerization in position allows nanofillers to swell in a monomer solution since the monomer's lower molecular solution will easily swell between layers and cause swelling. The resulting mixture is polymerized, whether it be with radiation, heat, diffusion of the initiator, or by the organic initiator. The monomer is a nanocomposite shaped between layers that is sub-polymerized, exfoliated or intercalated [41]. Synthesis is equivalent to the Situ Template. In the presence of polymer chains, the clay layers are synthesized. Polymer matrix and clay layers are often dissolved in an aqueous solution by relaxing a high-temperature gel. The polymer chains are in barrier layers, and the polymer chains are elevated with high temperatures for nucleation and clay layers. The only downside is that synthesis allows polymers to decompose at high temperatures. The methodology of obtaining conductive polymers by *in situ* polymerization method is shown in Fig. (**3**).

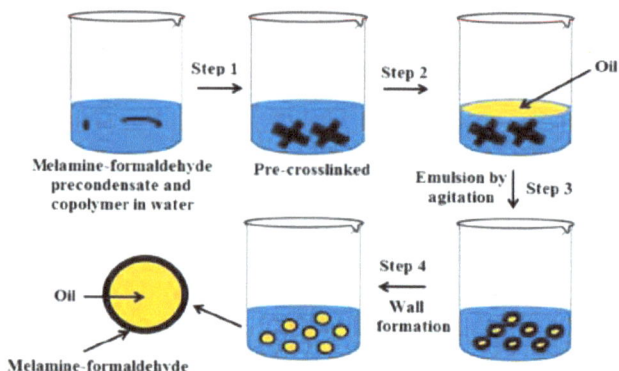

Fig. (3). Preparation of conductive polymers by *in Situ* Polymerization Method [42].

Melt Intercalation Method

Melts are a popular process widely used on the land. In this process, nanofillers are combined in the polymer matrix at melting temperature. A permanent or under shear polymer and nanofibers mixture is needed for this approach. The method is compatible with existing industrial processes such as injection and extrusion, enabling the use of polymers not suitable for *in situ* polymerization or intercalation solutions [43]. The method of melting is the same and the melt intercalation process is shown in Fig. (4). The process consists of melting polymers or pellets into a solution with viscus, and nanofillers are combined with high diffusion temperatures in order to form a high shear rate polymer solution. Compression, injection molding or fiber processing technique may produce the final form.

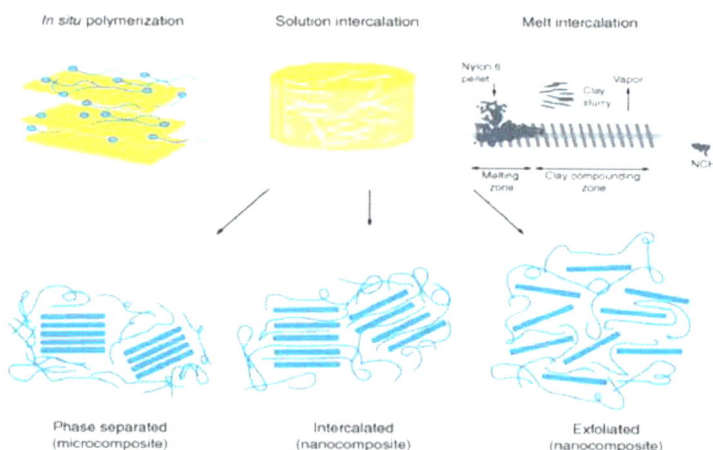

Fig. (4). Melt intercalation process for the preparation of nanocomposite [44].

ADVANTAGES OF NANOCOMPOSITES

A large range of materials, where one of the dimensions belongs to a nano range, is defined by a nanocomposite. In certain cases, nanocomposites are stronger than typical composites. Nanocomposites have the following benefits.

A small number of nanofiller materials compared to conventional combinations that require a high microparticle concentration can help improve the properties of the matrix materials in the nano-composites [45].

The added nanocomposites are much lighter in weight than standard compounds because of the small percentage of nano-fillers. Nanomaterials with size-dependent properties are substantially greater than standard composites in terms of thermal, chemical, mechanical, optical, magnet and electrical characteristics [46].

APPLICATIONS OF NANOCOMPOSITES

From the above, it is clear that the advantages of nanocomposites include enhanced characteristics, reduced solid waste as well as increased production capacities, in particular in packaging applications. All new materials are created and popular devices such as fuel cells, sensors and covers performance in nanocomposites are promising. While nanocomposites in the industry are still very little used, the research in the coming years has already begun and is expected to be turned into an industry. Similarly, the car industry is one of the leading applications and has an impact because of enhanced features, such as ecology, protection and comfort. Information on the industrial use and future advances of automotive nanocomposites (including nanocomposites based on the CNTs) is now available [47]. Lightweight boards also indicate that composites made of metal or plastic nanocomposites with enough refinement are of low density and very high strength (carbon fibers of 150 GPa strength and weight per the fifth stain). Two-phasis heterogeneous nanodielectrics are often commonly used in electrical and electrical applications; they are usually called dielectric nanocomposites. Nanocomposites of metals and ceramics are expected to have a significant impact on many industries, including aerospace, electronics and the military, while nanocomposites of polymer are expected to produce large effects in battery cathodes, microelectronics, nonlinear optics, and sensors. Enhanced features include major increases in breakage strength (around twice) and strength (around a half time), wear changes over time, even at very low weights, improved resistance to high temperature and crack, higher thermal heat temperature toughness, and higher hardness values than existing stains and alloys [48]. This comes mainly from nanosized reinforcements that lead to sufficient product morphology. Potential developments in electrical, magnetic, electronic, mechanical and energy transfer instruments proposed by field researchers, catalysts and sensors. A wide range of polymer nana composites was developed to meet the basic requirements of isolation, semiconduction or metal nanoparticles in some applications [49]. The existing and future application of nanocomposites in various fields is shown in Table **2**. Recently, some PLS nanocomposites have been available on the consumer market as ablatives, biodegradable high-performance composites, as well as in the e-packing and food industries. These include nylon 6, polypropylene, semicrystalline nylon for packaging containers and fuels systems, epoxy coating elements and higher voltage insulation, watercraft structure saturated polyester and external advertisement panels, polyolefin fire resistant lines, power containers and energy systems.

Table 2. Existing and future applications of nanocomposites.

S. No	Broad Area of Application	Components
1	Thermal	Fuselage structures, graphene platelets
2	Aerospace	Structural health monitoring (SHM) of aircrafts
3	Biomedical	Drug delivery, biosensors
4	Electrical	Electroluminescent devices, electro catalysis, batteries
5	Marine	Gratings, ducts, shafts, piping

CONCLUDING REMARKS

Nanotechnologies involve the study and management of materials with diameters ranging from 1 to 100 nanometers. As the scale of nanomaterials reduces, 'nanoeffects' produce certain rare and exotic qualities. The nanotechnology field has recently become one of the most well-known areas of study and innovation. It involves polymer science. Compounds of polymers are formed by combining a polymer with synthetic materials or natural inorganic fillers. In order to strengthen the properties of polymer composites, filler materials are used. Polymer nanocomposites, both in industry and academia, have demonstrated significant improvement in their properties relative to traditional micro composites in recent years. Nano-composites in polymers include nano-filler materials that cause nanocomposite properties to be "nano-effects". Nanomaterials are, in this sense, the most appropriate material to fulfill the current demands of the science community. Compared with traditional composites and monolithic counterparts, nanocomposites provide enhanced performance and, in many industries, polymer nanocomposites have been used and their applications have evolved dramatically.

CONSENT FOR PUBLICATION

Not applicable.

CONFLICT OF INTEREST

The author declares no conflict of interest, financial or otherwise.

ACKNOWLEDGEMENTS

Declared none.

REFERENCES

[1]   Zaferani, S. Hooshmand Introduction of polymer-based nanocomposites. In: *Polymer-based nanocomposites for energy and environmental applications*; Woodhead Publishing, **2018**; pp. 1-25.
        [http://dx.doi.org/10.1016/B978-0-08-102262-7.00001-5]

[2]   Hussain, C.M.; Mishra, A.K., Eds. *New Polymer Nanocomposites for Environmental Remediation*; Elsevier, **2018**.

[3]   Pal, K., Ed. *Hybrid nanocomposites: fundamentals, synthesis, and applications*; CRC Press, **2019**.
        [http://dx.doi.org/10.1201/9780429000966]

[4]   Noh, H.S.; Jung, J. Synthesis of organic–inorganic hybrid nanocomposites *via* a simple two-phase ligands exchange. *Sci. Adv. Mater.,* **2020**, *12*(3), 326-332.
        [http://dx.doi.org/10.1166/sam.2020.3644]

[5]   Li, Z.; Qi, X.; Gong, X.; Xie, R.; Deng, C.; Zhong, W. Carbon nanotubes/Ni and chain-like carbon nanospheres/Ni nanocomposites: selective production and their microwave absorption performances. *J. Mater. Sci. Mater. Electron.,* **2020**, 1-10.

[6]   Li, Q.; Li, J.; Tran, D.; Luo, C.; Gao, Y.; Yu, C.; Xuan, F. Engineering of carbon nanotube/polydimethylsiloxane nanocomposites with enhanced sensitivity for wearable motion sensors. *J. Mater. Chem. C Mater. Opt. Electron. Devices,* **2017**, *5*(42), 11092-11099.
        [http://dx.doi.org/10.1039/C7TC03434B]

[7]   Hou, G.; Tao, W.; Liu, J.; Gao, Y.; Zhang, L.; Li, Y. Tailoring the dispersion of nanoparticles and the mechanical behavior of polymer nanocomposites by designing the chain architecture. *Phys. Chem. Chem. Phys.,* **2017**, *19*(47), 32024-32037.
        [http://dx.doi.org/10.1039/C7CP06199D] [PMID: 29181472]

[8]   Nisar, M.; Maria da Graca, S.B. Luiz CP da Silva Filho, Julian Geshev, Nara R. de Souza Basso, and Griselda Barrera Galland. Polypropylene nanocomposites with electrical and magnetic properties. *J. Appl. Polym. Sci.,* **2018**, *135*(42), 46820.
        [http://dx.doi.org/10.1002/app.46820]

[9]   Parthasarathy, V.; Nakandhrakumar, R.S.; Mahalakshmi, S.; Sundaresan, B. Structural, optical, thermal and non-isothermal decomposition behavior of PMMA nanocomposites. *J. Inorg. Organomet. Polym. Mater.,* **2020**, *30*(8), 2998-3013.
        [http://dx.doi.org/10.1007/s10904-020-01453-5]

[10]   Ramazanov, M.A.; Babayev, Y. Preparation and structure of nanocomposites based on zinc sulfide in polyvinylchloride. *J. Non-Oxide Glasses,* **2018**, *10*, 1-6.

[11]   Eckert, A.; Abbasi, M.; Mang, T.; Saalwächter, K.; Walther, A. Structure, Mechanical Properties, and Dynamics of Polyethylenoxide/Nanoclay Nacre-Mimetic Nanocomposites. *Macromolecules,* **2020**, *53*(5), 1716-1725.
        [http://dx.doi.org/10.1021/acs.macromol.9b01931]

[12]   Guo, Yongqiang; Ruan, Kunpeng; Yang, Xutong; Ma, Tengbo; Kong, Jie; Wu, Nannan; Zhang, Jiaoxia; Gu, Junwei; Guo, Zhanhu Constructing fully carbon-based fillers with a hierarchical structure to fabricate highly thermally conductive polyimide nanocomposites. *J.Mat. Che. C 7*, **2019**, *23*, 7035-7044.

[13]   Jiang, J.; Shen, Z.; Cai, X.; Qian, J.; Dan, Z.; Lin, Y.; Liu, B.; Nan, C-W.; Chen, L.; Shen, Y. Polymer nanocomposites with interpenetrating gradient structure exhibiting ultrahigh discharge efficiency and energy density. *Adv. Energy Mater.*, **2019**, *9*(15), 1803411.
       [http://dx.doi.org/10.1002/aenm.201803411]

[14]   Murugesan, P.; Moses, J.A.; Anandharamakrishnan, C. J. A. Moses, and C. Anandharamakrishnan. Photocatalytic disinfection efficiency of 2D structure graphitic carbon nitride-based nanocomposites: a review. *J. Mater. Sci.*, **2019**, *54*(19), 12206-12235.
       [http://dx.doi.org/10.1007/s10853-019-03695-2]

[15]   Cheng, S.; Carroll, B.; Bocharova, V.; Carrillo, J.M.; Sumpter, B.G.; Sokolov, A.P. Focus: Structure and dynamics of the interfacial layer in polymer nanocomposites with attractive interactions. *J. Chem. Phys.*, **2017**, *146*(20), 203201.
       [http://dx.doi.org/10.1063/1.4978504] [PMID: 28571333]

[16]   Zare, Y.; Rhee, K.Y.; Hui, D. Influences of nanoparticles aggregation/agglomeration on the interfacial/interphase and tensile properties of nanocomposites. *Compos., Part B Eng.*, **2017**, *122*, 41-46.
       [http://dx.doi.org/10.1016/j.compositesb.2017.04.008]

[17]   Almessiere, M.A.; Trukhanov, A.V.; Slimani, Y.; You, K.Y.; Trukhanov, S.V.; Trukhanova, E.L.; Esa, F.; Sadaqati, A.; Chaudhary, K.; Zdorovets, M.; Baykal, A. Correlation between composition and electrodynamics properties in nanocomposites based on hard/soft ferrimagnetics with strong exchange coupling. *Nanomaterials (Basel)*, **2019**, *9*(2), 202.
       [http://dx.doi.org/10.3390/nano9020202] [PMID: 30720737]

[18]   Cobos, M.; De-La-Pinta, I.; Quindós, G.; Fernández, M.J.; Fernández, M.D. Synthesis, physical, mechanical and antibacterial properties of nanocomposites based on poly (vinyl alcohol)/graphene oxide–silver nanoparticles. *Polymers (Basel)*, **2020**, *12*(3), 723.
       [http://dx.doi.org/10.3390/polym12030723] [PMID: 32214025]

[19]   Rodríguez-García, S.; Santiago, R.; López-Díaz, D.; Merchán, M.D.; Velázquez, M.M.; Fierro, J.L.G.; Palomar, J. Role of the structure of graphene oxide sheets on the CO_2 adsorption properties of nanocomposites based on graphene oxide and polyaniline or Fe3O4-nanoparticles. *ACS Sustain. Chem.& Eng.*, **2019**, *7*, 12464-12473.
       [http://dx.doi.org/10.1021/acssuschemeng.9b02035]

[20]   Dilova, T.; Atanasova, G.; Og Dikovska, A.; Nedyalkov, N.N. The effect of light irradiation on the gas-sensing properties of nanocomposites based on ZnO and Ag nanoparticles. *Appl. Surf. Sci.*, **2020**, *505*, 144625.
       [http://dx.doi.org/10.1016/j.apsusc.2019.144625]

[21]   Yang, B.J.; Jang, J.; Eem, S-H.; Kim, S.Y. Ji-un Jang, Seung-Hyun Eem, and Seong Yun Kim. A probabilistic micromechanical modeling for electrical properties of nanocomposites with multi-walled carbon nanotube morphology. *Compos., Part A Appl. Sci. Manuf.*, **2017**, *92*, 108-117.
       [http://dx.doi.org/10.1016/j.compositesa.2016.11.009]

[22]   Wang, Y.; Zhang, Y.; Zhao, H.; Li, X.; Huang, Y.; Schadler, L.S.; Chen, W.; Catherine Brinson, L. Identifying interphase properties in polymer nanocomposites using adaptive optimization. *Compos. Sci. Technol.*, **2018**, *162*, 146-155.
       [http://dx.doi.org/10.1016/j.compscitech.2018.04.017]

[23]   Vahabi, H.; Movahedifar, E.; Ganjali, M.R.; Saeb, M.R. Polymer nanocomposites from the flame retardancy viewpoint: A comprehensive classification of nanoparticle performance using the flame retardancy index. In: *Handbook of Poly Nanocomp for Industrial App*; Elsevier, **2021**; pp. 61-146.

[http://dx.doi.org/10.1016/B978-0-12-821497-8.00003-4]

[24]   Varghese, N.; Francis, T.; Shelly, M.; Nair, A.B. Nanocomposites of polymer matrices: Nanoscale processing. In: *Nanoscale Processing*; Elsevier, **2021**; pp. 383-406.
         [http://dx.doi.org/10.1016/B978-0-12-820569-3.00014-1]

[25]   Saleh, T.A. *Nanomaterials: Classification, properties, and environmental toxicities*; Envi. Tech & Inn, **2020**, p. 101067.

[26]   Valapa, R.B.; Loganathan, S. *G. Pugazhenthi, Sabu Thomas, and T. O. Varghese. An overview of polymer–clay nanocomposites*; Clay-Polymer Nanocomposites, **2017**, pp. 29-81.
         [http://dx.doi.org/10.1016/B978-0-323-46153-5.00002-1]

[27]   Bramhill, J.; Ross, S.; Ross, G. Bioactive nanocomposites for tissue repair and regeneration: a review. International j. of envi. Res. and pub. *Health,* **2017**, *14*, 66.

[28]   Kumar, S.; Nehra, M.; Dilbaghi, N. K. Tankeshwar, and Ki-Hyun Kim. Recent advances and remaining challenges for polymeric nanocomposites in healthcare applications. Progress in Poly. *Sci,* **2018**, *80*, 1-38.

[29]   Azarniya, A.; Sovizi, S.; Azarniya, A.; Rahmani Taji Boyuk, M.R.; Varol, T.; Nithyadharseni, P.; Madaah Hosseini, H.R.; Ramakrishna, S.; Reddy, M.V. Physicomechanical properties of spark plasma sintered carbon nanotube-containing ceramic matrix nanocomposites. *Nanoscale,* **2017**, *9*(35), 12779-12820.
         [http://dx.doi.org/10.1039/C7NR01878A] [PMID: 28832057]

[30]   Rathod, V.T.; Kumar, J.S.; Jain, A. Polymer and ceramic nanocomposites for aerospace applications. *Appl. Nanosci.,* **2017**, *7*(8), 519-548.
         [http://dx.doi.org/10.1007/s13204-017-0592-9]

[31]   Tabandeh-Khorshid, M.; Kumar, A.; Omrani, E.; Kim, C.; Rohatgi, P. Synthesis, characterization, and properties of graphene reinforced metal-matrix nanocomposites. *Compos., Part B Eng.,* **2020**, *183*, 107664.
         [http://dx.doi.org/10.1016/j.compositesb.2019.107664]

[32]   Hassanzadeh-Aghdam, M.K.; Mahmoodi, M.J. A comprehensive analysis of mechanical characteristics of carbon nanotube-metal matrix nanocomposites. *Mater. Sci. Eng. A,* **2017**, *701*, 34-44.
         [http://dx.doi.org/10.1016/j.msea.2017.06.066]

[33]   Bachmaier, A.; Katzensteiner, A.; Wurster, S.; Aristizabal, K.; Suarez, S.; Pippan, R. Thermal stabilization of metal matrix nanocomposites by nanocarbon reinforcements. *Scr. Mater.,* **2020**, *186*, 202-207.
         [http://dx.doi.org/10.1016/j.scriptamat.2020.05.014]

[34]   Camargo, P.H.C.; Satyanarayana, K.G.; Wypych, F. Nanocomposites: synthesis, structure, properties and new application opportunities. *Mater. Res.,* **2009**, *12*(1), 1-39.
         [http://dx.doi.org/10.1590/S1516-14392009000100002]

[35]   Hasnidawani, J.N.; Azlina, H.N.; Norita, H.; Bonnia, N.N.; Ratim, S.; Ali, E.S. Synthesis of ZnO nanostructures using sol-gel method. *Procedia Chem.,* **2016**, *19*, 211-216.
         [http://dx.doi.org/10.1016/j.proche.2016.03.095]

[36]   Azlina, H.N.; Hasnidawani, J.N.; Norita, H.; Surip, S.N. Synthesis of SiO2 nanostructures using sol-gel method. *Acta Phys. Pol. A,* **2016**, *129*(4), 842-844.
         [http://dx.doi.org/10.12693/APhysPolA.129.842]

[37]   Jaworek, A.; Krupa, A.; Lackowski, M.; Sobczyk, A.T.; Czech, T.; Ramakrishna, S.; Sundarrajan, S.; Pliszka, D. Nanocomposite fabric formation by electrospinning and electrospraying technologies. *J. Electrost.,* **2009**, *67*(2-3), 435-438.
         [http://dx.doi.org/10.1016/j.elstat.2008.12.019]

[38]   Mahapatra, A.; Mishra, B.G.; Hota, G. Electrospun Fe2O3-Al2O3 nanocomposite fibers as efficient adsorbent for removal of heavy metal ions from aqueous solution. *J. Hazard. Mater.,* **2013**, *258-259*,

116-123.
[http://dx.doi.org/10.1016/j.jhazmat.2013.04.045] [PMID: 23708454]

[39]   Fathollahipour, S.; Abouei Mehrizi, A.; Ghaee, A.; Koosha, M. Electrospinning of PVA/chitosan nanocomposite nanofibers containing gelatin nanoparticles as a dual drug delivery system. *J. Biomed. Mater. Res. A,* **2015**, *103*(12), 3852-3862.
[http://dx.doi.org/10.1002/jbm.a.35529] [PMID: 26112829]

[40]   Czech, T.; Ramakrishna, S.; Sundarrajan, S. Electrospinning and electrospraying techniques for nanocomposite non-woven fabric production. *Fibres Text. East. Eur.,* **2009**, *17*, 77-81.

[41]   Zapata, P.A.; Tamayo, L.; Páez, M.; Cerda, E.; Azócar, I.; Rabagliati, F.M. Nanocomposites based on polyethylene and nanosilver particles produced by metallocenic *in situ* polymerization: synthesis, characterization, and antimicrobial behavior. *Eur. Polym. J.,* **2011**, *47*(8), 1541-1549.
[http://dx.doi.org/10.1016/j.eurpolymj.2011.05.008]

[42]   Shukla, V. Review of electromagnetic interference shielding materials fabricated by iron ingredients. *Nanoscale Adv.,* **2019**, *1*(5), 1640-1671.
[http://dx.doi.org/10.1039/C9NA00108E]

[43]   Zhang, G.; Wu, T.; Lin, W.; Tan, Y.; Chen, R.; Huang, Z.; Yin, X.; Qu, J. Preparation of polymer/clay nanocomposites *via* melt intercalation under continuous elongation flow. *Compos. Sci. Technol.,* **2017**, *145*, 157-164.
[http://dx.doi.org/10.1016/j.compscitech.2017.04.005]

[44]   Bharimalla, A.K.; Deshmukh, S.P.; Vigneshwaran, N.; Patil, P.G.; Prasad, V. Nanocellulose-polymer composites for applications in food packaging: Current status, future prospects and challenges. *Polym. Plast. Technol. Eng.,* **2017**, *56*(8), 805-823.
[http://dx.doi.org/10.1080/03602559.2016.1233281]

[45]   Kim, I.Y.; Jo, Y.K.; Lee, J.M.; Wang, L.; Hwang, S-J. Unique advantages of exfoliated 2D nanosheets for tailoring the functionalities of nanocomposites. J.phy chem. *J. Phys. Chem. Lett.,* **2014**, *5*(23), 4149-4161.
[http://dx.doi.org/10.1021/jz502038g] [PMID: 26278947]

[46]   Olad, A. Polymer/clay nanocomposites. In: *Advances in diverse industrial applications of nanocomposites*; IntechOpen, **2011**.
[http://dx.doi.org/10.5772/15657]

[47]   Okpala, C.C. The benefits and applications of nanocomposites. *Int. J. Adv. Eng. Technol.,* **2014**, *12*, 18.

[48]   Pandey, S. Turning to nanotechnology for water pollution control: applications of nanocomposites. *Focus On Medical Sciences Journal,* **2016**, *2*, 2016.

[49]   Koo, J.H. *Polymer nanocomposites: processing, characterization, and applications*; McGraw-Hill Education, **2019**.

CHAPTER 2

Nanocomposites: A Boon To Material Sciences

Sachin Saxena[1,*], **Nazia Siddiqui**[1] and **Manju Srivastava**[1]

[1] *Dayalbagh Educational Institute, Dayalbagh-282005, India*

Abstract: The imperfections of microstructures and monolithic in different realms of material sciences have been completely engulfed by the improved characteristics and excellent properties of nanocomposites. Their multiphase components with nano dimensions provide these structures with much superiority over conventional composites. This paper is a brief review about nanocomposites and their classification. Based on the dimensionality of particle size these can be grouped into one-two- and three-dimensional nanomaterial derived composites while if the number of components form the basis, they can be classified into binary, ternary and quaternary nanocomposites. This work also discusses and focuses on the computational analysis of nanocomposites with designing and energy calculation studies, based on DFT, and other mathematical tools and models. The role of metal organic framework-based nanocomposites in sensor fabrication and quantification of different redox system have also been listed.

Keywords: Nanocomposites, Binary, Ternary, Quaternary, Metal organic frameworks.

INTRODUCTION

This decade has observed a detailed review and research reports in the field of nanocomposites, either from prosthetics [1, 2], food and heath corporations [3 - 5], chemical industries [6], or in highly sensitive and specific electrochemical applications [7 - 10]. The composites and in case, one or more phases are in nanoscale, the nanocomposites, have many superior properties than their precursors.

This blending of properties of single components provides extended range of high mechanical strength, chemical inertness, thermal stability, surface to volume ratio and better optical properties with much insignificant wear and tear losses *etc* [11]. It can be stated, that the reinforced material (phase) distribution or dispersion in the continuous phase or the matrix determine the performance of the nanocompos-

* **Corresponding Author S. Saxena**: Dayalbagh Educational Institute, Dayalbagh-282005, India; Email: sachinusic@gmail.com

Manorama Singh, Vijai K Rai and Ankita Rai (Eds.)

ite [12]. In other words, the homogenous distribution of the particles in the matrix is the key to the properties found in the nanocomposite. If the distribution of particles is homogenous, a sufficient amount of reinforcement will occur, this will lead to better interfacial interactions, but this is rarely found. Due to this, inferior properties of the nanocomposite may also be observed.

The properties of the nanocomposites are governed by essential factors that are needed to be optimized. These factors are mainly

- characteristic of matrix
- structure
- composition (the filler content), and
- interfacial interactions.

High aspect ratio, adhesion and reinforcement component makes the composite with enhanced principal properties. These properties rely on the structure of nanocomposite which is dependent upon the surface to volume ratio of the reinforced phase. Further, these peculiar properties of the nanocomposite are the outcome of the different interfacial interactions [13].

The nanomaterials used are basically nanofillers, nanoclays and nanoparticles. The characteristics of matrix and fillers highly influence the properties of nanocomposites. Literature reports the decrease in matrix stiffness leads to an increase in reinforcement property while the purity of filler material affects the nanocomposite applicability, for *e.g.*, in glassy polymer the matrix stiffness and reinforcement observed is moderate, while in elastomers very large increase is observed, in case of impure filler nanomaterial decrease in stability and even discoloration is observed. Heavy metal contamination and impurity affect the exfoliation and deteriorate the properties of the nanocomposite [14].

CLASSIFICATION

As per the studied classifications, nanocomposites are classified on the basis of polymer and non-polymer nanocomposites [15]. Further, if the criteria basis is reinforcement material, they can be classified into carbon nanotubes, polymer, noble metal and metal oxide-based nanocomposites. If matrix material is concerned, the nanocomposites may further be grouped into ceramic, polymer and metal matrix nanocomposites, respectively [16]. Detailed work has been accounted comprehensively in books and research articles, regarding the classification of nanocomposites. Here we can further classify these nanocomposites on the basis of number of components:

- Binary nanocomposites
- Ternary nanocomposites; and
- Quaternary nanocomposites

If the dispersion of nanofillers is heterogeneous, then according to the dimensionality and particle size, nanocomposites can be divided into 3 groups:

- One dimensional
- Two dimensional; and
- Three dimensional

When the size in all three dimensions is in nanoscale (<100nm), such particles include TiO_2, $CaCO_3$, SiO_2, oligomeric silsesquioxane *etc.*

Nanofibers and nanotubes lie in the two-dimension range. It is also found in the micrometer range.

Layered silicates have larger dimensions but one of its dimensions is found to be almost 1nm thick [13].

Based on the components and their properties, nanocomposites can be summarized as below:

Binary Nanocomposite

Production of advanced nanocomposites nowadays, has resolved many complex problems in real-life applications. One such application is decontamination of wastewater (removal of inorganic contaminants, degradation of dyes, *etc.*) using spinel ferrite MFe_2O_4 based binary nanocomposite [17]. Further, the metal oxides binary nanocomposites have been found to detect complex organic compounds and serve as better catalytic nanocomposites. They enhance the electron transfer mechanism by serving acidic and basic components in a single entity, due to which a high adsorption phenomenon occurs [18 - 20]. The adsorption rate is enhanced by the presence of nanofillers or nanomaterials used during its preparation. Recently, Hareesh etal. 2020 have summarized in a review that the graphitic carbon nitride binary nanocomposites can be used as energy storage materials, owing to high supercapacitor performance [21]. Ce and Cd binary metal oxide nanocomposites exhibited good antibacterial, antimicrobial activity and have reported absorption spectra of binary metal oxide nanocomposite [22].

Ternary Nanocomposite

Other than binary nanocomposites, ternary nanocomposites have shown greater performances in application like supercapacitance [23], photocatalytic mineralization [24], and photoelectrochemical sensing [25], enhanced thermoelectric properties [26], improved dielectric constant *etc* [27]. The synergistic effects of nanometer range particle and individual component property of precursor plays pivotal role in getting high fracture toughness and ductility, which actually depends upon interfacial compatibilization and interactions [28]. The role of carbon material, metal oxide and conducting polymer based ternary nanocomposites is observed in the case of graphene/PEDOT/MnO_2 where enhanced specific capacitance percentage was obtained [29]. PEDOT: PSS coated $NiFe_2O_4$/reduced graphene oxide tertiary nanocomposite proved to be effective energy storage systems from its high specific capacitance energy density values [30]. An orderly bead chain Cu_2O/Mn_3O_4/NiO nanocomposite has been synthesized by using electrostatic spinning technology which improved the supercapacitance of the electrode material [31]. Au/CuS/TiO_2 based non-enzymatic biosensor exhibited highly sensitive photoelectrochemical sensing for the electrochemical determination of glucose [25]. Reduced graphene oxide/double walled carbon nanotube, nacre-based synthesized ternary bioinspired nanocomposites have shown excellent strength and stiffness with high fatigue resistant capabilities [32]. Further, these nanocomposites have applications in bioremediation supplementing the solutions for solving pollution problems for *e.g.*, g-C_3N_4/Co_3O_4/Ag_2O ternary heterojunction nanocomposite shows excellent photodegradation efficiency towards rhodamine B [33]. Aqueous cephalexin degradation and toxicity evaluation work has proven the role of solar active TiO_2/WO_3/CQDs heterojunction nanocomposite material as a potential material for decontamination of water [24].

Quaternary Nanocomposites

Synthesis of lightweight graphene, Fe_3O_4@Fe core/ZnO nanomaterial-based quaternary nanocomposite showed a remarkable increase in electromagnetic absorption properties [34]. Reduction of Cr(VI) to Cr (III) was studied utilizing the Alg/CMC/MnO_2/g-C_3N_4/Ca nanocomposite, which was prepared and used as a sensing material and fabricated sensor with a low detection limit of 0.63ng/ml [35]. Further, quaternary ferrite magnetic nanocomposite, in the presence of solar irradiation mineralized almost 80% of the dyes found in wastewater. Therefore, the process is helpful in decontamination of wastewater [36]. Catalyzing the photoreduction process of carbon dioxide to methane has been found effective when nanocomposite TiO_2/CdS/reduced graphene oxide/Pt was used [37]. Studies involving hydrogen evolution revealed the property of Pd-Pt/graphene-TiO_2

nanocomposite to act as photocatalyst. Pt and Pd nanoparticles are spread over graphene sheets and provide large surface area and increases efficiency of the system [38]. Chalcogenide/graphene based quaternary nanocomposite further revealed the increased photocatalytic and charge storage phenomenon occurrence where modified solvothermal process was used for the synthesis [39]. Controlled synthesis is observed in case of CuO@Ni/polyaniline/MWCNT, where the nanocomposite was tested to be a good electrode material and exhibited high charge storage capacity *i.e.* supercapacitance [40]. Fe_2O_3-Sm_2O_3-ZnO-P_2O_5 nanocomposite is another material synthesized, whose optical properties and electrical conducting mechanism has also been studied in this class [41].

Many more research reports have shown significant improvement in the material property such as, specific capacitance *etc.* which is found better in case of ternary and quaternary nanocomposites when compared to those of individual or binary nanocomposites.

COMPUTATIONAL STUDIES ON NANOCOMPOSITES

Nanocomposites are matter that assimilates nanosized particles into a forge of standard material. Inclusion of nanoparticles leads to significant enhancement in various properties such as, electrical conductivity [42], mechanical strength [43], flame retardancy [44], chemical resistance [45], optical clarity [46] *etc.* At present, nanocomposites are found to exhibit diverse applications [47], like gas sensor devices, photocatalysts, capacitors for computer chips, *etc.* Numerous experimental and theoretical approaches are appearing and redefining the method of synthesis, analysis and cost control designing of nanocomposites. This section is mainly devoted to reviewing computational studies of different nanocomposites.

A number of computational investigations were carried out to determine the adsorption behaviour of different gases on nanocomposites in view of its applicability as gas sensor devices. For instance, Abbasi *et al.* [48] performed DFT calculations to determine the adsorption energies of O_3 molecules on the undoped and N-doped TiO_2/MoS_2 nanocomposites. Similarly, they also investigate the adsorption properties of undoped and N-doped TiO_2/Au nanocomposites for NO_2 and H_2S molecules [49], respectively. The properties which were explored include structural parameters, electronic parameters in terms of density of states, molecular orbitals and allocation of spin densities and Mullikan population analysis. The result of these studies implies that the N-doped composites strengthen the adsorption process, which makes them potential candidates for gas sensors. Grand canonical Monte Carlo (GCMC) simulations to reveal enhanced hydrogen storage on Li-doped fullerene pillared graphene nanocomposites were carried out by Baykasoglu research group [50]. One of the DFT studies of polyaniline/ZnO nanocomposites predicted its photocatalytic activity for the reduction of methylene blue dye [51]. Semiempirical PM6 quantum mechanical study reveals more reactivity and thermal stability of Graphene/Calcium Oxide nanocomposite as compared to graphene alone [52]. Also, there are many reports on combined experimental and computational studies on nanocomposites, such as, synthesis, structural characterization and DFT analysis of g-C_3N_4/ZnS nanocomposites which exhibits photocatalytic performance [53], polylactic acid/silver-NP nanocomposite having antimicrobial activity, which was further scrutinized using both experimental and computational approaches [54], analysis of optical properties of polycarbonate/TiO_2/ZnO nanocomposite by employing experimental and DFT methods [55] and results for graphene oxide-epoxy polymer nanocomposite was validated by applying theoretical methods [56]. Summary of a few computational studies on different nanocomposites is presented in Table **1**.

Table 1. Summary of computational studies on different nanocomposites.

Nanocomposite	Computational Method Used	Properties Studied	Application	Refs
Polyaniline (PANI)/(ZnO)n Nanocomposites (n = 7–12)	DFT /B3LYP/6-311G (d,p)	Binding energies, Natural bond orbital (NBO) analysis, Electrostatic potential (ESP), density of states, nonlinear-optical properties (NLO)	Photocatalyst	[51]
TiO_2/Au Nanocomposites	DFT/GGA/PBE	Structural properties (bond lengths, angles), adsorption energies and electronic properties	Gas sensor devices	[49b]
TiO_2/Au Nanocomposites	DFT/GGA/PBE	Structural properties (bond lengths, bond angles), adsorption energies, mulliken population analysis and electronic properties	Gas sensor devices	[49a]
TiO_2/MoS_2 Nanocomposites	DFT/GGA/PBE	Structural properties (bond lengths, angles), adsorption energies and electronic properties	Gas sensor devices	[48]
Graphene/Calcium Oxide Nanocomposite	Semiempirical PM6 method	Ionization potential, charge distribution, TDM and thermal parameters	Electrochemical Sensors	[52]

(Table 1) cont.....

Nanocomposite	Computational Method Used	Properties Studied	Application	Refs
Graphene Oxide/Epoxy Polymer Nanocomposite	DFT/GGA/PBE	Structural properties, adsorption energies, interfacial interaction	Adhesive, electronic encapsulates	[56]
g-C₃N₄/ZnS Nanocomposite	DFT/GGA/PBE	Electrostatic potential, electronic structure, density of state	Photocatalyst	[53]
Polycarbonate/TiO₂/ZnO Nanocomposite	DFT	Band gap	Photo- catalysis and antibacterial materials	[55]
Lithium Doped Fullerene Pillared Graphene (Li-FPGNs) Nanocomposites	GCMC simulations	Structural properties, hydrogen physisorption	Light- weight Hydrogen storage devices	[50]
Polylactic Acid/silver-NP (PLA/AgNPs) Nanocomposite	Molecular dynamics (MD) simulation	-	Food packaging	[54]

Abbreviations: DFT= Density functional theory, B3LYP = Becke, 3-parameter, Lee–Yang–Parr, GGA = Generalized gradient approximation, PBE = Perdew–Burke–Ernzerhof, GCMC = Grand canonical monte carlo.

METAL ORGANIC FRAMEWORK-BASED NANOCOMPOSITES

Metal organic framework (MOF) based nanocomposites are the futuristic nanomaterials or nanocomposites possessing wide range of applications. They are formed by the self-assembly process where the linkage is formed by the organic ligands. Fig. (**1**) shows schematic drawings of some well celebrated MOFs

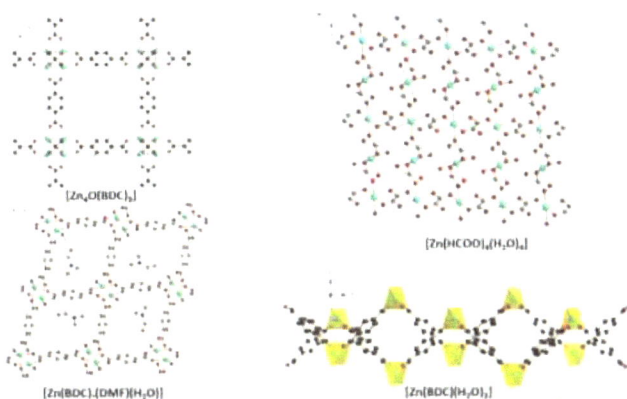

[Zn₄O(BDC)₃]

[Zn(HCOO)₄(H₂O)₄]

[Zn(BDC)₂(DMF)(H₂O)]

[Zn(BDC)(H₂O)₂]

Fig. (1). Different structures of well celebrated porous MOFs.

They are the coordination polymers comprising metal ion/clusters and organic ligands coordinating to form one, two or three-dimensional networks. MOFs possess several unique features such as high and tuneable porosity, tailorable chemistry with diverse structures, thermal and chemical stability. These properties with the existence of electrically conductive and semi conductive MOFs is the prime motivation behind using MOFs electrochemical sensing applications. The performances of MOFs based electrochemical sensors have attracted researchers to develop new electroanalytical methods for the determination of redox systems with high sensitivity, selectivity, easy operation and trace determinations. Some of the MOF based sensors have been reported in Table **2**.

Table 2. Electrochemical sensing of analyte systems based on MOF based sensors.

Modified Sensor	Method of Modification	Analyte	Limit of Detection	Ref
AP-Ni-MOF/CPE	Drop coating	Hydrogen peroxide	0.9 µM	[57]
MIP/PC/GCE	Drop-casting	Lidocaine	67 fM	[58]
Cu MOF/AuNPs GCE	Drop-casting	Paracetamol	1.1 pM	[59]
MOF-199/GCE	Drop-coating	Caffeine and Paracetamol	1.2 µM and 1.3 µM	[60]
EPC/GCE	Drop-coating	Chloramphenicol	2.9 nM	[61]
MPA-CdTe QDs	Drop-coating	6-mercapto-purine	0.15 µM	[62]
G-ZIF-8/GCE	Drop-coating	Dopamine	1.0 µM	[63]
MOF-ERGO-n/GCE	Electrodeposition	Catechol and Hydroquinone	0.1 µM	[64]
UiO-66/MC-3/GCE	Drop coating	Hydroquinone, Catechol and Resorcinol	0.056 µM, 0.072 µM and 3.51 µM	[65]
EPC/GCE	Drop-coating	Chloramphenicol	2.9 nM	[66]
$CuCo_2O_4$/PrGO-n/GCE	Drop coating	Glucose	0.15 µM	[67]
Nafion/MSA/ PANI/GCE	Drop-coating	Carcinoembryonic antigen and Alpha-fetoprotein	50 ng L^{-1} and 200 ng L^{-1}	[68]
Cu- MOF/AB/GCE	Drop-coating	Hydrogen peroxide	14 nM	[69]
HKUST-CNF	Drop-coating	Diclofenac and Ibuprofen	3.2 µg/L^{-1} and 6.1µg/L^{-1}	[70]
Ni(II)-MOFs/GCE	Drop-coating	Hydrogen peroxide	2.1 µM	[71]
$Cu_3(BTC)_2$–CPE	Drop-coating	2,4- dichlorophenol	9 nM	[72]
AgNPs/MIL-101/GCE	Drop-coating	Tryptophan	0.14 µM	[73]

(Table 2) cont.....

Modified Sensor	Method of Modification	Analyte	Limit of Detection	Ref
GC/Au-MOF-5	Drop-coating	Nitrite and Nitrobenzene	1.0 µM and 15.3 µM	[74]
Fe_3O_4@ZIF-8/GCE	Drop casting	Dopamine	6.67 nM	[75]
Nafion/C/Al-MIL-53 $(OH)_2$	Drop-coating	Dopamine	8 nM	[76]
Co-MOF/GCE	Drop-coating	Hydrogen peroxide	3.76 µM	[77]
Cu-MOF	Drop-coating	Bisphenol A	13 nM	[78]
Au-SH-SiO_2@Cu-MOF/GCE	Drop-coating	Hydrazine	0.01 µM	[79]
Au-SH-SiO_2@Cu-MOF/GCE	Drop-coating	L-cysteine	8 nM	[80]
Fc-MOF-GE	Drop-coating	Acetaminophen	6.4 nM	[81]

CONCLUDING REMARKS

The acceptable properties of the nanocomposite depend on the percentage homogeneity, or the heterogeneity of the reinforced nanomaterial or nanofiller in the matrix phase or the continuous phase. If one or more component lies in the range of nanometer (<100nm), then the composite is referred to as nanocomposite. Nanocomposites have many classifications, but they can also be divided and classified according to the number of components present in the nanocomposite as binary, ternary and quaternary nanocomposites. Effective and efficient computational models have been studied nowadays, to find the different energy values and functionality of the nanocomposites. For *e.g.*, adsorptive behavior of nanocomposites has been reported using cost-effective designs and contribution calculations connecting the role of nanosized material to the property of the system. Further, new advanced nanocomposites like MOFs are being used for the determination of numerous analyte solution systems. They with different functional materials form highly sensitive and selective electrochemical probes with much lower limits of detection. This area of nanocomposites has played a pivotal role in future technological interventions and still has a lot of scope for the future prospects of life.

CONSENT FOR PUBLICATION

Not applicable.

CONFLICT OF INTEREST

The author declares no conflict of interest, financial or otherwise.

ACKNOWLEDGEMENTS

Declared none.

REFERENCES

[1]   Hasnain, M. S.; Ahmad, S.A; Minhaj, M.A.; Ara, T.J.; Nayak, A.K. Nanocomposite materials for prosthetic devices, Applications of Nanocomposites Materials in Orthopaedics. *Woodhead Publioshing series in biomaterials,* **2019**, 127-144.

[2]   Cavaco, A.; A., Ramalho; Pais, S.; Duraes, L. CNT-polydimethylsiloxane nanocomposites for prosthesis interfaces. 10th International conference on composite science and technology (ICCST/10). , **2015**.

[3]   Vasile, C. Polymer nanocomposites and nanocoating in Food packaging: A review. *Materials (Basel),* **2018**, *11*(10), 1834.
[http://dx.doi.org/10.3390/ma11101834] [PMID: 30261658]

[4]   Jawad, S.; Sarfaraz, T.G.; Nygaard, J.N.; Petterson, M.K. Nanocomposites for food packaging application: An overview. *Nanomaterials (Basel),* **2021**, *11*, 1-27.

[5]   Feldman, D. Polymer nanocomposites in medicine. *J. Macromol. Sci. Part A Pure Appl. Chem.,* **2016**, *53*(1), 55-62.
[http://dx.doi.org/10.1080/10601325.2016.1110459]

[6]   Camargo, P.H.C.; Satyanarayana, K.G.; Wypych, F. Nanocomposites: synthesis, structure, properties and new application opportunities. *Mater. Res.,* **2009**, *12*(1), 1-39.
[http://dx.doi.org/10.1590/S1516-14392009000100002]

[7]   Lu, L.; Zhu, Z.; Hu, X. Multivariate nanocomposites for electrochemical sensing in the application of food. *Trends Analyt. Chem.,* **2019**, *118*, 759-769.
[http://dx.doi.org/10.1016/j.trac.2019.07.010]

[8]   Munonde, T.S.; Nomngongo, P.N. Nanocomposites for electrochemical sensors & their applications on the detection of trace metals in environmental water samples. *Sensors (Basel),* **2020**, *21*(1), 131.
[http://dx.doi.org/10.3390/s21010131] [PMID: 33379201]

[9]   Chen, T.W.; Ramachandran, R.; Chen, S.M.; Anushya, G.; Ramachandran, K. Graphene & Pervoskite -based nanocomposites for both electrochemical and gas sensors applications: An overview. *Sensors (Basel),* **2020**, *20*(23), 6755.
[http://dx.doi.org/10.3390/s20236755]

[10]   Kumar, M.R.A.; Abebe, B. Nagaswarupa, H.P.; Murthy, H.C.A.; Kumar, C.R.R.; Sabir, F.K. Enhanced photocatalytic & electrochemical performance of TiO_2-Fe_2O_3 nanocomposite: Its application in dye decolorization & as supercapacitors. *Sci. Rep.,* **2020**, *10*, 1249.
[http://dx.doi.org/10.1038/s41598-020-58110-7] [PMID: 31988344]

[11]   Miklicanin, E.O.; Badnjevic, A.; Kazlagic, A.; Hajlovac, M. Nanocomposites: a brief review. *Health Technol. (Berl.),* **2020**, *10*(1), 51-59.
[http://dx.doi.org/10.1007/s12553-019-00380-x]

[12]   Said, R.A.M.; Hasan, M.A.; Abdelzaher, A.M.; Abdel-Raoof, A.M. Review-Insights into the developments of nanocomposites for processing & application as sensing materials. *J. Electrochem. Soc.,* **2020**, *167*(3), 037549.
[http://dx.doi.org/10.1149/1945-7111/ab697b]

[13]   Kumar, S.K.; Krishnamoorti, R. Nanocomposites: structure, phase behavior, and properties. *Annu. Rev. Chem. Biomol. Eng.,* **2010**, *1*(1), 37-58.
[http://dx.doi.org/10.1146/annurev-chembioeng-073009-100856] [PMID: 22432572]

[14]   Hari, J.D.; Pukanszky, B. *Nanocomposites: Preparation, structure & properties, Applied Plastics*

Engineering Handbook; Kutz, M., Ed.; Elsevier Inc.: Waltham, MA, USA, **2011**, pp. 109-142.
[http://dx.doi.org/10.1016/B978-1-4377-3514-7.10008-X]

[15]   Cangialosi, D.; Boucher, V.M.; Alegria, A.; Colmenero, J. Physical ageing in polymers & polymer nanocomposites: recent results & open questions. *Soft Matter,* **2013**, *9*(36), 8619-8630.
[http://dx.doi.org/10.1039/c3sm51077h]

[16]   Jordan, J.; Jacob, K.I.; Tannenbaum, R.; Sharaf, M.A.; Jasiuk, I. Experimental trends in polymer nanocomposites: a review. *Mater. Sci. Eng. A,* **2005**, *393*(1-2), 1-11.
[http://dx.doi.org/10.1016/j.msea.2004.09.044]

[17]   Suresh, R.; Rajendran, S.; Kumar, P.S.; Vo, D.N.; Cornejo-Ponce, L. Recent advancements of spinel ferrite based binary nanocomposite photocatalysts in wastewater treatment. *Chemosphere,* **2021**, *274*, 129734.
[http://dx.doi.org/10.1016/j.chemosphere.2021.129734] [PMID: 33548641]

[18]   Gerasimov, G.N.; Gromov, V.F.; Illegbuse, O.J.; Trakhttenberg, L.I. the mechanism of sensory phenomenon in binary metal-oxide nanocomposites. *Sens. Actuators B Chem.,* **2017**, *240*, 613-624.
[http://dx.doi.org/10.1016/j.snb.2016.09.007]

[19]   Riyadh, S.M.; Khalil, K.D.; Ali, H.B. Structural properties & catalytic activity of binary poly(vinyl alcohol)/Al_2O_3 nanocomposite film for synthesis of thiazoles. *Catalysts,* **2020**, *10*(1), 1-9.
[http://dx.doi.org/10.3390/catal10010100]

[20]   Majumdar, D. Review on current progress of MnO_2-based ternary nanocomposites as supercapacitor applications. *ChemElectroChem,* **2021**, *8*(2), 291-336.
[http://dx.doi.org/10.1002/celc.202001371]

[21]   Ashritha, M.G.; Hareesh, K. A review on graphitic carbon nitride based binary nanocomposites as supercapacitors. *J. Energy Storage,* **2020**, *32*, 101840.
[http://dx.doi.org/10.1016/j.est.2020.101840]

[22]   Magdalane, C.M.; Kaviyarasu, K.; Vijaya, J.J.; Siddhardha, B.; Jeyaraj, B. Photocatalytic activity of binary metal oxide nanocomposites of CeO2/CdO nanospheres: Investigation of optical and antimicrobial activity. *J. Photochem. Photobiol. B,* **2016**, *163*, 77-86.
[http://dx.doi.org/10.1016/j.jphotobiol.2016.08.013] [PMID: 27541568]

[23]   Huisheng, P.; Xuemei, S.; Wei, W.; Fang, X. Energy storage devices based on polymers In: *Polymer materials for energy & electronic application*; H, Peng; X, Sun; W, Weng; X, Fang, Eds.; Academia Press, **2017**; pp. 197-242.

[24]   Sun, X.; He, W.; Yang, T.; Ji, H.; Lui, W.; Lei, J.; Lui, Y.; Cai, Z. Ternary TiO_2/WO_3/CQDs nanocomposites for enhanced photocatalytic mineralization of aqueous cephalexin: Degradation mechanism and toxicity evaluation. *Chem. Eng. J.,* **2021**, *412*, 128679.
[http://dx.doi.org/10.1016/j.cej.2021.128679]

[25]   Wang, Y.; Bai, L.; Wang, Y.; Qin, D.; Shan, D.; Lu, X. Ternary nanocomposites of Au/CuS/TiO_2 for an ultrasensitive photoelectrochemical non-enzymatic glucose sensor. *Analyst (Lond.),* **2018**, *143*(7), 1699-1704.
[http://dx.doi.org/10.1039/C8AN00187A] [PMID: 29521385]

[26]   Wang, Y.; Wu, S.; Yin, Q.; Jiang, B.; Mo, S. Poly(3,4-ethylenedioxythiophene)/polypyrrole/carbon nanoparticle ternary nanocomposite film with enhanced thermoelectric properties. *Polymer (Guildf.),* **2021**, *212*, 123131.
[http://dx.doi.org/10.1016/j.polymer.2020.123131]

[27]   Li, H.; Ren, H.; Ai, D.; Han, Z.; Liu, Y.; Yao, B.; Wang, Q. Ternary polymer nanocomposites with concurrently enhanced dielectric constant & breakdown strength for high temperature electrostatic capacitors. *Infomat.,* **2020**, *2*(2), 389-400.
[http://dx.doi.org/10.1002/inf2.12043]

[28]   Maguire, S.M.; Krook, N.M.; Kulshreshtha, A.; Bilchak, C.R.; Brosnan, R.; Pana, A.M.; Rannou, P.;

Marechal, M.; Ohno, K.; Jayaraman, A.; Composto, R.J. Interfacial Compatibilization in ternary polymer nanocomposite: Comparing theory and experiments. *Macromolecules,* **2021**, *54*(2), 797-811.
[http://dx.doi.org/10.1021/acs.macromol.0c02345]

[29]   Moussa, M.; Shi, G.; Wu, H.; Zhao, Z.; Voelcker, N.H.; Losic, D.; Jun, M. Development of flexible supercapacitors using an inexpensive graphene/PEDOT/MnO$_2$ sponge composite. *Mater. Des.,* **2017**, *125*, 1-10.
[http://dx.doi.org/10.1016/j.matdes.2017.03.075]

[30]   Hareesh, K.; Shateesh, B.; Joshi, R.P.; Dahiwale, S.S.; Bhorasker, V.N.; Haram, S.K.; Dhole, S.D. PEDOT:PSS wrapped NiFe$_2$O$_4$/rGO tertiary nanocomposite for the supercapacitor applications. *Electrochim. Acta,* **2016**, *201*, 106-116.
[http://dx.doi.org/10.1016/j.electacta.2016.03.205]

[31]   Su, L.; Zhang, C.; Shu, L.; Huang, L.; Li, J.; Qin, H. Orderly arranged bead chain Cu$_2$O-Mn$_3$O$_4$-NiO ternary nanocomposites with high specific capacitance for supercapacitors. *Nano,* **2020**, *6*(6), 2050082.
[http://dx.doi.org/10.1142/S1793292020500824]

[32]   Gong, S.; Cui, W.; Zhang, Q.; Cao, A.; Jiang, L.; Cheng, Q. Integrated ternary bioinspired nanocomposites *via* synergistic toughening of reduced graphene oxide & double walled carbon nanotubes. *ACS Nano,* **2015**, *9*(12), 11568-11573.
[http://dx.doi.org/10.1021/acsnano.5b05252] [PMID: 26469807]

[33]   Xu, Q.; Zhao, P.; Shi, Y.K.; Li, J.S.; You, W.S.; Zhang, L.C.; Sang, X.J. Preparation of a g-C$_3$N$_3$/Co$_3$O$_4$/Ag$_2$O ternary heterojunction nanocomposite & its photocatalytic activity & mechanism. *New J. Chem.,* **2020**, *44*(16), 6261-6268.
[http://dx.doi.org/10.1039/D0NJ01122C]

[34]   Ren, Y.L.; Wu, H.Y.; Lu, M.M.; Chen, Y.J.; Zhu, C.L.; Gao, P.; Cao, M.S.; Li, C.Y.; Ouyang, Q.Y. Quaternary nanocomposites consisting of graphene, Fe3O4@Fe core@shell, and ZnO nanoparticles: synthesis and excellent electromagnetic absorption properties. *ACS Appl. Mater. Interfaces,* **2012**, *4*(12), 6436-6442.
[http://dx.doi.org/10.1021/am3021697] [PMID: 23176086]

[35]   Bano, M.; Khan, I.; Ahirwar, D.; Khan, F. Synthesis of quaternary nanocomposites for the catalytic reduction of Cr(IV) to Cr(III) & its sensing. *React. Funct. Polym.,* **2020**, *150*, 104545.
[http://dx.doi.org/10.1016/j.reactfunctpolym.2020.104545]

[36]   Ciocarlan, R.G.; Seftel, E.M.; Mertens, M.; Pui, A.; Mazaj, M.; Tusar, N.N.; Cool, P. Novel magnetic nanocomposites containing quaternary ferrites systems Co 0.5 Zn 0.25 M0.25 Fe$_2$O$_4$ (M= Ni, Cu, Mn, Mg) & TiO$_2$-anatase phase as photocatalysts for wastewater remediation under solar light irradiation. *Mater. Sci. Eng. B,* **2018**, *230*, 1-7.
[http://dx.doi.org/10.1016/j.mseb.2017.12.030]

[37]   Benedetti, J.E.; Bernardo, D.R.; Morais, A.; Nogueira, A.F. Synthesis & characterization of a quaternary nanocomposite based on TiO$_2$/CdS/rGO/Pt & its application in the photoreduction of CO$_2$ to methane under visible light. *RSC Advances,* **2015**, *43*(43), 33914-33922.
[http://dx.doi.org/10.1039/C4RA15605F]

[38]   Ye, S.; Oh, W.C. Novel synthesis of quaternary nanocomposites based on chemical vapour grown graphene for photocatalytic hydrogen evaluation. *Fuller. Nanotub. Carbon Nanostruct.,* **2016**, *24*(8), 487-493.
[http://dx.doi.org/10.1080/1536383X.2016.1188283]

[39]   Sarkar, S.; Howli, P.; Das, B.; Das, N.S.; Samanta, M.; Das, G.C.; Chattopadhyay, K.K. Novel quaternary chalcogenide/reduced graphene oxide based asymmetric supercapacitor with high energy density. *ACS Appl. Mater. Interfaces,* **2017**, *9*(27), 22652-22664.
[http://dx.doi.org/10.1021/acsami.7b00437] [PMID: 28616963]

[40]   Chakraborty, I.K.; Chakraborty, N.; Senapati, A.; Chakraborty, A.K. CuO@NiO/Polyaniline/

MWCNT/nanocomposites as high-performance electrode for supercapacitors. *J. Phys. Chem. C,* **2018**, *122*(48), 27180-27190.
[http://dx.doi.org/10.1021/acs.jpcc.8b08091]

[41]   Singh, Y.B.; Biswas, D.; Shah, S.K.; Shaw, S.; Mondal, R.; Das, A.S.; Kabi, S.; Singh, S.S. Investigation of optical properties and electrical conductive mechanism of Fe_2O_3-Sm_2O_3-ZnO-P_2O_5 quaternary glass nanocomposite systems. *Materialia (Oxf.),* **2021**, *15*, 100963.
[http://dx.doi.org/10.1016/j.mtla.2020.100963]

[42]   Ma, P.C.; Liu, M.Y.; Zhang, H.; Wang, S.Q.; Wang, R.; Wang, K.; Wong, Y.K.; Tang, B.Z.; Hong, S.H.; Paik, K.W.; Kim, J.K. Enhanced electrical conductivity of nanocomposites containing hybrid fillers of carbon nanotubes and carbon black. *ACS Appl. Mater. Interfaces,* **2009**, *1*(5), 1090-1096.
[http://dx.doi.org/10.1021/am9000503] [PMID: 20355896]

[43]   Hong, S.H. Enhanced mechanical properties of boron nitride nanosheet/copper nanocomposites *via* a molecular-level mixing process. *Compos., Part B Eng.,* **2020**, *195*, 108088-108098.
[http://dx.doi.org/10.1016/j.compositesb.2020.108088]

[44]   Lopez-Cuesta, J.M. *Flame-retardant polymer nanocomposites*; Advances in Polymer Nanocomposites, **2012**, pp. 540-566.

[45]   Roy, D.V. Chemical resistance/thermal and mechanical properties of unsaturated polyester-based nanocomposites. *Appl. Nanosci.,* **2014**, *4*(2), 233-240.
[http://dx.doi.org/10.1007/s13204-013-0193-1]

[46]   Joseph, H.K. *Optical Properties of Polymer Nanocomposites*; Cambridge University Press, **2017**, pp. 550-565.

[47]   Sonawane, G. H.; Patil, S. P.; Sonawane, S. H. Nanocomposites and Its Applications. *Applications of Nanomaterials,* **2018**, 1-22.
[http://dx.doi.org/10.1016/B978-0-08-101971-9.00001-6]

[48]   Abbasi, A.; Sardroodi, J.J. Density functional theory (DFT) study of O_3 molecules adsorbed on nitrogen-doped TiO_2/MoS_2 nanocomposites: applications to gas sensor devices. *J. Iran. Chem. Soc.,* **2017**, *14*(12), 2615-2626.
[http://dx.doi.org/10.1007/s13738-017-1196-8]

[49]   (a) Abbasi, A. TiO_2/Gold nanocomposite as an extremely sensitive molecule sensor for NO_2 detection: A DFT study. *J. Water Environ. Nanotechnol.,* **2016**, *1*, 55-62 .(b) Abbasi, A.; Sardroodi, J. J. Adsorption of H_2S molecule on TiO_2/Au nanocomposites: A density functional theory study. *Nanochem. Res,* **2017**, *2*, 1-7.

[50]   Baykasoglu, C. Li-doped fullerene pillared graphene nanocomposites for enhancing hydrogen storage: A computational study. *Comput. Mater. Sci.,* **2021**, *186*, 110023-110031.
[http://dx.doi.org/10.1016/j.commatsci.2020.110023]

[51]   Singh, T.; Sharma, D. A DFT study of polyaniline/ZnO nanocomposite as a photocatalyst for the reduction of methylene blue dye. *J. Mol. Liq.,* **2019**, *293*, 111528-111541.
[http://dx.doi.org/10.1016/j.molliq.2019.111528]

[52]   Al-Bagawia, A.H.; Salama, E.E. Computational Study on the Electronic Properties of Graphene/Calcium Oxide Nanocomposite. *Egypt. J. Chem.,* **2021**, *64*, 407-412.

[53]   Mukherjee, B. Experimental Investigation with DFT Analysis Towards a Promising Recyclable Photocatalyst from g-C_3N_4/ZnS Nanocomposite. *ChemistrySelect,* **2020**, *5*(31), 9736-9744.
[http://dx.doi.org/10.1002/slct.202002785]

[54]   Sirotkin, N.A.; Gurina, D.L.; Khlyustova, A.V.; Costerin, D.Y.; Naumova, I.K.; Titov, V.A. Agafonov.Experimental and computational investigation of polylactic acid/silver-NP nanocomposite with antimicrobial activity prepared by plasma in liquid. *Plasma Process. Polym.,* **2020**, 1-17.

[55]   Eskandari, M.; Najafi, L.M.; Malekfar, R.; Taboada, P. Investigation of Optical Properties of Polycarbonate/TiO_2/ZnO Nanocomposite: Experimental and DFT Calculations. J Inorg. *J. Inorg.*

Organomet. Polym. Mater., **2020**, *30*(12), 5283-5292.
[http://dx.doi.org/10.1007/s10904-020-01644-0]

[56]   Abhishek, K.P.; Sanjay, R.D. Validation of experimental results for graphene oxide-epoxy polymer nanocomposite through computational analysis. *J. Polym. Sci.,* **2020**, 1-16.

[57]   Manan, N.S.A.; Sherino, B.; Mohamad, S.; Halim, S.N.A. Electrochemical detection of hydrogen peroxide on a new microporous Ni–metal organic framework material-carbon paste electrode. *Sens. Actuators B Chem.,* **2018**, *254*, 1148-1156.
[http://dx.doi.org/10.1016/j.snb.2017.08.002]

[58]   Zhang, J.; Liu, J.; Zhang, Y.; Yu, F.; Wang, F.; Peng, Z.; Li, Y. Voltammetric lidocaine sensor by using a glassy carbon electrode modified with porous carbon prepared from a MOF, and with a molecularly imprinted polymer. *Mikrochim. Acta,* **2018**, *87*(1), 185.
[http://dx.doi.org/10.1007/s00604-017-2551-2] [PMID: 29594562]

[59]   Shil, Y.; Zhang, Y. Wang1, Y.; Huang1, H.; Ma, J. Amperometric Sensing of Paracetamol Using a Glassy Carbon Electrode Modified with a Composite of Water–Stable Metal−Organic Framework and Gold Nanoparticles. *Int. J. Electrochem. Sci.,* **2018**, *13*, 7643-7654.

[60]   Minh, T.T.; Phong, N.H.; Duc, H.V.; Khieu, D.Q. Microwave synthesis and voltammetric simultaneous determination of paracetamol and caffeine using an MOF-199-based electrode. *J. Mater. Sci.,* **2018**, *53*(4), 2453-2471.
[http://dx.doi.org/10.1007/s10853-017-1715-0]

[61]   Xiao, L.; Xu, R.; Yuan, Q.; Wang, F. Highly sensitive electrochemical sensor for chloramphenicol based on MOF derived exfoliated porous carbon. *Talanta,* **2017**, *167*, 39-43.
[http://dx.doi.org/10.1016/j.talanta.2017.01.078] [PMID: 28340736]

[62]   Jin, M.; Mou, Z.L.; Zhang, R.L.; Liang, S.S.; Zhang, Z.Q. An efficient ratiometric fluorescence sensor based on metal-organic frameworks and quantum dots for highly selective detection of 6-mercaptopurine. *Biosens. Bioelectron.,* **2017**, *91*, 162-168.
[http://dx.doi.org/10.1016/j.bios.2016.12.022] [PMID: 28006684]

[63]   Zheng, Y.Y.; Li, C.X.; Ding, X.T.; Yang, Q.; Qi, Y.M.; Zhang, H.M.; Qu, L.T. Detection of dopamine at graphene-ZIF-8 nanocomposite modified electrode. *Chin. Chem. Lett.,* **2017**, *28*(7), 1473-1478.
[http://dx.doi.org/10.1016/j.cclet.2017.03.014]

[64]   Chen, Q.; Li, X.; Min, X.; Jian, D.C.; Li, Y.; Peng, Z.X.; Cai, W.; Zhang, C. Determination of catechol and hydroquinone with high sensitivity using MOF-graphene composites modified electrode. *J. Electroanal. Chem. (Lausanne),* **2017**, *789*, 114-122.
[http://dx.doi.org/10.1016/j.jelechem.2017.02.033]

[65]   Deng, M.; Lin, S.; Bo, X.; Guo, L. Simultaneous and sensitive electrochemical detection of dihydroxybenzene isomers with UiO-66 metal-organic framework/mesoporous carbon. *Talanta,* **2017**, *174*, 527-538.
[http://dx.doi.org/10.1016/j.talanta.2017.06.061] [PMID: 28738619]

[66]   Xiao, L.; Xu, R.; Yuan, Q.; Wang, F. Highly sensitive electrochemical sensor for chloramphenicol based on MOF derived exfoliated porous carbon. *Talanta,* **2017**, *167*, 39-43.
[http://dx.doi.org/10.1016/j.talanta.2017.01.078] [PMID: 28340736]

[67]   Yang, J.; Ye, H.; Zhang, Z.; Zhao, F.; Zeng, B. Metal–organic framework derived hollow polyhedron $CuCo_2O_4$ functionalized porous graphene for sensitive glucose sensing Sensor. *Sens. Actuators B Chem.,* **2017**, *242*, 728-735.
[http://dx.doi.org/10.1016/j.snb.2016.11.122]

[68]   Zhang, P.; Huang, H.; Wang, N.; Li, H.; Shen, D.; Ma, H.; Zhang, P. Duplex voltammetric immunoassay for the cancer biomarkers carcinoembryonic antigen and alpha-fetoprotein by using metal-organic framework probes and a glassy carbon electrode modified with thiolated polyaniline nanofibers. *Mikrochim. Acta,* **2017**, *184*(10), 4037-4045.
[http://dx.doi.org/10.1007/s00604-017-2437-3]

[69]   Meng, W.; Xu, S.; Dai, L.; Li, Y.; Zhu, J.; Wang, L. An enhanced sensitivity towards H_2O_2 reduction based on a novel Cu metal–organic framework and acetylene black modified electrode. *Electrochim. Acta,* **2017**, *230*, 324-332.
[http://dx.doi.org/10.1016/j.electacta.2017.02.017]

[70]   Motoc, S.; Manea, F.; Iacob, A.; Joaristi, A. M.; Gascon, J.; Pop, A.; Schoonman, J. *Electrochemical Selective and Simultaneous Detectionof Diclofenac and Ibuprofen in Aqueous SolutionUsing HKUST-1 Metal-Organic Framework-CarbonNanofiber Composite ElectrodeSensors,* **2016**, *16*, 1719-1730.

[71]   Wang, M.Q.; Zhang, Y.; Bao, S.J.; Yu, Y.N.; Ye, C. Ni (II)-based metal-organic framework anchored on carbon nanotubes for highly sensitive non-enzymatic hydrogen peroxide sensing. *Electrochim. Acta,* **2016**, *190*, 365-370.
[http://dx.doi.org/10.1016/j.electacta.2015.12.199]

[72]   Dong, S.; Suo, G.; Lia, N.; Chen, Z.; Peng, L.; Fu, Y.; Yang, Q.; Huang, T. A simple strategy to fabricate highly sensitive 2,4-dichlorophenol electrochemical sensor based on metal organic framework $Cu_3(BTC)_2$ Sensor. *Sens. Actuators B Chem.,* **2016**, *222*, 972-979.
[http://dx.doi.org/10.1016/j.snb.2015.09.035]

[73]   Peng, Z.; Jiang, Z.; Huang, X.; Li, Y. A novel electrochemical sensor of tryptophan based on silver nanoparticles/metal–organic framework composite modified glassy carbon electrode. *RSC Advances,* **2016**, *6*(17), 13742-13748.
[http://dx.doi.org/10.1039/C5RA25251B]

[74]   Yadav, D.K.; Ganesan, V.; Sonkar, P.K.; Gupta, R.; Rastogi, P.K. Electrochemical investigation of gold nanoparticles incorporated zinc-based metal-organic framework for selective recognition of nitrite and nitrobenzene. *Electrochim. Acta,* **2016**, *200*, 276-282.
[http://dx.doi.org/10.1016/j.electacta.2016.03.092]

[75]   Wang, Y.; Zhang, Y.; Hou, C.; Liu, M. Magnetic Fe_3O_4@MOFs decorated graphene nanocomposites as novel electrochemical sensor for ultrasensitive detection of dopamine. *RSC Advances,* **2015**, *5*(119), 98260-98268.
[http://dx.doi.org/10.1039/C5RA20996J]

[76]   Wang, Y.; Ge, H.; Ye, G.; Chen, H.; Hu, X. Carbon functionalized metal organic framework/Nafion composites as novel electrode materials for ultrasensitive determination of dopamine. *J. Mater. Chem. B Mater. Biol. Med.,* **2015**, *3*(18), 3747-3753.
[http://dx.doi.org/10.1039/C4TB01869A] [PMID: 32262849]

[77]   Yang, L.; Xu, C.; Ye, W.; Liu, W. An electrochemical sensor for H_2O_2 based on a new Co-meta--organic framework modified electrode. *Sens. Actuators B Chem.,* **2015**, *215*, 489-496.
[http://dx.doi.org/10.1016/j.snb.2015.03.104]

[78]   Wang, X.; Lu, X.; Wu, L.; Chen, J. 3D metal-organic framework as highly efficient biosensing platform for ultrasensitive and rapid detection of bisphenol A. *J. Biosens. Bioelectron.,* **2015**, *65*, 295-301.
[http://dx.doi.org/10.1016/j.bios.2014.10.010] [PMID: 25461172]

[79]   Hosseini, H.; Ahmar, H.; Dehghani, A.; Bagheri, A.; Fakhari, A.R.; Amini, M.M. Au-SH-SiO2 nanoparticles supported on metal-organic framework (Au-SH-SiO2@ Cu-MOF) as a sensor for electrocatalytic oxidation and determination of hydrazine. *Electrochim. Acta,* **2013**, *88*, 301-309.
[http://dx.doi.org/10.1016/j.electacta.2012.10.064]

[80]   Hosseini, H.; Ahmar, H.; Dehghani, A.; Bagheri, A.; Tadjarodi, A.; Fakhari, A.R. A novel electrochemical sensor based on metal-organic framework for electro-catalytic oxidation of L-cysteine. *Biosens. Bioelectron.,* **2013**, *42*, 426-429.
[http://dx.doi.org/10.1016/j.bios.2012.09.062] [PMID: 23228494]

[81]   Chang, Z.; Gao, N.; Li, Y.; He, X. Preparation of ferrocene immobilized metal–organic-framework modified electrode for the determination of acetaminophen. *Anal. Methods,* **2012**, *4*(12), 4037-4041.
[http://dx.doi.org/10.1039/c2ay26061a]

CHAPTER 3

Bimetallic-Carbon Based Composites for Electrochemical Sensors

S. Keerthana[1]**, A. Rajapriya**[1] and **N. Ponpandian**[1,*]

[1] *Department of Nanoscience & Technology, Bharathiar University, Coimbatore 641 046, Tamilnadu, India*

Abstract: The robust, sensitive, and selective finding of various biomolecules and environmental factors by potential nanostructures holds much promise for accurate electrochemical sensors. However, to be competitive, present electrochemical sensor technologies need noteworthy developments, particularly in specificity output rate, and long-lasting steadiness in complex biological environments. Bimetallic carbon nanocomposites are newly emerging materials with fascinating physicochemical properties and are very prospective in the innovative point-of-care study of various healthcare issues. Particularly, the multidimensionality of bimetallic carbon composites and their structural, optical, electronic, and electrocatalytic properties are suitable for the design of various electrochemical sensing devices. This chapter summarizes the sensing applications of bimetallic @C and its modern advances in the detection of different analytes. The chapter begins with a brief introduction to the advancement of bimetallic @C based electrochemical sensors followed by the discussion of the structure and properties of the bimetallic @C nanocomposites. We also discuss in detail the utilization of these bimetallic @C nanocomposites with graphene, MWCNTs, CQDs, and g-C$_3$N$_4$ for their worthwhile application in electrochemical sensors. Finally, the chapter concludes with a positive outlook on the use of bimetallic @C nanocomposites for day-to-day life and clinical applications based on the present growth.

Keywords: Bimetallic, Biomolecules, Biosensors, Carbon nanomaterials, Electrochemical sensors, Graphene, Sensitivity, Selectivity.

INTRODUCTION

Due to the present advancement and demands in the field of medical diagnosis and several environmental applications, the development of fast, sensitive, and selective analytical techniques are required for the sensing of health factors as the clinical investigations in laboratories are expensive and time-consuming processes in the hospital point-of-care settings [1]. A variety of detection techniques are

---

* **Corresponding Author N. Ponpandian**: Department of Nanoscience & Technology, Bharathiar University, Coimbatore 641 046, Tamilnadu, India; Email: ponpandian@buc.edu.in

Manorama Singh, Vijai K Rai and Ankita Rai (Eds.)

available such as chromatography, chemiluminescence, and spectroscopic techniques [2, 3]. Among them, the technique of electrochemical analytical techniques has many outstanding properties like accuracy, reliability, sensitivity, and specificity over the other methods. Moreover, electrochemical sensors have gained massive attention in the field of pharmaceutical, biological, environmental, and food industries.

These electrochemical devices should also have additional features like high flexibility, non-invasive detection, wearability, biocompatibility, lightweight, ease of fabrication, and cost-effectiveness [4, 5]. The analytical techniques to study the electrochemical characteristics include cyclic voltammetry, differential pulse voltammetry, and square-wave amperometric techniques. All of them are effective techniques to study and determine the electrochemical behavior of prominent electrodes developed for electrochemical sensors. These electrochemical processes can be influenced by many factors, such as the nature of the analyte, structure of the electrode material, type of the electrode, and the choice of the electrolyte [6]. Among all factors, the electrode material plays a crucial part and the researchers are particularly driven to prepare an effective electrode material for high-quality electrochemical sensors. Generally, numerous nanomaterials, including metal-nanoparticles (NPs), metal oxides, polymer nanocomposites, various types of 2D nanomaterials transition metal dichalcogenides, black phosphorous, layered double hydroxides, and graphene, *etc*.) and the carbon derivatives are widely exploited and investigated as the electrode material for electrochemical detection. This chapter comprises an analysis of electrochemical sensors employing the composites of bimetallic NPs and carbon-based materials, which can improve both the sensitivity and selectivity of the sensors. Carbon nanomaterials and their allotropes like carbon nanotubes (CNTs), graphene oxide (GO), carbon-based quantum dots (CQDs), crystalline diamond, and fullerenes [7] exhibit outstanding electrochemical properties, which have resulted in their employment in a wide range of applications. Currently several bottom-up and top-down techniques are available to prepare carbon structures with tuned uniformity, size, and morphology by varying reaction parameters. Hence, the above-mentioned diverse carbon nanomaterials are deemed to be wonder materials in the electrochemical field due to their excellent properties like electrical conductivity, high mechanical strength, higher electron transfer kinetics, and large surface area [8]. NPs of noble metals and non-noble metals have been studied for several decades and the precious metals are shown great interest owing to their greater surface area and number of corner atoms and edges, which greatly enhance their catalytic activity [9]. Bimetallic NPs are significantly unique from their monometallic counterparts and have attained great attention from both technical as well as scientific points of view owing to their distinctive catalytic properties [10]. By choosing the suitable combination of metal NPs and their sizes, we can tune the properties and hence, the performance of the nanostructures. However, pure metals (mainly Pt or Au) have poor sensitivity and selectivity. Bimetallic NPs composed of two dissimilar metals consume greater attention than monometallic NPs from both technical as well as scientific points of view. Some approaches have been adopted for the synthesis of metal NPs like chemical reduction using reducing agents like sodium borohydride, citrate, hydrazine, and glucose [11]. The synergy of carbon materials with the appropriate addition of metal NPs can inhibit the aggregation of composites and so, improves the catalytic activity because of the interaction and surface reactivity between the support and the supported electrocatalysts. Therefore, many pieces of research have been accomplished on the preparation of electrochemical sensors with highly efficient bimetallic-carbon nanocomposites [12]. As a result, it is deemed that combining the carbon materials and bimetallic NPs to construct an effective electrochemical sensing platform is a noble alternative to detect complex systems with interfering compounds.

STRUCTURE AND PROPERTIES OF BIMETALLIC NPS AND CARBON NANOMATERIALS

Carbon-based materials such as derivatives of graphene, carbon nanotubes, and carbon quantum dots are shown in Fig. (**1a**). Graphene, the greatly explored carbon material is a layered sheet of sp^2-bonded carbon atoms. Graphene is synthesized by various top-down and bottom-up approaches [13]. The family of graphene-related materials includes structural or chemical derivatives of graphene, such as a few layers of graphene. Another derivative is graphene oxide (GO), which is an oxygenated derivative of 2D graphene, and reduced graphene oxide (rGO), a product of the reduction of GO. CNT is an allotrope of carbon exhibiting a narrow cylindrical structure, composed of sp^2 hybridized carbon hexagon network rolled up with at least one end-capped with fullerene molecule. The sp^2 hybridization builds a stacking of layers with van der Waals force combined with weak out-of-plane and strong in-plane force. The general classification of CNTs includes single-walled carbon nanotube (SWCNT) and multi-walled carbon nanotube (MWCNT). SWCNT is known to be a single layer of rolled carbon network with a diameter ranging from 1-2 nm and MWCNTs can be presumed as a few graphene sheets rolled into hollow cylindrical tubes with a high aspect ratio. MWCNTs have numerous advantages, such as excellent thermal conductivity, higher electrical conductivity, and chemical stability. Due to their excellent physicochemical characteristics, MWCNTs can effectively serve as electrodes for electrochemical analysis of different chemicals, food preservatives, clinical and environmental agents. MWCNTs based electrochemical sensors show enhanced sensitivity, a lower limit of detection because of their high aspect ratio and rapid electrode kinetics [14]. GQDs, a novel class of carbon nanomaterials, are a new kind of fluorescent material that has been used mainly in a variety of metal ion sensing approaches. One-pot synthesis, cost-effectiveness, suitable optoelectronic, and biocompatibility properties are the major benefits of CQDs. The g-C$_3$N$_4$ is an analog of graphite and is a 2D semiconductor with van-der Waals layered semiconductor. The structure of g-C$_3$N$_4$ can be held as an N-substituted graphite network formed by sp^2 hybridization of carbon and nitrogen atoms [15]. The steady bonding between carbon and nitrogen atoms, and strong Vander Waal forces between the layers, make it stable in acidic and alkaline environments and at high temperatures. Among various N-doped carbon materials, the g-C$_3$N$_4$ has been a potential candidate due to its unique electronic structure, high chemical stability, electrocatalytic and photocatalytic properties [16, 17]. As a final point, due to the excellent properties of bimetallic-nanocomposites has great attention in the field of electrochemical sensors. Many factors lead to the development of effective sensing of biomolecules and other chemicals. The major fields in which bimetallic-carbon nanocomposites are developed are depicted in Fig. (**2**). Metallic NPs are nanomaterials engineered in nanometer size. Still, several noble (Ru, Rh, Pt, Ag, Au, and Pd) and non-noble (3d transition metals- Cu, Fe, Co, and Ni) metal NPs are used to prepare novel metallic composites for the development of the diagnosis applications. Monometallic NPs consist of single metal, and this constituted metal atom determines the properties of these metallic NPs. Bimetallic nanomaterials are categorized into two types, *i.e.*, mixed and separated ones, which can be categorized according to their configuration of atoms. Notably, these structures can be further categorized into sub-cluster structures, core-shell structures, and multishell core-shell structures [18], as shown in Fig. (**1b**).

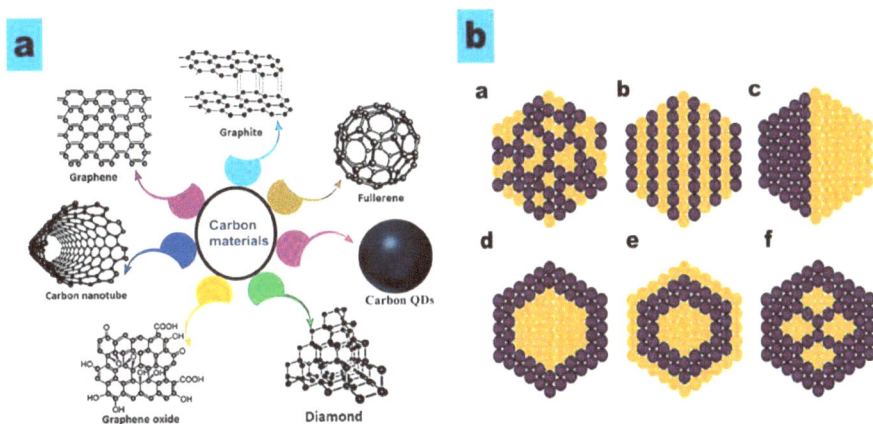

Fig. (1). (a) Various carbon nanostructures and **(b)** Bimetallic nanostructures [**(A)** alloyed, **(B)** intermetallic, **(C)** subclusters, **(D)** core–shell, **(E)** multishell core–shell, and **(F)** multiple core materials coated by a single shell. Yellow and purple spheres represent two different kinds of metal atoms] [18].

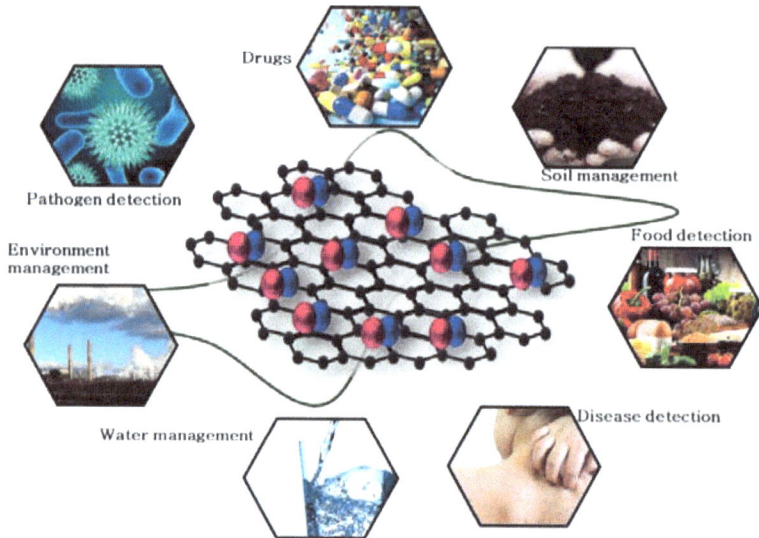

Fig. (2). Bimetallic-carbon nanocomposites for electrochemical sensors towards various biomolecules, environmental and health factors [18].

Usually, electrochemical sensors comprise two basic elements, **(a)** a molecular recognition system which is the most significant part of a sensor, and **(b)** a physicochemical transducer system which is a component that converts the chemical or biological response into a signal that can be detected by modern electrical instrumentations is shown in Scheme. **(1)**.

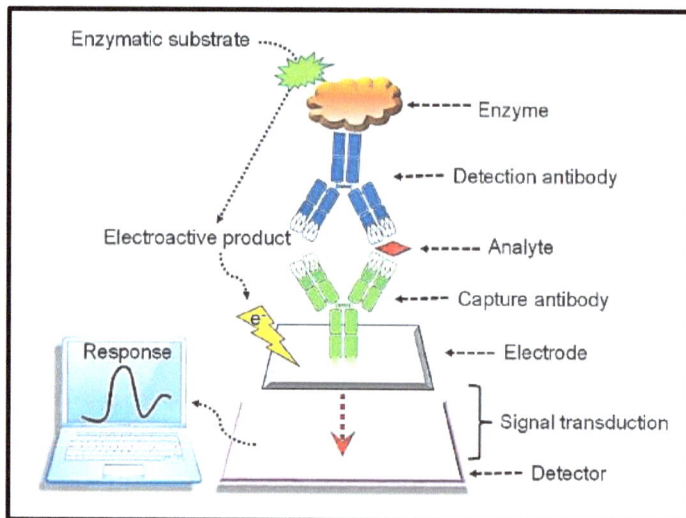

Scheme 1. Analytical principle of electrochemical sensors [32].

BIMETALLIC-2D GRAPHENE-BASED COMPOSITE FOR ELECTROCHEMICAL SENSORS

Generally, graphene is a widely used efficient composite material due to its easy modification or integration with other nanomaterials. Due to the poor sensitivity of monometallic NPs, Yang *et al.* fabricated GO with bimetallic nano-clusters using the noble bi-metals Pt and Au. In this report, the Pt-Au bimetals incorporated GO were prepared by chemical and physical reactions. The contrived sensors are applied to the electrochemical analysis of dopamine and uric acid due to their synergistic effect of Au-Pt-GO. The prepared nanocomposites show the electrochemical oxidation towards dopamine [19]. Similarly, Zhao *et al.* fabricated the electrochemical sensor by AuPt bimetallic nanoparticles-graphene nanocomposites were obtained by electrochemical co-reduction of graphene oxide (GO), $HAuCl_4$ and H_2PtCl_6 [20]. This fabricated sensor exposed to the detection of dopamine in the presence of uric acid and ascorbic acid and their DPV responses are shown in Fig. (**3a**). In another report, Yu *et al.* synthesized bimetallic of Pt-Ni/Poly fabricated by dropping nanocomposite of Pt-Ni-PDA/rGO on a surface of GCE [21]. Similarly, another bimetallic Pt-Co alloy/rGO composite enhanced the electrochemical sensors for the detection of AA, DA, and UA [22]. The structural and functional properties of Pt-Co alloy/rGO were confirmed by X-Ray diffraction pattern and Raman analysis (Fig. **3b**). In this approach, the electro-catalytic properties of bimetallic-carbon nanocomposites were boosted by the synergistic effect of both bimetallic alloy nanoparticles and the r-GO. Mahmoudi *et al.* fabricated a high-performance electrochemical sensors for the detection of bisphenol A by a novel template-based graphene nanoribbons with protein capped Au-Cu for. This proposed template-based sensor showed a higher sensitivity with a lower low detection limit and selectivity among the several biomolecules and ions.Their linear range and LOD are 0.01 to 70 µmol. L^{-1} and 0.004 µmol. L^{-1} [23]. Similarly, a bimetallic Pd-Cu functionalized with MIP graphene was developed for the electrochemical sensing of another contaminant amaranth. The electrochemical behavior towards amaranth was studied by CV and DPV. This system shows better conductivity and linear regression with the LOD value of 2 nM. Additionally, this Pd-Cu-graphene nanocomposite exhibits satisfactory recoveries in the detection of soft drinks [24]. A simple electrochemical approach was intended for the analysis of glycoprotein existing on the apical surfaces of epithelial cells. For this study, Au-Pt NPs anchored on the carboxylated GO were deposited on the FTO electrode for signal amplification. The prepared sensing platform used biotin tagged aptamer as a recognition element [23]. Au-Pt bimetallic NPs supported N doped graphene were developed and utilized for the EC detection of nitride. Nitride is a food additive to prevent bacterial growth. However, the excess amount of nitride leads to several health problems. The study proved that the N-doped graphene facilitates higher electron transfer kinetics [24].

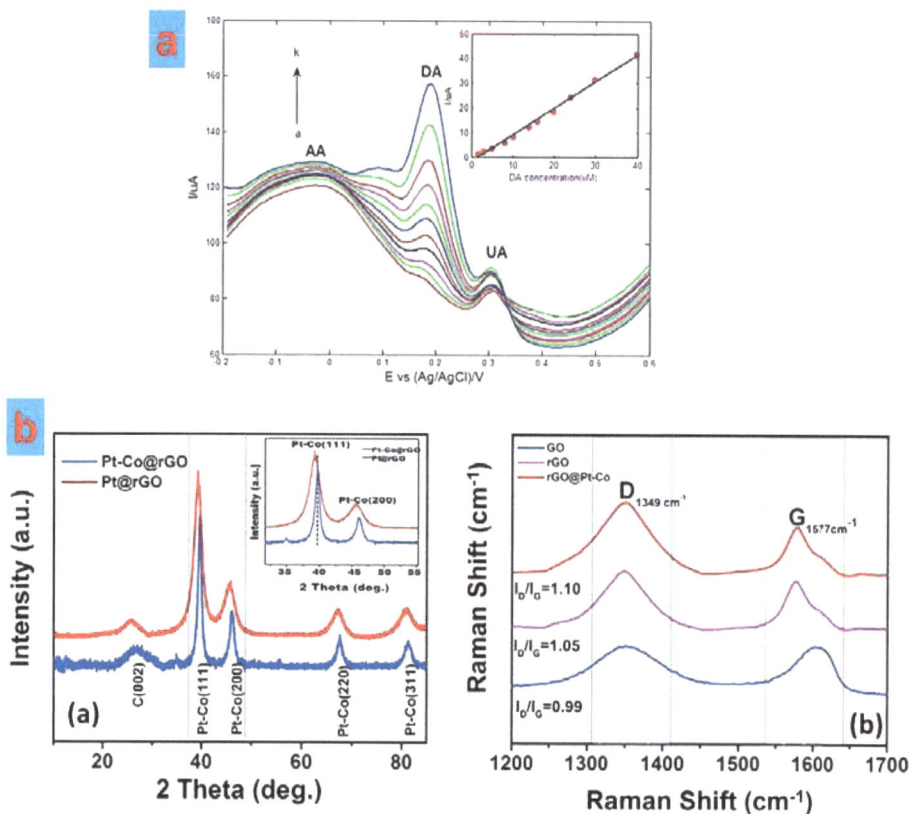

Fig. (3). (a) DPV response of various AuPt /GO nanocomposites towards DA in the presence of uric acid and ascorbic acid (20), **(b)** X-ray diffraction pattern and Raman analysis of Pt-Co alloy NPs/rGO [22].

BIMETALLIC-CNT BASED COMPOSITE FOR ELECTROCHEMICAL SENSORS

Bimetallic-CNT-based composites are in the current scope of catalysis and electrocatalysis and are of high importance. Subsequently, the addition of another metal brings about many deliberated variations in particle size, shape, morphology, and catalytic activity. A research group developed the electrochemical amperometric sensor based on PtM (where M=Pd, Ir)/MWCNT for the rapid and accurate determination of hydrogen peroxide. Developed PtPd/MWCNT/GCE exhibited a low detection limit of about 1.2 μM along with specificity among various interfering compounds [25]. XRD patterns of PtPd/MWCNT confirms the structural properties of MWCNT along with the metal composites. Recently, a real-time, reliable biosensor was constructed for the detection of neurotransmitters and reactive oxygen species using another bimetallic AgAu decorated MWCNT composite. Due to the high conductivity of AgAu/MWCNTs, the developed sensor provided an outstanding electrochemical activity that can quantify DA and H_2O_2 to the LOD of 0.23 nM and 0.59 nM with a broad dynamic range [26]. However, the major drawback of this biosensor was its high cost owing to the utilization of a high amount of noble metals. To switch this problem in an optimizing way, increasing the activity of nanomaterials and decreasing the concentration of noble metals is an effective approach. A group developed the electrochemical sensor for BPA by using AuPd incorporated MWCNT [4].The most common metals for catalytic purposes are Au, Pt, Ag, and Pd owing to their inherent physical and chemical properties. While these metals are outstanding electrocatalysts, their expensiveness creates them unsuitable for day-to-day health care analysis. To overcome these problems, the transition metals such as Mn, Fe, Cu, Ni, Zn, and Co are being explored for their use as electrocatalysts due to their abundancy, cost-effectiveness, and the improved electron capacity nature with MWCNT. A group developed a screen printed ECS by preparing NiFe anchored f-MWCNT for the detection of nitrofurantoin [27]. The spherical structure of NiFe anchored on the tubular-shaped f-MWCNT is confirmed by the HRTEM images (Fig. **4b**) (i-ii). This structure also reflected the d-spacing values according to their XRD results and displayed a high sensitivity towards DPV with the LOD 0.03 µM (Fig. **4b**) (iii).

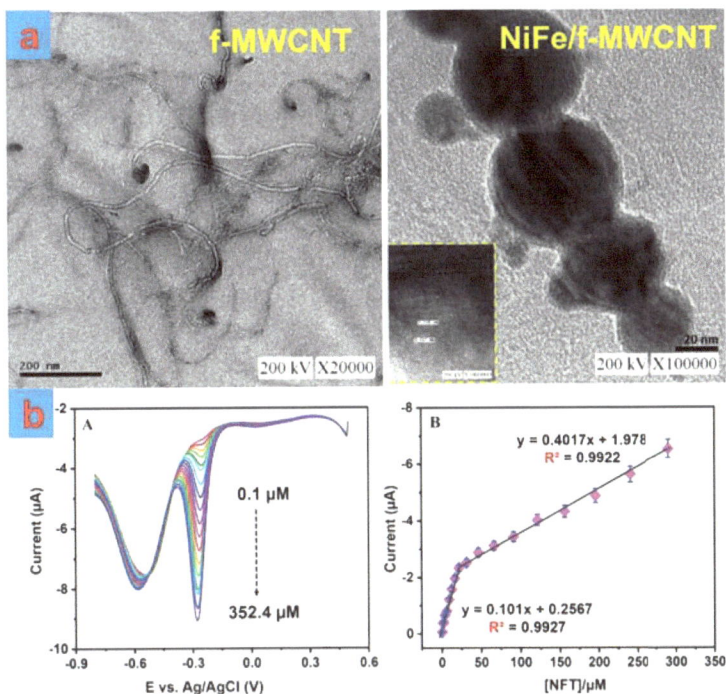

Fig. (4). (a) HRTEM images of NiFe/f-MWCNT and **(b)** their DPV response towards nitrofurantoin for the detection of LOD [27].

BIMETALLIC/CARBON QDS BASED COMPOSITES FOR ELECTROCHEMICAL SENSORS

As discussed above, the optoelectronic properties of CQDs can lead to the development of efficient electrochemical sensors. Typically, N-doped GQDs are synthesized by infrared assisted pyrolysis method on ITO substrates and were used for the sensing of Hg^+ ions. These N/GQDs have a particle size of 4.5 nm and exhibit a lower LOD of Hg^+ ions about 10 ppb with a buildup time of 32 s. Therefore the introduction of electron-rich heteroatoms of nitrogen (N) into the GQDs induces higher catalytic activity and also their p-type (or) n-type conductivity [28]. Hence, the heteroatoms (B, S, F, and P) doping into the system of GQDs lead to the improvement of higher quantum yield, optical and electronic properties. Similarly, fluorescent probes were developed for the sensitive and selective detection of quercetin using S doped GQDs. The development of GQDs with the functionalization of monometallic and bimetallic NPs exhibits outstanding properties and leads to the development of the sensors of improved version [29]. An effective approach was established for the "on-off" photoelectrochemical platforms for the detection of caffeic acid using QDs and Au/NGQDs prepared by hydrothermal and calcination methods. In this work, NGQDs exhibited unique quantum effects and biocompatibility and AuNPs lead to the enhancement of the sensitivity of the sensors due to their plasmonic functionalities and their catalytic properties [30]. Another effective approach of molecularly imprinted polymer-based (MIP) electrochemical sensor was developed for the detection of DA and chlorpromazine by N-GOQDs supported the noble metal atom and metal sulfide of Au/NiS_2 [31]. Moreover, amperometric immunosensors are usually the most preferred for the sensing of biomolecules, owing to their sensitivity and versatility. The general behavior of signal generated by GQDs towards the electrochemical measurements [32] is shown in Fig. 5(**a-e**).

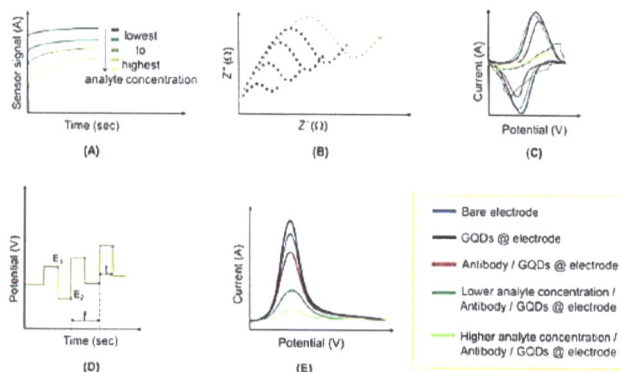

Fig. (5). (a-e) The general behavior of signal generated by GQDs towards electrochemical sensors [32].

A group developed an electrochemical sensor for the detection of BPA, by immobilizing the aptamer on the GCE which was modified by N, S doped GQDs. In this research, mainly AuNPs were decorated for the enhancement of the sensitivity of the sensor by improving the specific surface area of the materials. Under the optimal conditions, the developed sensor shows lower LOD and recovery values about 0.03 µM and 82.6 to 107%, respectively. The density of states and their sensing mechanisms were also investigated and are shown in Fig. (**6 a-b**). BPA can be precisely recognized and taken by the probe, which creates a closed canal of the probe, and hence additional interference ions cannot be sensed. BPA can be detected from the variation in the output signals [32].

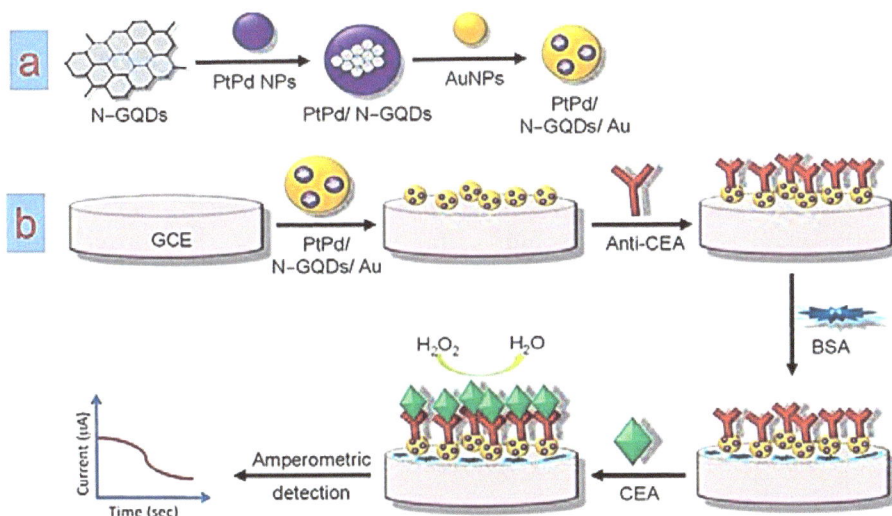

Fig. (**6**). **a)** Synthesis of PtPd/N-GQDs **b)** Sensor set-up for the detection of CEA immunosensors [32].

Table 1. Various Bimetallic-carbon nanocomposites for the various electrochemical sensors.

S. No	Bimetallic-carbon Nanocomposites	Method of Synthesis	Analyte	LOD	Ref
1	Cu–Ag	Hydrothermal process	glucose	0.08 µM	[33]
2	Pd-Ag	Fungal extracted aqueous method	Uric acid	5.54 nM	[34]
3	Au-Pd /reduced graphene oxide	Electrochemical co-reduction	Hydroquinone, catechol, resorcinol	0.5 µM 0.6 µM 0.8 µM	[35]
4	Pd-Au/PEDOT/ graphene	One-pot approach	Hydrazine	0.37 nM	[36]
5	Pd-Ag/rGO	Modified hummer's and reduction method	Hydrogen peroxidase	0.7 µM	[13]
6	Ni-CO/MWCNT	Chemical vapor deposition	glucose	1.2 µM	[37]

(Table 1) cont.....

S. No	Bimetallic-carbon Nanocomposites	Method of Synthesis	Analyte	LOD	Ref
7	Pt-Au/MWCNT	Electrochemical synthesis	Nitride	0.09 µM	[12]
8	Ag-N/graphene QDs	Hydrothermal and calcination	Caffeic acid	0.03 µM	[30]
9	Au/N, S-GQDs	hydrothermal route	Bisphenol A	0.03 µM	[38]

BIMETALLIC/G-C$_3$N$_4$ BASED COMPOSITE FOR ELECTROCHEMICAL SENSORS

A quick and effective electrochemical sensor was developed for the detection of vanillin. In this work, lamellar g-C$_3$N$_4$ sheets were synthesized by simple pyrolysis of melamine. The g-C$_3$N$_4$ modified GCE exhibited excellent sensing performance. Due to the enhanced condition, the prepared electrochemical sensor exhibited a low detection limit of 4 nM [39]. It is an effective approach to the use of carbon-metallic nanostructures, because of their unique features such as fast electron transfer, excellent biocompatibility, relatively wide potential window, and feasibility for surface functionalization. Owing to this, a group developed a successful sensor for non-enzymatic detection of H$_2$O$_2$ by the doping of metallic ions (Fe^{2+}, CO^{2+}, Ni^{2+}, or Cu^{2+}) into the g-C$_3$N$_4$ [40]. Another novel non-enzymatic electrochemical glucose sensor was fabricated by Pt/ZnO modified g-C$_3$N$_4$. This sensor shows a wide linear range of 0.25–110 mM with a lower limit of detection of 0.1 mM. This Pt/ZnO modified g-C$_3$N$_4$shows the four times reusability and to prepare a strip in a cost-effective manner [41].

CONCLUDING REMARKS AND CHALLENGES

In conclusion, the field of bimetallic carbon nanocomposites based electro-chemical sensors is developing rapidly with the creation of bimetallic-carbon materials with new kind of structures. Carbon materials and bimetallic NPs have numerous advantages for developing electrochemical sensors, including rapid electron transfer kinetics and lower limit of detection, sensitivity, and selectivity. However, many new materials are still being developed; it will be significant to compare the nanomaterials with one another to consider which ones are the most beneficial for the fabrication of electrochemical sensors. However, controlled synthesis of bimetallic-carbon nanocomposites with appropriate structures is still tough to realize by currently developed synthesis methods. Therefore, the synthesis of bimetallic-carbon nanocomposites with required structural and morphological characteristics in a highly controllable manner is still challenged. The alternative challenge is the development of electrochemical sensors with stability, durability and reproducibility for better potential applications.

CONSENT FOR PUBLICATION

Not applicable.

CONFLICT OF INTEREST

The author declares no conflict of interest, financial or otherwise.

ACKNOWLEDGEMENTS

The authors thank the DST-FIST, DST-PURSE and UGC-SAP, Government of India, for financial support to build the instrumental facilities.

REFERENCES

[1] Wang, Y.; Xu, H.; Zhang, J.; Li, G. Electrochemical sensors for clinic analysis. *Sensors (Basel),* **2008**, *8*(4), 2043-2081.
[http://dx.doi.org/10.3390/s8042043] [PMID: 27879810]

[2] Huy, G.D.; Jin, N.; Yin, B.C.; Ye, B.C. A novel separation and enrichment method of 17β-estradiol using aptamer-anchored microbeads. *Bioprocess Biosyst. Eng.,* **2011**, *34*(2), 189-195.
[http://dx.doi.org/10.1007/s00449-010-0460-4] [PMID: 20734205]

[3] Blanco, E.; Hristova, L.; Martínez-Moro, R.; Vázquez, L.; Ellis, G.J.; Sánchez, L.; del Pozo, M.; Petit-Domínguez, M.D.; Casero, E.; Quintana, C. A 2D tungsten disulphide/diamond nanoparticles hybrid for an electrochemical sensor development towards the simultaneous determination of sunset yellow and quinoline yellow. *Sens. Actuators B Chem.,* **2020**, *324*(August), 128731.
[http://dx.doi.org/10.1016/j.snb.2020.128731]

[4] Mo, F.; Xie, J.; Wu, T.; Liu, M.; Zhang, Y.; Yao, S. A sensitive electrochemical sensor for bisphenol A on the basis of the AuPd incorporated carboxylic multi-walled carbon nanotubes. *Food Chem.,* **2019**, *292*, 253-259.
[http://dx.doi.org/10.1016/j.foodchem.2019.04.034] [PMID: 31054673]

[5] Xiao, T.; Huang, J.; Wang, D.; Meng, T.; Yang, X. Au and Au-Based nanomaterials: Synthesis and recent progress in electrochemical sensor applications. *Talanta,* **2020**, *206*, 120210.
[http://dx.doi.org/10.1016/j.talanta.2019.120210] [PMID: 31514855]

[6] Power, A., B.; Gorey, S.; Chandra, JC. Carbon nanomaterials and their application to electrochemical sensors. **2011**.

[7] Yang, C.; Denno, M.E.; Pyakurel, P.; Venton, B.J. Recent trends in carbon nanomaterial-based electrochemical sensors for biomolecules: A review. *Anal. Chim. Acta,* **2015**, *887*, 17-37.
[http://dx.doi.org/10.1016/j.aca.2015.05.049] [PMID: 26320782]

[8] Naqvi, S.T.R.; Shirinfar, B.; Hussain, D.; Majeed, S.; Ashiq, M.N.; Aslam, Y.; Ahmed, N. Electrochemical Sensing of Ascorbic Acid, Hydrogen Peroxide and Glucose by Bimetallic (Fe, Ni)–CNTs Composite Modified Electrode. *Electroanalysis,* **2019**, *31*(5), 851-857.
[http://dx.doi.org/10.1002/elan.201800768]

[9] Dou, N.; Zhang, S.; Qu, J. Simultaneous detection of acetaminophen and 4-aminophenol with an electrochemical sensor based on silver-palladium bimetal nanoparticles and reduced graphene oxide. *RSC Advances,* **2019**, *9*(54), 31440-31446.
[http://dx.doi.org/10.1039/C9RA05987C]

[10] Zhang, C.; Li, F.; Huang, S.; Li, M.; Guo, T.; Mo, C.; Pang, X.; Chen, L.; Li, X. *In-situ* facile preparation of highly efficient copper/nickel bimetallic nanocatalyst on chemically grafted carbon nanotubes for nonenzymatic sensing of glucose. *J. Colloid Interface Sci.,* **2019**, *557*, 825-836.

[http://dx.doi.org/10.1016/j.jcis.2019.09.076] [PMID: 31580978]

[11]   Rick, J.; Tsai, M.C.; Hwang, B.J. Biosensors incorporating bimetallic nanoparticles. *Nanomaterials (Basel)*, **2015**, *6*(1), 1-30.
       [http://dx.doi.org/10.3390/nano6010005] [PMID: 28344262]

[12]   Yang, C.Y.; Chen, S.M.; Palanisamy, S.; Thirumalraj, B.; Liu, X. Electrochemical synthesis of PtAu bimetallic nanoparticles on multiwalled carbon nanotubes and application for amperometric determination of nitrite. *Int. J. Electrochem. Sci.*, **2016**, *11*(5), 4027-4036.
       [http://dx.doi.org/10.20964/110452]

[13]   Guler, M.; Turkoglu, V.; Bulut, A.; Zahmakiran, M. Electrochemical sensing of hydrogen peroxide using Pd@Ag bimetallic nanoparticles decorated functionalized reduced graphene oxide. *Electrochim. Acta*, **2018**, *263*, 118-126.
       [http://dx.doi.org/10.1016/j.electacta.2018.01.048]

[14]   Vashist, S.K.; Zheng, D.; Al-Rubeaan, K.; Luong, J.H.T.; Sheu, F.S. Advances in carbon nanotube based electrochemical sensors for bioanalytical applications. *Biotechnol. Adv.*, **2011**, *29*(2), 169-188.
       [http://dx.doi.org/10.1016/j.biotechadv.2010.10.002] [PMID: 21034805]

[15]   Tan, C.; Cao, X.; Wu, X.J.; He, Q.; Yang, J.; Zhang, X.; Chen, J.; Zhao, W.; Han, S.; Nam, G.H.; Sindoro, M.; Zhang, H. Recent Advances in Ultrathin Two-Dimensional Nanomaterials. *Chem. Rev.*, **2017**, *117*(9), 6225-6331.
       [http://dx.doi.org/10.1021/acs.chemrev.6b00558] [PMID: 28306244]

[16]   Song, Y; Qiao, J; Li, W; Ma, C; Chen, S; Li, H Bimetallic PtCu nanoparticles supported on molybdenum disulfide–functionalized graphitic carbon nitride for the detection of carcinoembryonic antigen. *Microchim Acta*, **2020**, *187*(9)
       [http://dx.doi.org/10.1007/s00604-020-04498-y]

[17]   Jiang, B.; Huang, A.; Wang, T.; Shao, Q.; Zhu, W.; Liao, F.; Cheng, Y.; Shao, M. Rhodium/graphitic-carbon-nitride composite electrocatalyst facilitates efficient hydrogen evolution in acidic and alkaline electrolytes. *J. Colloid Interface Sci.*, **2020**, *571*, 30-37.
       [http://dx.doi.org/10.1016/j.jcis.2020.03.022] [PMID: 32179306]

[18]   Arora, N.; Thangavelu, K.; Karanikolos, G.N. Bimetallic Nanoparticles for Antimicrobial Applications. *Front Chem.*, **2020**, *8*, 412.
       [http://dx.doi.org/10.3389/fchem.2020.00412] [PMID: 32671014]

[19]   Liu, Y.; She, P.; Gong, J.; Wu, W.; Xu, S.; Li, J.; Zhao, K.; Deng, A. A novel sensor based on electrodeposited Au-Pt bimetallic nano-clusters decorated on graphene oxide (GO)-electrochemically reduced GO for sensitive detection of dopamine and uric acid. *Sens. Actuators B Chem.*, **2015**, *221*, 221.
       [http://dx.doi.org/10.1016/j.snb.2015.07.086]

[20]   Zhao, Z.; Zhang, M.; Chen, X.; Li, Y.; Wang, J. Electrochemical co-reduction synthesis of AuPt bimetallic nanoparticles-graphene nanocomposites for selective detection of dopamine in the presence of ascorbic acid and uric acid. *Sensors (Basel)*, **2015**, *15*(7), 16614-16631.
       [http://dx.doi.org/10.3390/s150716614] [PMID: 26184200]

[21]   Shen, Y.; Sheng, Q.; Zheng, J. A high-performance electrochemical dopamine sensor based on a platinum-nickel bimetallic decorated poly(dopamine)-functionalized reduced graphene oxide nanocomposite. *Anal. Methods*, **2017**, *9*(31), 4566-4573.
       [http://dx.doi.org/10.1039/C7AY00717E]

[22]   Demirkan, B.; Bozkurt, S.; Şavk, A.; Cellat, K.; Gülbağca, F.; Nas, M.S.; Alma, M.H.; Sen, F. Composites of Bimetallic Platinum-Cobalt Alloy Nanoparticles and Reduced Graphene Oxide for Electrochemical Determination of Ascorbic Acid, Dopamine, and Uric Acid. *Sci. Rep.*, **2019**, *9*(1), 12258.
       [http://dx.doi.org/10.1038/s41598-019-48802-0] [PMID: 31439896]

[23]   Mahmoudi, E.; Hajian, A.; Rezaei, M.; Afkhami, A.; Amine, A.; Bagheri, H. A novel platform based

on graphene nanoribbons/protein capped Au-Cu bimetallic nanoclusters: Application to the sensitive electrochemical determination of bisphenol A. *Microchem. J., 2019, 145*, 242-251.
[http://dx.doi.org/10.1016/j.microc.2018.10.044]

[24]   Li, L.; Zheng, H.; Guo, L.; Qu, L.; Yu, L. A sensitive and selective molecularly imprinted electrochemical sensor based on Pd-Cu bimetallic alloy functionalized graphene for detection of amaranth in soft drink. *Talanta,* **2019**, *197*, 68-76.
[http://dx.doi.org/10.1016/j.talanta.2019.01.009] [PMID: 30771990]

[25]   Chen, K.J.; Chandrasekara Pillai, K.; Rick, J.; Pan, C.J.; Wang, S.H.; Liu, C.C.; Hwang, B.J. Bimetallic PtM (M=Pd, Ir) nanoparticle decorated multi-walled carbon nanotube enzyme-free, mediator-less amperometric sensor for H_2O_2. *Biosens. Bioelectron.,* **2012**, *33*(1), 120-127.
[http://dx.doi.org/10.1016/j.bios.2011.12.037] [PMID: 22236778]

[26]   Balasubramanian, P.; He, S.; Jansirani, A.; Peng, H-P.; Huang, L-L.; Deng, H-H.; Chen, W. Bimetallic AgAu decorated MWCNTs enable robust nonenzyme electrochemical sensors for *in-situ* quantification of dopamine and H2O2 biomarkers expelled from PC-12 cells. *J. Electroanal. Chem. (Lausanne), 2020, 878,* 114554.
[http://dx.doi.org/10.1016/j.jelechem.2020.114554]

[27]   Hwa, K.Y.; Sharma, T.S.K. Nano assembly of NiFe spheres anchored on f-MWCNT for electrocatalytic reduction and sensing of nitrofurantoin in biological samples. *Sci. Rep., 2020, 10*(1), 12256.
[http://dx.doi.org/10.1038/s41598-020-69125-5] [PMID: 32704113]

[28]   Fu, C.C.; Te Hsieh, C.; Juang, R.S.; Gu, S.; Ashraf Gandomi, Y.; Kelly, R.E. Electrochemical sensing of mercury ions in electrolyte solutions by nitrogen-doped graphene quantum dot electrodes at ultralow concentrations. *J. Mol. Liq., 2020, 302,* 112593.
[http://dx.doi.org/10.1016/j.molliq.2020.112593]

[29]   Kadian, S.; Manik, G. Sulfur doped graphene quantum dots as a potential sensitive fluorescent probe for the detection of quercetin. *Food Chem., 2020, 317,* 126457.
[http://dx.doi.org/10.1016/j.foodchem.2020.126457] [PMID: 32106009]

[30]   Zhao, Q.; Zhou, L.; Li, X.; He, J.; Huang, W.; Cai, Y.; Wang, J.; Chen, T.; Du, Y.; Yao, Y. Au–nitrogen-doped graphene quantum dot composites as "on–off" nanosensors for sensitive photo-electrochemical detection of caffeic acid. *Nanomaterials (Basel), 2020, 10*(10), 1-12.
[http://dx.doi.org/10.3390/nano10101972] [PMID: 33027974]

[31]   Lu, Z.; Li, Y.; Liu, T.; Wang, G.; Sun, M.; Jiang, Y.; He, H.; Wang, Y.; Zou, P.; Wang, X.; Zhao, Q.; Rao, H. A dual-template imprinted polymer electrochemical sensor based on AuNPs and nitrogen-doped graphene oxide quantum dots coated on NiS2/biomass carbon for simultaneous determination of dopamine and chlorpromazine. *Chem. Eng. J., 2020, 389,* 124417.
[http://dx.doi.org/10.1016/j.cej.2020.124417]

[32]   Mansuriya, B.D.; Altintas, Z. Graphene Quantum Dot-Based Electrochemical Immunosensors for Biomedical Applications. *Materials (Basel), 2019, 13*(1), E96.
[http://dx.doi.org/10.3390/ma13010096] [PMID: 31878102]

[33]   Li, H.; Guo, C.Y.; Xu, C.L. A highly sensitive non-enzymatic glucose sensor based on bimetallic Cu-Ag superstructures. *Biosens. Bioelectron.,* **2015**, *63*, 339-346.
[http://dx.doi.org/10.1016/j.bios.2014.07.061] [PMID: 25113052]

[34]   Mallikarjuna, K.; Veera Manohara Reddy, Y.; Sravani, B.; Madhavi, G.; Kim, H.; Agarwal, S.; Gupta, V.K. Simple synthesis of biogenic Pd–Ag bimetallic nanostructures for an ultra-sensitive electrochemical sensor for sensitive determination of uric acid. *J. Electroanal. Chem. (Lausanne),* **2018**, *822*, 163-170.
[http://dx.doi.org/10.1016/j.jelechem.2018.05.019]

[35]   Chen, Y.; Liu, X.; Zhang, S.; Yang, L.; Liu, M.; Zhang, Y.; Yao, S. Ultrasensitive and simultaneous detection of hydroquinone, catechol and resorcinol based on the electrochemical co-reduction prepared

Au-Pd nanoflower/reduced graphene oxide nanocomposite. *Electrochim. Acta,* **2017**, *231*, 677-685.
[http://dx.doi.org/10.1016/j.electacta.2017.02.060]

[36]   Liu, Z.; Lu, B.; Gao, Y.; Yang, T.; Yue, R.; Xu, J.; Gao, L. Facile one-pot preparation of Pd-Au/PEDOT/graphene nanocomposites and their high electrochemical sensing performance for caffeic acid detection. *RSC Advances,* **2016**, *6*(92), 89157-89166.
[http://dx.doi.org/10.1039/C6RA16488A]

[37]   Ramachandran, K.; Raj Kumar, T.; Babu, K.J.; Gnana Kumar, G. Ni-Co bimetal nanowires filled multiwalled carbon nanotubes for the highly sensitive and selective non-enzymatic glucose sensor applications. *Sci. Rep.,* **2016**, *6*(1), 36583.
[http://dx.doi.org/10.1038/srep36583] [PMID: 27833123]

[38]   Yao, J.; Li, Y.; Xie, M.; Yang, Q.; Liu, T. The electrochemical behaviors and kinetics of AuNPs/N, S-GQDs composite electrode: A novel label-free amplified BPA aptasensor with extreme sensitivity and selectivity. *J. Mol. Liq.,* **2020**, *320*, 114384.
[http://dx.doi.org/10.1016/j.molliq.2020.114384]

[39]   Fu, L.; Xie, K.; Wu, D.; Wang, A.; Zhang, H.; Ji, Z. Electrochemical determination of vanillin in food samples by using pyrolyzed graphitic carbon nitride. *Mater. Chem. Phys.,* **2020**, *242*, 242.
[http://dx.doi.org/10.1016/j.matchemphys.2019.122462]

[40]   Sheng, Z.M.; Gan, Z.Z.; Huang, H.; Niu, R.L.; Han, Z.W.; Jia, R.P. M-Nx (M = Fe, Co, Ni, Cu) doped graphitic nanocages with High specific surface Area for non-enzymatic electrochemical detection of H2O2. *Sens. Actuators B Chem.,* **2020**, *305*, 127550.
[http://dx.doi.org/10.1016/j.snb.2019.127550]

[41]   Imran, H.; Vaishali, K.; Antony Francy, S.; Manikandan, P.N.; Dharuman, V. Platinum and zinc oxide modified carbon nitride electrode as non-enzymatic highly selective and reusable electrochemical diabetic sensor in human blood. *Bioelectrochemistry,* **2021**, *137*, 107645.
[http://dx.doi.org/10.1016/j.bioelechem.2020.107645] [PMID: 32916428]

Current and Future Developments, 2022, Vol. 2, 45-61   **45**

CHAPTER 4

Two-dimensional Graphene-based Nanocomposites for Electrochemical Sensor

J. Debbarma[1] and **M. Saha**[1,*]

¹ Department of Chemistry, National Institute of Technology Agartala, 799046, Tripura, India

Abstract: This chapter reviews the usage of graphene nanostructures in the fabrication of electrochemical sensors. In current ages, graphene derivatives have attracted a great deal of attention due to their outstanding electrical, mechanical, and thermal properties, making them one of the most popular choices to develop the electrodes of a sensor. In addition, the high effective surface area, electrocatalytic activity, excellent electrical conductivity, high porosity and adsorption capability, fascinating their electrochemical properties, which turn them as potential candidates for electrochemical applications, particularly sensing. This chapter deals with an overview of the work done on graphene-based nanocomposites in very recent years. It explains the properties of graphene-based nanocomposites for their usage as electrochemical sensors.

Keywords: Electrochemical sensor, Graphene, Nanocomposite.

INTRODUCTION

The remarkable properties of carbon family members, including fullerenes, carbon nanotubes and graphene, have stimulated a broad range of researches in various applications. All the carbon nanomaterials, especially graphene is the latest family member, have attracted considerable attention, because of their unique structure and properties. Moreover, the large surface area, high electrical and electrochemical activities make them suitable candidates for developing energy-related devices (supercapacitors and batteries) as well as sensors. Electrochemical sensors have proved to be powerful analytical tools which have great potential to improve both selectivity and sensitivity through tuned signal amplifications. This technique also possesses outstanding properties such as low cost, fast response time, instrumental simplicity, the possibility of miniaturization, and integration in portable devices.

---

* **Corresponding Author Mitali Saha:** Department of Chemistry, National Institute of Technology Agartala, 799046, Tripura, India; E-mail: mitalichem71@gmail.com

Manorama Singh, Vijai K Rai and Ankita Rai (Eds.)

In addition, another distinguishing feature of electrochemical sensors is their capability of detecting a wide range of compounds from neutral molecules, organic, inorganic, ionic and metal ions. The combination of these fascinating properties with graphene and its derivatives has been emerged in powerful electrochemical sensing platforms based on so called "graphene modified electrodes" which are being used for the electroanalysis of a large range of analyte species. During the past decade, the modified electrodes based on graphene were applied for the determination of various compounds and have been extensively applied as the modifier electrodes because of their chemical stability, antifouling resistance as well as large adsorption ability towards various analytes. This chapter represents an overview of the current state of researches on the application of electrochemical sensors with various electrode modification approaches using graphene. The electrochemical sensors based on graphene and its composites for the determination of diverse target analytes such as drugs, pharmaceutical formulations, biological species, neurotransmitters, biomarkers and heavy metals, in very recent years have been discussed in detail.

ELECTROCHEMICAL SENSING OF VARIOUS ANALYTES USING GRAPHENE BASED NANOCOMPOSITES

In recent years, graphene and its derivatives have provided a most promising platform for the sensing of various electrochemical analytes. More recently, FGP/AuNC nanocomposite was prepared *via* ultrasonic method based on the fact that AuNC facilitated to reduce the agglomeration of graphene materials and also improves the catalytic activity of nanocomposite as the synergistic effect of the AuNC and derived graphene materials enhances their physical and chemical properties, thus providing a novel technology for real-time monitoring of heavy metal pollution in food [1]. The electrochemical detection of amikacin sulphate was examined by utilizing a nanocomposite of silver nanoparticles and reduced graphene oxide (AgNPs/rGO) on a nickel foam (NiF) electrode. The practical viability of the developed sensor was evaluated in spiked human urine samples under validated conditions and the recovery rate was found to be in the range of 99%–102%, which indicates the sensor's effectiveness in the detection of amikacin sulphate [2]. Fe_2O_3 NPs is a promising catalyst for rapid electron transfer between the electrode and the active site of the redox system. Hence, a one-pot hydrothermal method was carried out for the green synthesis of ɤ-Fe_2O_3 (~40 nm) decorated with N-rGO hydrogel which exhibits electrochemical detection performance of 4-Nitrophenol under neutral pH [3]. It is well known that; metal or metal oxide nanoparticles exhibit potential catalytic performances for sensing various analytes. Among various metal/metal oxide nanoparticles, ZnO acts as a promising candidate for the electrochemical detection of bio analytes due to its significant features such as large loading capacity, easy fabrication, biocompatibility, highly physicochemical stability and electrocatalytic activity. Owing to these properties, (rGO)-g-CN/ZnO nanocomposites were developed for the electrochemical analysis of sunset yellow in real samples [4]. Conductive polymers possess unique physical and chemical properties, which enhance their electrocatalytic activities towards dopamine and ascorbic acid [5]. Polymers such as polyaniline-co-thionine based GO/P(ANI-co-THI) composite modified electrode was successfully prepared *via* a simple one-step electrochemical deposition method. A nanocomposite consisting of polyaniline-GO-iron tungsten nitride was developed and then applied for accurate detection of 4-NP in the aqueous samples through a real-time and repeatable approach, making it a fantastic alternative for the development of precise sensors and electronic devices [6]. A sensor was fabricated based on a modified screen-printed carbon electrode (SPCE) for Rohypnol measurements. The surface of SPCE was modified with amine-functionalized graphene oxide sheets reinforced through Cu nanoparticles (SPCE/AGO-Cu). It provides an accurate and precise assay pathway of Rohypnol in the fruit juice [7]. Biochar and reduced graphene oxide (rGO) were prepared as a modifier electrode for the construction of an electrochemical sensor for sensing carbendazim (CBZ, methyl benzimidazol-2-yl carbamate) in different samples. The modified electrode showed to be able to preconcentrate CBZ and presented the highest analytical response in comparison to the unmodified electrode [8]. In 2020, graphene oxide based $CaTiO_3NPs$ nanocomposite was prepared by an *in-situ* sonochemical process, which applied towards electrocatalytic and electrochemical sensing of chemotherapeutic drug flutamide. This bimetal ($CaTiO_3$) enhances the electrocatalytic performance towards detection of flutamide drug [9]. A simple modified glassy carbon electrode (PVP-GR/GCE) based on the polyvinylpyrrolidone (PVP)-graphene film was also reported in the same year. This modified electrode was used for the determination of dopamine, uric acid, and ascorbic acid in real samples. It was established that PVP was allowed to make the carbon atoms less prone to aggregation in graphene nanosheets. Hence, the dispersion and stability of PVP-GR nanocomposite improved compared with pure GR [10]. An electrochemical sensor was developed with a metal-organic gel (MOG) consisting of $Fe_3Cl(H_2O)_2O$ clusters interconnected with 5-aminoisophthalic acid (NH_2-ip) in the presence of isophthalate functionalized graphene (IG) for sensing of dopamine. Here, IG acted as a structure directing templating agent which also enhanced conductivity of the material [11]. Electrochemical sensor based on rGO-Ag@SiO_2 nanocomposite was established to analysis etidronic acid (EA) using the di erential pulse voltammetric (DPV) technique. The real-time application of the rGO-Ag@SiO_2/Au PCB for EA detection was investigated using EA-based pharmaceutical samples. Recovery percentages were found to be 96.2% to 102.9% [12]. It was found that, copper has high electrochemical activity, low resistance and excellent catalytic performance compare to other metals. Therefore copper-aluminum metal hydroxide doped graphene nanoprobe electrode (labeled as CuAl–LDH/Gr NC) was developed for detection of trace glyphosate, where Cu-based nanomaterials have showed a high affinity for glyphosate [13]. Ag nanoparticles are recommended for the alloying with Pt due to their electrochemical stability, biocompatibility, abundance, low cost and sustainable electrocatalytic activity which could lead to an enhancement in the electrocatalytic activity of the bimetallic nanoparticles. Thus, a nanocomposite of platinum–silver bimetallic nanoparticles supported on graphene (Pt–Ag/Gr) was synthesized and electrochemical performance of the Pt–Ag/Gr/GCE towards the detection of dopamine (DA) was analysed using voltammetric techniques [14]. It was established that when metal oxide nanoparticles can enter into the inner regions of the GO sheets which enhances the surface area of GO as well as promote the electron transfer towards active sites. Metal oxide semiconductors materials have become very useful for the development of sensors due to high sensitivity, good stability, smaller size and simplicity of integration. The development of a highly sensitive and reproducible electrochemical catechol sensor based on nickel molybdate graphene oxide ($NiMoO_4$/GO) nanocomposite was fabricated by ultrasonication method for monitoring of catechol [15]. Due to biocompatibility and low-cost advantages, TiO_2 is highly preferred for the development of biosensors or electrochemical sensors. MoS_2 has satisfactory long-term durability in biosensing studies due to its stability in laboratory conditions. Therefore, a MoS_2-TiO_2/rGO/SPE composite was designed on screen printed electrode (SPE) for selective and sensitive detection of paracetamol (acetaminophen, Ac) [16]. Nanocomposite of functionalized graphene oxide decorated tin oxide nanoparticles (APTMS-GO@SnO_2) was fabricated *via* refluxing route for the determination of eugenol. Here, GO has been modified by SnO_2 nanoparticles to increase the rate of electron transfer and conductivity. SnO_2 is n-type of semiconductor and has been well loaded into the 2D graphene structure to increase the higher charge carrier mobility [17]. Polyaniline emeraldine salt (PANI-ES) and its composite with graphene are evaluated for sensing ability towards chemical warfare agents like hydrogen cyanide and cyanogen chloride [18]. Binary metal oxides (La_2O_3/SnO_2) decorated with reduced graphene oxide (rGO) nanocomposite was prepared by ultrasound process in an environmentally benign solvent. In contrast to other modified and unmodified electrodes, this modified La_2O_3/SnO_2/rGO nanocomposite exhibits better electrochemical activity towards the oxidation of methyl nicotinate with a higher anodic current [19]. Generally, it was found that, gold nanoparticles (AuNPs) are used to increase the sensitivity of electrochemical sensors as it supplies great conductivity and catalytic properties. On the other hand, rGO is a suitable material for decorating electrochemical sensors because of its excellent electron transfer capability and large surface area. Recently, rGO/AuNPs nanocomposite was also developed for the direct analysis of L-tryptophan (Try). In this fabricated electrode, E. tereticornis leaves was used as an environmentally friendly reducing agent for the development of electrode [20]. Bi_2S_3 and Bi_2S_3/rGO nanocomposite were employed for the sensing of hexanitrodiphenylamine using microwave assisted one-pot tandem reaction. The electrode results show better electrochemical behaviour of Bi_2S_3/rGO nanocomposite as compared to Bi_2S_3 alone [21]. It has been found that, poly (L-Cys) can provide plentiful active sites for binding analytes, show high sensitivity and selectivity, tunable conductivity and enriched electrocatalytic activity as electrode material. Therefore, a high potential electrochemical sensor was fabricated by modification of the GCE with TiO_2/CuO-N doped rGO and a layer of poly (L-Cys) was used for trace analysis of flunitrazepam [22]. The modified glassy carbon electrode (GCE) coated with methylcellulose/graphene oxide/iron oxide nano hydrogel (MC-GO-Fe_3O_4) was designed for analysis of uric acid content in urine. The electrode exhibits a diffusion-controlled process with a two-electron, two-proton transfer reaction [23]. Layered nanocomposite of zinc sulfide nanospheres (ZnS NPs) encapsulated on reduced graphene oxide (rGO) hybrid was fabricated by an ultrasonic bath for the electrochemical determination of caffeic acid (CA) in various food samples. When compared to other modified electrodes, this hybrid modified electrode (GCE) exhibits an excellent electrocatalytic performance towards caffeic acid determination. This is due to the fact that ZnS has a different crystal structure; zinc blende and wurtzite which influence excellent electrochemical properties in sensing [24]. Due to high electron transport and electrochemical activity, ZnO was used to develop graphene (GR) and ZnO nanorods (GR-ZnO/GCE) based sensor for the detection of sulfamethoxazole (SMX) and trimethoprim (TMP), simultaneously [25]. An electrochemically prepared graphene oxide modified pencil graphite electrode (EGO-PGE) have been reported for the electrochemical determination of sunset yellow. Cyclic and differential pulse voltammetry measurements of developed modified electrode was investigated at pH 9 using phosphate buffer solution [26]. It was investigated that Pt-based modified graphene oxide (GO) not only maximizes the availability of nano-sized electro-active surface area for electron transfer but also cause better mass transport of reactants to the electro-catalyst [27]. The Pt nanoparticles/over-oxidized poly-pyrrole nanofiber/reduced graphene oxide (PtNPs/OPPy/RGO) nanocomposite was electrochemically synthesized and used for analyzing the electrocatalytic activities toward the simultaneous determination of dopamine and 5-Hydroxy tryptamine. It has been well established that graphene/metal oxide nanocomposites have shown more attractive properties instead of single graphene or nano metal oxide. Based on the excellent electrochemical properties of nanoneedles (NN)-CuO/N-rGO as well as the specific complexation of natural substance L-cysteine with Pb2+, the L-Cys/NN-CuO/N-rGO, was designed for the determination of trace Pb^{2+} at pH 5.0 [28]. Another graphene and CuO nanoparticles have been fabricated for simultaneous determination of two anti-cancer drugs, raloxifene and tamoxifen [29]. The MIP was decorated on the surface of zeolitic imidazolate framework (ZIF-8) and reduced graphene oxide (rGO) composite modified with glassy carbon electrode (GCE) to fabricate GCE/rGO/ZIF-8/MIP electrode. The sensor exhibits high recognition ability for rutin in the optimal conditions due to the synergetic effects of excellent electrical conductivity of rGO, high surface area of ZIF-8 and high recognition accuracy of imprinted approach [30]. A hybrid binary composite made of 2D hexagonal boron nitride nanosheets doped graphene film *via* a scalable top-down technique for the electrochemical detection of nicotine has been established. The structure of boron nitride is similar to graphene as it exhibited outstanding physical, chemical, mechanical and thermal properties towards nicotine determination [31]. A nonenzymatic graphene stabilized nickel sulphide with modified glassy carbon (NiS/GO/MGCE) electrode was fabricated for electrochemical detection of urea. Here, NiS was used as electron distribution on the compound, which helps in selective interaction with analytes [32]. The electrochemical oxidation of NADH was investigated on chemically produced reduced graphene oxide (CRGO) modified glassy carbon electrode (GCE). The electro-oxidation is found to be an irreversible process and the voltammograms exhibited low potential for NADH oxidation on CRGO/GCE compared to GO/GCE [33]. One label-free biosensor, reduced graphene oxide/Au nanoparticles (RGO/AuNPs) has been designed for the detection of glycated albumin. The sensor was modified by anti-GA aptamer, where aptasensor regeneration takes place more easily. Therefore, aptamers can be developed against a wide range of targets, including non-immunogenic or toxic components or even ions [34]. It is well known that the sensing performances can be improved by the increased electrochemical active sites along with the fast electron and mass transport, *via* microstructure engineering of the nanomaterials. 3D nanoporous PtSi (NP-PtSi) alloy was fabricated using dealloying ternary PtSiAl alloy ribbons. A new composite material was developed by combining the nanoporous morphology of PtSi and graphene (GR), which was applied to modify the surface of a glassy carbon electrode (GCE). The prepared modified electrodes showed an excellent electrocatalytic activity towards the electrooxidation of bisphenol A [35]. An electrochemical sensor determination of dopamine (DA) has been developed using the polymerized L-tryptophan (p-L-Trp) and graphene (GN) composite as the electrode. p-L-Trp is an essential amino acid for the biosynthesis of proteins, which is an aromatic heterocyclic nonpolar α-amino acid, plays an important role in the quality and weight of the human body [36]. It is well known that, noble metal NPs and mixed metal oxide have been studied as electrocatalysts for the reduction of 4-NP. A sensor was developed based on templated silver nanoparticles (AgNPs) on reduced graphene oxide (rGO) nanosheets (Ag-rGO) for the detection of 4-Nitrophenol (referred to as 4-NP) and also utilized as an electrocatalyst [37]. Another, Ag nanoparticles-based polydopamine-reduced graphene oxide modified glassy carbon electrode (AgNPs-PDA-GR/GCE), was developed for recognition of 2,4,6-trichlorophenol [38]. Determination of uric acid was carried out by enzyme biosensor based on screen-printed electrode, which was developed modified chitosan-graphene composite cryogel coated with Prussian blue. Amperometric detection of UA catalyzed by uricase was based on the change in cathodic current of PB at a potential of 0.00 V in a flow injection system and developed biosensor achieved recoveries ranging from 98±2 to 102±5% [39]. Development of an accurate detection system over a wide range of Crt concentrations in both blood and urine is medically essential. Hence, a nonenzymatic electrochemical probe, carbon dots doped tungstic anhydride embedded on graphene oxide nanopanels (CDs/WO$_3$@GO), have been decorated for picomolar-level Crt detection in blood and urine with a wide linear range (0.2-112.0 nM) [40]. A chemical treatment involving the sequential DMF/HNO$_3$/NaBH$_4$ reagents to fabrication rGO within 3D-printed PLA electrodes has been reported. The fabricated rGO-PLA electrode was applied for serotonin determination in synthetic urine using DPV with notable sensing properties. Moreover, the electrode surface was used to fabricate a catechol biosensor, which was capable of determining the analytes in natural water samples with acceptable analytical features [41]. In combination with graphene oxide nanosheets using a crossing membrane metal electrode sensor system was established at low volume (<5μL) for detection of lactate in appropriate ranges of perspired human sweat. The ability of the developed sensor was done by measuring impedance changes associated with lactate binding to the lactate oxidase at the GO nanosheets interface using electrochemical impedance spectroscopy [42].

ELECTROCHEMICAL SENSING OF GLUCOSE AND HYDROGEN PEROXIDE USING GRAPHENE-BASED NANOCOMPOSITES

Accurate and real time determination of glucose level in blood sugar is of great significance to the treatment of diabetes. Therefore, a variety of electrochemical sensors for glucose determination have been developed, which are mainly divided into two categories: enzymatic and nonenzymatic. Three-dimensional porous graphene aerogel and glucose oxidase (GOx) based on enzymatic electrochemical microfluidic biosensor was developed for glucose detection. The porous structure of this fabricated 3D graphene aerogel greatly increased the specific surface area, which resulted in more GOx that could be modified on graphene, and thus, the sensitivity of the sensor increased [43]. A flexible electrochemical sensor has been designed with poly (3, 4-ethylene dioxythiophene)-poly (styrene sulfonate)

(PEDOT: PSS) modified 3D stable porous laser-induced graphene (LIG) for the detection of glucose and pH. PEDOT/PSS was spray-coated on the LIG to increase electrode robustness and deliver uniform electrical conductivity [44]. A tapered single-mode fiber (SMF) structure coated with GO/AuNPs was proposed for the detection of glucose. The GO/AuNPs of the electrode increase the biocompatibility of the sensor [45]. Titanium dioxide, carboxylic functionalized graphene, and cross-linked with APTES nanocomposite were decorated for glucose determination. The silanization of titanium dioxide nanoparticles of this electrode and an increase in electron shuttle was proven feasible, when the composite was able to achieve about 30% higher current than non-silanized material [46]. ZnO nanowires (NWs) are known to be highly sensitive to glucose due to their strong electron transfer and adsorption, as well as stability and high surface area. Therefore, the sensor was modified by incorporation of graphene nanoplates (GNPs) beneath the ZnO NWs for glucose sensing with high proficiency [47]. Chitosan-glucose oxidase (GOx) nanocomposite was successfully immobilized onto the LIG/PtNPs electrode to fabricate a sweat glucose biosensor. The as-prepared LIG/PtNPs electrode exhibited a high sensitivity since porous laser-induced graphene (LIG) is an attractive and promising carbon material for electrochemical applications, thus it can immobilize various proteins, such as enzymes, antibodies, and receptors [48]. An integrated flexible and reusable graphene-based field-effect transistor (GFET) nanosensor was fabricated for the detection of glucose using pyrene-1-boronic acid (PBA) as the receptor. The graphene surface functionalized with PBA binds with glucose to induce a change in the carrier concentration in the graphene, which is used to analysis of glucose concentration from human bodily fluids (*e.g.*, sweat) [49]. Another biosensor AuNDs-rGO/Nafion/GCE have been developed for glucose sensing. Since nafion, a perfluoro sulfonated polymer, bears a polar side chain, can act as a cation exchanger with a preference for hydrophobic cations; hence it was used in the fabrication of electrodes [50]. A nonenzymatic sensor was constructed with Cu_2O/CuO hollow microspheres (CuxOHM) and reduced graphene oxide (rGO) nanosheets to form CuxOHM/RGO hybrid by a simple one-step hydrothermal method for glucose sensor application. In comparison with pristine CuxOHM and rGO, the prepared hybrid glucose sensor showed significantly enhanced electrocatalytic activity [51]. A well designed $ZnFe_2O_4/\alpha\text{-}Fe_2O_3$/ZFGr nanocomposite was fabricated by exploiting the electrocatalytic property of $ZnFe_2O_4/\alpha\text{-}Fe_2O_3$ and the electrical conductivity of graphene. This nanocomposite, contributes to the electrochemical glucose sensing property due to the reduction in the probable agglomerations of $ZnFe_2O_4/\alpha\text{-}Fe_2O_3$ magnetic nanoparticles and physiochemical properties of graphene [52]. Conifer-like copper-carboxylated graphene electrode modified with gold has been developed for glucose sensors. The presence of carboxylated graphene (G-COOH) on the electrode can improve the dispersion of graphene in water and maintain good conductivity [53]. A nonenzymatic sensor based on NiO-N-doped carbon/reduced graphene oxide (NiO-NC/rGO) material was prepared for the detection of glucose. Here, rGO/NC exhibits good electronic transmission performance towards glucose detection. NC has the advantage of simplicity and diversity, and the unique property of rGO *i.e.*, two-dimensional structure and single-atom thickness were able to make high-performance electrode material for nonenzymatic analysis of glucose [54]. On account of multiple valence state, the different metal center can give metal-organic framework (MOF) with good electrochemical. Different metal-coordinated two-dimensional (2D) MOF with electrochemical exfoliated graphene (EG) was prepared successfully by a simple and economical method for the nonenzymatic glucose sensor [55]. Cobalt oxides designed MoS_2/rGO electrode was fabricated using one-pot hydrothermal method for determination of glucose. Screen printed sensor exhibited good conductivity and high catalytic activity towards glucose sensing since first-row transition metals *i.e.*, Co, which achieves satisfactory electrochemical performance by increasing the specific surface area and providing many highly accessible active sites on their surfaces [56]. PANI is a phenyl-based polymer having chemically flexible –NH group in polymer chain flanked either side by phenylene ring and has gained much more attention in recent years. It showed various properties such as protonation, deprotonation and other physiochemical properties. The $PANI_S$ based rGO sensor was fabricated and it exhibited better sensitivity, selectivity and low detection limit towards an enzyme-free glucose detection [57]. The FET sensor using silver nanoparticles (AgNPs) and graphene has been fabricated for the detection of various glucose concentration levels. The behavioral characteristics of FET sensor towards glucose detection were found to offer a potential device for further real-time clinical applications [58]. Reagent less electrochemical sensor based on 3D graphene was prepared for glucose detection. Prussian blue, a well-known electrochemical indicator, was co-deposited with AuNPs on 3DG electrode by a simple one-step electrodeposition method to increase the catalytical activity towards glucose detection [59]. Well-dispersed poly (cysteine)-$Ni(OH)_2$ nanocomposites on graphene modified electrodes have also been reported for the determination of glucose. The interconversion of $Ni(OH)_2$/NiOOH redox couple on GCE/GN/Pcys-$Ni(OH)_2$ electrode showed good electrochemical responses towards glucose detection under strong alkaline conditions [60]. Silver nanoparticles decorated polyaniline (PANI) with reduced graphene oxide (rGO) as Ag-PANI/rGO nanocomposites have been designed *via* simple electrochemical method at room temperature. As compared to bare PANI and PANI/rGO, the fabricated nanocomposite exhibited excellent electrochemical characteristics and sensing performances towards glucose detection [61]. Transition metal nanoparticles supported on 3D-GR electrodes were fabricated using a universal laser-assistant growth method. With the porous nature and catalytic property of metal nanoparticles, metal nanoparticle/3D graphene shows high direct oxidation activity towards the glucose in alkaline solution [62]. An abnormal H_2O_2 concentration in the body could lead to some diseases; hence health monitoring, sensitive and accurate analysis of H_2O_2 in the body is of great importance. For the developed electrochemical sensors, the MnO_2-GNSs composites were synthesized by a facile one-step hydrothermal method and applied for the determination of H_2O_2. In comparison with bare GCE, GNSs/GCE and c-MnO_2/GNSs/GCE, as well as MnO_2-GNSs/GCE, showed higher electrocatalytic activities toward H_2O_2 owing to the synergic effects of the catalytic activity of MnO_2 and the electrical conductivity of GNS [63]. Crumpled graphene structures electrode designed by manganese ferrite ($MnFe_2O_4$) nanoparticles was established for two different applications; as an electrochemical sensor for hydrogen peroxide and electrochemical supercapacitors. The samples were produced using an aerosol-assisted capillary compression process, and the crumpled paper ball-like shape of the nanocomposites was adjusted to incorporate increasing amounts of manganese ferrite nanoparticles [64]. Real-time monitoring of H_2O_2 released from cells was detected by electrochemical sensor based on reduced graphene oxide supported dumbbell shaped $CuCo_2O_4$, which was fabricated *via* a facile hydrothermal procedure. The 3D dumbbell-shaped structure consisting of 2D nanosheets layers with large surface area and high conductivity can enhance the electrocatalytic performance of the composites [65]. Compared with extensively-used chemically-exfoliated graphene, physically-exfoliated graphene nanosheets (GN) were used as the substrate material, coupled with high-activity FeOOH nanoparticles for rapid electrochemical sensing of H_2O_2. Relative to pristine GN in the prepared electrode, the electrochemical active area and electron transfer kinetics are greatly increased with the introduction of high-activity of FeOOH nanoparticles [66]. The fabrication of a copper-based metal-organic framework (Cu-MOF) incorporated with graphene oxide (Cu-BT--MOF/GO) (BTC=1,3,5-benzene-tricarboxylate and GO=graphene oxide) *via* the reflux method have been reported towards the detection of hydrogen peroxide in milk. In the comparison of Cu-MOF/ERGO/ITO electrode-based H_2O_2 detection with other, this MOF-based electrochemical sensor exhibited its superior performance in terms of sensitivity, the limit of detection, and linear range [67]. An effective strategy was developed for constructing a robust enzymatic bio-sensing platform based on one-step electrodeposition of 0D enzyme-loaded bio-polymeric nanoparticles (NPs) with 2D graphene oxide nanosheets. The fabricated enzymatic sensor exhibited excellent sensing performance towards H_2O_2 detection, which is a synergistic result from both the design of materials and the hierarchically 0D-2D nanostructure [68]. Bi-functional amperometric sensor based on GOD/Pt-LEPG was decorated for sensitive electrochemical assay of glucose and H_2O_2 at near-neutral pH conditions. This unique structure of LEPG was functionalized with Pt NPs by CV electrodeposition to form Pt-LEPG, which exhibited good conductivity and excellent sensitivity for H_2O_2 sensing. Therefore, it was immobilized glucose oxidase (GOD) and effectively electro-catalyze glucose oxidation in near-neutral pH conditions [69].

CONCLUDING REMARKS

In current ages, electrochemical sensors play a leading role towards the determination of various analytes species. The development of highly sensitive electrochemical sensors is still greatly in demand. This chapter focuses on the recent progress on graphene-based nanocomposites towards sensing various analyte species such as biological species, biomarkers, neurotransmitters, heavy metals, drugs and pharmaceutical formulations. It also deals with the design, fabrication and development of different electrodes with functionalized graphene, heteroatom incorporated graphene and graphene-based nanocomposites for the electrochemical sensor.

CONSENT FOR PUBLICATION

Not applicable.

CONFLICT OF INTEREST

The author declares no conflict of interest, financial or otherwise.

ACKNOWLEDGEMENTS

The authors are grateful to the Director of NIT Agartala.

REFERENCES

[1]   Tan, Z.; Wu, W.; Feng, C.; Wu, H.; Zhang, Z. Simultaneous determination of heavy metals by an electrochemical method based on a nanocomposite consisting of fluorinated graphene and gold nanocage. *Mikrochim. Acta,* **2020**, *187*(7), 414.
[http://dx.doi.org/10.1007/s00604-020-04393-6] [PMID: 32602018]

[2]   Sharma, N.; Selvam, S.P.; Yun, K. Electrochemical detection of amikacin sulphate using reduced graphene oxide and silver nanoparticles nanocomposite. *Appl. Surf. Sci.,* **2020**, *512*, 145742.
[http://dx.doi.org/10.1016/j.apsusc.2020.145742]

[3]   Ramu, A.G.; Salla, S.; Gopi, S.; Silambarasan, P.; Yang, D.J.; Song, M.J.; Ali, H.M.; Salem, M.Z.M.; Choi, D. Surface-tuned hierarchical ɤ-Fe₂O₃-N-rGO nanohydrogel for efficient catalytic removal and electrochemical sensing of toxic nitro compounds. *Chemosphere,* **2021**, *268*, 128853.
[http://dx.doi.org/10.1016/j.chemosphere.2020.128853] [PMID: 33187664]

[4]   Ezhil Vilian, A.T.; Kang, S.M.; Yeong Oh, S.; Woo Oh, C.; Umapathi, R.; Suk Huh, Y.; Han, Y.K. A simple strategy for the synthesis of flower-like textures of Au-ZnO anchored carbon nanocomposite towards the high-performance electrochemical sensing of sunset yellow. *Food Chem.,* **2020**, *323*, 126848.

[http://dx.doi.org/10.1016/j.foodchem.2020.126848] [PMID: 32330645]

[5]   Song, N.N.; Wang, Y.Z.; Yang, X.Y.; Zong, H.L.; Chen, Y.X.; Ma, Z.; Chen, C.X. A novel electrochemical biosensor for the determination of dopamine and ascorbic acid based on graphene oxide/poly (aniline-co-thionine) nanocomposite. *J. Electroanal. Chem. (Lausanne)*, **2020**, *873*, 114352.
[http://dx.doi.org/10.1016/j.jelechem.2020.114352]

[6]   Hashemi, S.A.; Mousavi, S.M.; Bahrani, S.; Ramakrishna, S. Integrated polyaniline with graphene oxide-iron tungsten nitride nanoflakes as ultrasensitive electrochemical sensor for precise detection of 4-nitrophenol within aquatic media. *J. Electroanal. Chem. (Lausanne)*, **2020**, *873*, 114406.
[http://dx.doi.org/10.1016/j.jelechem.2020.114406]

[7]   Mohammadnia, M.S.; Naghian, E.; Ghalkhani, M.; Nosratzehi, F.; Adib, K.; Zahedi, M.M.; Nasrabadi, M.R.; Ahmadi, F. Fabrication of a new electrochemical sensor based on screen-printed carbon electrode/amine-functionalized graphene oxide-Cu nanoparticles for Rohypnol direct determination in drink sample. *J. Electroanal. Chem. (Lausanne)*, **2021**, *880*, 114764.
[http://dx.doi.org/10.1016/j.jelechem.2020.114764]

[8]   Sant'Anna, M.V.S.; Carvalho, S.W.M.M.; Gevaerd, A.; Silva, J.O.S.; Santos, E.; Carregosa, I.S.C.; Wisniewski, A., Jr; Marcolino-Junior, L.H.; Bergamini, M.F.; Sussuchi, E.M. Electrochemical sensor based on biochar and reduced graphene oxide nanocomposite for carbendazim determination. *Talanta*, **2020**, *220*, 121334.
[http://dx.doi.org/10.1016/j.talanta.2020.121334] [PMID: 32928384]

[9]   Tseng, T.W.; Rajaji, U.; Chen, T.W.; Chen, S.M.; Huang, Y.C.; Mani, V.; Irudaya Jothi, A. Sonochemical synthesis and fabrication of perovskite type calcium titanate interfacial nanostructure supported on graphene oxide sheets as a highly efficient electrocatalyst for electrochemical detection of chemotherapeutic drug. *Ultrason. Sonochem.*, **2020**, *69*, 105242.
[http://dx.doi.org/10.1016/j.ultsonch.2020.105242] [PMID: 32673961]

[10]   Wu, Y.; Deng, P.; Tian, Y.; Feng, J.; Xiao, J.; Li, J.; Liu, J.; Li, G.; He, Q. Simultaneous and sensitive determination of ascorbic acid, dopamine and uric acid *via* an electrochemical sensor based on PVP-graphene composite. *J. Nanobiotechnology*, **2020**, *18*(1), 112.
[http://dx.doi.org/10.1186/s12951-020-00672-9] [PMID: 32778119]

[11]   Vermisoglou, E.C.; Jakubec, P.; Malina, O.; Kupka, V.; Schneemann, A.; Fischer, R.A.; Zbořil, R.; Jayaramulu, K.; Otyepka, M. Hierarchical Porous Graphene-Iron Carbide Hybrid Derived From Functionalized Graphene-Based Metal-Organic Gel as Efficient Electrochemical Dopamine Sensor. *Front Chem.*, **2020**, *8*, 544.
[http://dx.doi.org/10.3389/fchem.2020.00544] [PMID: 32850616]

[12]   Panneer Selvam, S.; Chinnadayyala, S.R.; Cho, S.; Yun, K. Differential pulse voltammetric electrochemical sensor for the detection of etidronic acid in pharmaceutical samples by using rGO-Ag@ SiO$_2$/Au PCB. *Nanomaterials (Basel)*, **2020**, *10*(7), 1368.
[http://dx.doi.org/10.3390/nano10071368] [PMID: 32674260]

[13]   Zhang, C.; Liang, X.; Lu, Y.; Li, H.; Xu, X. Performance of CuAl-LDH/Gr Nanocomposite-Based Electrochemical Sensor with Regard to Trace Glyphosate Detection in Water. *Sensors (Basel)*, **2020**, *20*(15), 4146.
[http://dx.doi.org/10.3390/s20154146] [PMID: 32722519]

[14]   Anuar, N.S.; Basirun, W.J.; Shalauddin, M.; Akhter, S. A dopamine electrochemical sensor based on a platinum–silver graphene nanocomposite modified electrode. *RSC Advances*, **2020**, *10*(29), 17336-17344.
[http://dx.doi.org/10.1039/C9RA11056A]

[15]   Boopathy, G.; Keerthi, M.; Chen, S.M.; Umapathy, M.J.; Kumar, B.N. Highly porous nickel molybdate@ graphene oxide nanocomposite for the ultrasensitive electrochemical detection of environmental toxic pollutant catechol. *Mater. Chem. Phys.*, **2020**, *239*, 121982.
[http://dx.doi.org/10.1016/j.matchemphys.2019.121982]

[16]   Demir, N.; Atacan, K.; Ozmen, M.; Bas, S.Z. Design of a new electrochemical sensing system based on MoS₂–TiO₂/reduced graphene oxide nanocomposite for the detection of paracetamol. *New J. Chem.,* **2020**, *44*(27), 11759-11767.
[http://dx.doi.org/10.1039/D0NJ02298E]

[17]   Fadillah, G.; Wicaksono, W.P.; Fatimah, I.; Saleh, T.A. A sensitive electrochemical sensor based on functionalized graphene oxide/SnO2 for the determination of eugenol. *Microchem. J.,* **2020**, *159*, 105353.
[http://dx.doi.org/10.1016/j.microc.2020.105353]

[18]   Farooqi, B.A.; Yar, M.; Ashraf, A.; Farooq, U.; Ayub, K. Graphene-polyaniline composite as superior electrochemical sensor for detection of cyano explosives. *Eur. Polym. J.,* **2020**, *138*, 109981.
[http://dx.doi.org/10.1016/j.eurpolymj.2020.109981]

[19]   Chen, T.W.; Arumugam, R.; Chen, S.M.; Altaf, M.; Manohardas, S.; Saeed Ali Abuhasil, M.; Ajmal Ali, M. Ultrasonic preparation and nanosheets supported binary metal oxide nanocomposite for the effective application towards the electrochemical sensor. *Ultrason. Sonochem.,* **2020**, *64*, 105007.
[http://dx.doi.org/10.1016/j.ultsonch.2020.105007] [PMID: 32092696]

[20]   Nazarpour, S.; Hajian, R.; Sabzvari, M.H. A novel nanocomposite electrochemical sensor based on green synthesis of reduced graphene oxide/gold nanoparticles modified screen printed electrode for determination of tryptophan using response surface methodology approach. *Microchem. J.,* **2020**, *154*, 104634.
[http://dx.doi.org/10.1016/j.microc.2020.104634]

[21]   Kumar, K.S.; Giribabu, K.; Suresh, R.; Manigandan, R.; Kumar, S.P.; Narayanan, V. Bismuth sulphide/reduced graphene oxide nanocomposites as an electrochemical sensing platform for hexanitrodiphenylamine. *Mater. Lett.,* **2021**, *283*, 128804.
[http://dx.doi.org/10.1016/j.matlet.2020.128804]

[22]   Sohouli, E.; Ghalkhani, M.; Rostami, M.; Rahimi-Nasrabadi, M.; Ahmadi, F. A noble electrochemical sensor based on TiO₂@CuO-N-rGO and poly (L-cysteine) nanocomposite applicable for trace analysis of flunitrazepam. *Mater. Sci. Eng. C,* **2020**, *117*, 111300.
[http://dx.doi.org/10.1016/j.msec.2020.111300] [PMID: 32919661]

[23]   Sohouli, E.; Khosrowshahi, E.M.; Radi, P.; Naghian, E.; Rahimi-Nasrabadi, M.; Ahmadi, F. Electrochemical sensor based on modified methylcellulose by graphene oxide and Fe₃O₄ nanoparticles: Application in the analysis of uric acid content in urine. *J. Electroanal. Chem. (Lausanne),* **2020**, *877*, 114503.
[http://dx.doi.org/10.1016/j.jelechem.2020.114503]

[24]   Vinoth, S.; Govindasamy, M.; Wang, S.F.; Anandaraj, S. Layered nanocomposite of zinc sulfide covered reduced graphene oxide and their implications for electrocatalytic applications. *Ultrason. Sonochem.,* **2020**, *64*, 105036.
[http://dx.doi.org/10.1016/j.ultsonch.2020.105036] [PMID: 32146333]

[25]   Yue, X.; Li, Z.; Zhao, S. A new electrochemical sensor for simultaneous detection of sulfamethoxazole and trimethoprim antibiotics based on graphene and ZnO nanorods modified glassy carbon electrode. *Microchem. J.,* **2020**, *159*, 105440.
[http://dx.doi.org/10.1016/j.microc.2020.105440]

[26]   Tahtaisleyen, S.; Gorduk, O.; Sahin, Y. Electrochemical determination of sunset yellow using an electrochemically prepared graphene oxide modified–pencil graphite electrode (EGO-PGE). *Anal. Lett.,* **2020**, 1-23.

[27]   Ghanbari, K.; Bonyadi, S. An electrochemical sensor based on Pt nanoparticles decorated over-oxidized polypyrrole/reduced graphene oxide nanocomposite for simultaneous determination of two neurotransmitters dopamine and 5-Hydroxy tryptamine in the presence of ascorbic acid. *IJPAC Int. J. Polym. Anal. Charact.,* **2020**, *25*(3), 105-125.
[http://dx.doi.org/10.1080/1023666X.2020.1766785]

[28]   Yang, S.; Liu, P.; Wang, Y.; Guo, Z.; Tan, R.; Qu, L. Electrochemical sensor using poly-(l-cysteine) functionalized CuO nanoneedles/N-doped reduced graphene oxide for detection of lead ions. *RSC Advances,* **2020**, *10*(31), 18526-18532.
[http://dx.doi.org/10.1039/D0RA03149F]

[29]   Fouladgar, M.; Karimi-Maleh, H.; Opoku, F.; Govender, P.P. Electrochemical anticancer drug sensor for determination of raloxifene in the presence of tamoxifen using graphene-CuO-polypyrrole nanocomposite structure modified pencil graphite electrode: Theoretical and experimental investigation. *J. Mol. Liq.,* **2020**, *311*, 113314.
[http://dx.doi.org/10.1016/j.molliq.2020.113314]

[30]   El Jaouhari, A.; Yan, L.; Zhu, J.; Zhao, D.; Zaved Hossain Khan, M.; Liu, X. Enhanced molecular imprinted electrochemical sensor based on zeolitic imidazolate framework/reduced graphene oxide for highly recognition of rutin. *Anal. Chim. Acta,* **2020**, *1106*, 103-114.
[http://dx.doi.org/10.1016/j.aca.2020.01.039] [PMID: 32145838]

[31]   Jerome, R.; Sundramoorthy, A.K. Preparation of hexagonal boron nitride doped graphene film modified sensor for selective electrochemical detection of nicotine in tobacco sample. *Anal. Chim. Acta,* **2020**, *1132*, 110-120.
[http://dx.doi.org/10.1016/j.aca.2020.07.060] [PMID: 32980101]

[32]   Naik, T.S.K.; Saravanan, S.; Saravana, K.S.; Pratiush, U.; Ramamurthy, P.C. A nonenzymatic urea sensor based on the nickel sulfide/graphene oxide modified glassy carbon electrode. *Mater. Chem. Phys.,* **2020**, *245*, 122798.
[http://dx.doi.org/10.1016/j.matchemphys.2020.122798]

[33]   Immanuel, S.; Sivasubramanian, R. Electrochemical studies of NADH oxidation on chemically reduced graphene oxide nanosheets modified glassy carbon electrode. *Mater. Chem. Phys.,* **2020**, *249*, 123015.
[http://dx.doi.org/10.1016/j.matchemphys.2020.123015]

[34]   Farzadfard, A.; Shayeh, J.S.; Habibi-Rezaei, M.; Omidi, M. Modification of reduced graphene/Au-aptamer to develop an electrochemical based aptasensor for measurement of glycated albumin. *Talanta,* **2020**, *211*, 120722.
[http://dx.doi.org/10.1016/j.talanta.2020.120722] [PMID: 32070572]

[35]   Zhang, S.; Shi, Y.; Wang, J.; Xiao, L.; Yang, X.; Cui, R.; Han, Z. Nanocomposites consisting of nanoporous platinum-silicon and graphene for electrochemical determination of bisphenol A. *Mikrochim. Acta,* **2020**, *187*(4), 241.
[http://dx.doi.org/10.1007/s00604-020-4219-6] [PMID: 32206895]

[36]   Gong, Q.J.; Han, H.X.; Wang, Y.D.; Yao, C.Z.; Yang, H.Y.; Qiao, J.L. An electrochemical sensor for dopamine detection based on the electrode of a poly-tryptophan-functionalized graphene composite. *N. Carbon Mater.,* **2020**, *35*(1), 34-41.
[http://dx.doi.org/10.1016/S1872-5805(20)60473-5]

[37]   Ahmad, N.; Al-Fatesh, A.S.; Wahab, R.; Alam, M.; Fakeeha, A.H. Synthesis of silver nanoparticles decorated on reduced graphene oxide nanosheets and their electrochemical sensing towards hazardous 4-nitrophenol. *J. Mater. Sci. Mater. Electron.,* **2020**, *31*(14), 11927-11937.
[http://dx.doi.org/10.1007/s10854-020-03747-3]

[38]   Wang, L.; Liu, Y.; Yang, R.; Li, J.; Qu, L. AgNPs–PDA–GR nanocomposites-based molecularly imprinted electrochemical sensor for highly recognition of 2, 4, 6-trichlorophenol. *Microchem. J.,* **2020**, *159*, 105567.
[http://dx.doi.org/10.1016/j.microc.2020.105567]

[39]   Jirakunakorn, R.; Khumngern, S.; Choosang, J.; Thavarungkul, P.; Kanatharana, P.; Numnuam, A. Uric acid enzyme biosensor based on a screen-printed electrode coated with Prussian blue and modified with chitosan-graphene composite cryogel. *Microchem. J.,* **2020**, *154*, 104624.
[http://dx.doi.org/10.1016/j.microc.2020.104624]

[40]   Ponnaiah, S.K.; Prakash, P. Carbon dots doped tungstic anhydride on graphene oxide nanopanels: A new picomolar-range creatinine selective enzymeless electrochemical sensor. *Mater. Sci. Eng. C,* **2020**, *113*, 111010.
[http://dx.doi.org/10.1016/j.msec.2020.111010] [PMID: 32487413]

[41]   Silva, V.A.O.P.; Fernandes-Junior, W.S.; Rocha, D.P.; Stefano, J.S.; Munoz, R.A.A.; Bonacin, J.A.; Janegitz, B.C. 3D-printed reduced graphene oxide/polylactic acid electrodes: A new prototyped platform for sensing and biosensing applications. *Biosens. Bioelectron.,* **2020**, *170*, 112684.
[http://dx.doi.org/10.1016/j.bios.2020.112684] [PMID: 33049481]

[42]   Lin, K.C.; Muthukumar, S.; Prasad, S. Flex-GO (Flexible graphene oxide) sensor for electrochemical monitoring lactate in low-volume passive perspired human sweat. *Talanta,* **2020**, *214*, 120810.
[http://dx.doi.org/10.1016/j.talanta.2020.120810] [PMID: 32278429]

[43]   Xu, J.; Xu, K.; Han, Y.; Wang, D.; Li, X.; Hu, T.; Yi, H.; Ni, Z. A 3D porous graphene aerogel@GOx based microfluidic biosensor for electrochemical glucose detection. *Analyst (Lond.),* **2020**, *145*(15), 5141-5147.
[http://dx.doi.org/10.1039/D0AN00681E] [PMID: 32573601]

[44]   Zahed, M.A.; Barman, S.C.; Das, P.S.; Sharifuzzaman, M.; Yoon, H.S.; Yoon, S.H.; Park, J.Y. Highly flexible and conductive poly (3, 4-ethylene dioxythiophene)-poly (styrene sulfonate) anchored 3-dimensional porous graphene network-based electrochemical biosensor for glucose and pH detection in human perspiration. *Biosens. Bioelectron.,* **2020**, *160*, 112220.
[http://dx.doi.org/10.1016/j.bios.2020.112220] [PMID: 32339151]

[45]   Yang, Q.; Zhu, G.; Singh, L.; Wang, Y.; Singh, R.; Zhang, B.; Zhang, X.; Kumar, S. Highly sensitive and selective sensor probe using glucose oxidase/gold nanoparticles/graphene oxide functionalized tapered optical fiber structure for detection of glucose. *Optik (Stuttg.),* **2020**, *208*, 164536.
[http://dx.doi.org/10.1016/j.ijleo.2020.164536]

[46]   Ognjanović, M.; Stanković, V.; Knežević, S.; Antić, B.; Vranješ-Djurić, S.; Stanković, D.M. TiO_2/APTES cross-linked to carboxylic graphene based impedimetric glucose biosensor. *Microchem. J.,* **2020**, *158*, 105150.
[http://dx.doi.org/10.1016/j.microc.2020.105150]

[47]   Rafiee, Z.; Mosahebfard, A.; Sheikhi, M.H. High-performance ZnO nanowires-based glucose biosensor modified by graphene nanoplates. *Mater. Sci. Semicond. Process.,* **2020**, *115*, 105116.
[http://dx.doi.org/10.1016/j.mssp.2020.105116]

[48]   Yoon, H.; Nah, J.; Kim, H.; Ko, S.; Sharifuzzaman, M.; Barman, S.C.; Xuan, X.; Kim, J.; Park, J.Y. A chemically modified laser-induced porous graphene based flexible and ultrasensitive electrochemical biosensor for sweat glucose detection. *Sens. Actuators B Chem.,* **2020**, *311*, 127866.
[http://dx.doi.org/10.1016/j.snb.2020.127866]

[49]   Huang, C.; Hao, Z.; Qi, T.; Pan, Y.; Zhao, X. An integrated flexible and reusable graphene field effect transistor nanosensor for monitoring glucose. *J. Materiomics.,* **2020**, *6*(2), 308-314.
[http://dx.doi.org/10.1016/j.jmat.2020.02.002]

[50]   Mei, L.; Zhang, Q.; Du, M.; Zeng, Z. Electrochemical biosensing platforms on the basis of reduced graphene oxide and its composites with Au nanodots. *Analyst (Lond.),* **2020**, *145*(10), 3749-3756.
[http://dx.doi.org/10.1039/C9AN02592H] [PMID: 32319461]

[51]   Zhang, F.; Huang, S.; Guo, Q.; Zhang, H.; Li, H.; Wang, Y.; Fu, J.; Wu, X.; Xu, L.; Wang, M. One-step hydrothermal synthesis of Cu_2O/CuO hollow microspheres/reduced graphene oxide hybrid with enhanced sensitivity for nonenzymatic glucose sensing. *Colloids Surf. A Physicochem. Eng. Asp.,* **2020**, *602*, 125076.
[http://dx.doi.org/10.1016/j.colsurfa.2020.125076]

[52]   Neravathu, D.; Paloly, A.R.; Sajan, P.; Satheesh, M.; Bushiri, M.J. Hybrid nanomaterial of $ZnFe_2O_4$/α-Fe_2O_3 implanted graphene for electrochemical glucose sensing application. *Diamond Related Materials,* **2020**, *106*, 107852.

[http://dx.doi.org/10.1016/j.diamond.2020.107852]

[53]   Wu, S.; Zhang, Y.; Liu, L.; Fan, W. Conifer-like copper-carboxylated graphene nanocomposites modified electrode for sensitive nonenzymatic glucose biosensing with very low limit of detection. *Mater. Lett.,* **2020**, *276*, 128253.
[http://dx.doi.org/10.1016/j.matlet.2020.128253]

[54]   Zhang, Y.; Liu, Y.Q.; Bai, Y.; Chu, W.Sh.J.; Sh, J. Confinement preparation of hierarchical NiO--doped carbon@ reduced graphene oxide microspheres for high-performance nonenzymatic detection of glucose. *Sens. Actuators B Chem.,* **2020**, *309*, 127779.
[http://dx.doi.org/10.1016/j.snb.2020.127779]

[55]   Liu, B.; Wang, X.; Liu, H.; Zhai, Y.; Li, L.; Wen, H. 2D MOF with electrochemical exfoliated graphene for nonenzymatic glucose sensing: Central metal sites and oxidation potentials. *Anal. Chim. Acta,* **2020**, *1122*, 9-19.
[http://dx.doi.org/10.1016/j.aca.2020.04.075] [PMID: 32503748]

[56]   Li, X.; Zhang, M.; Hu, Y.; Xu, J.; Sun, D.; Hu, T.; Ni, Z. Screen-printed electrochemical biosensor based on a ternary Co@ MoS$_2$/rGO functionalized electrode for high-performance nonenzymatic glucose sensing. *Biomed. Microdevices,* **2020**, *22*(1), 1-8.
[http://dx.doi.org/10.1007/s10544-020-0472-z]

[57]   Kailasa, S.; Reddy, R.K.K.; Reddy, M.S.B.; Rani, B.G.; Maseed, H.; Sathyavathi, R.; Rao, K.V. High sensitive polyaniline nanosheets (PANINS)@ rGO as nonenzymatic glucose sensor. *J. Mater. Sci. Mater. Electron.,* **2020**, *31*(4), 2926-2937.
[http://dx.doi.org/10.1007/s10854-019-02837-1]

[58]   Archana, R.; Sreeja, B.S.; Nagarajan, K.K.; Radha, S.; BalajiBhargav, P.; Balaji, C.; Padmalaya, G. BalajiBhargav, P.; Balaji, C.; Padmalaya, G.; Development of highly sensitive Ag NPs decorated graphene FET sensor for detection of glucose concentration. *J. Inorg. Organomet. Polym. Mater.,* **2020**, *30*(9), 3818-3825.
[http://dx.doi.org/10.1007/s10904-020-01541-6]

[59]   Liu, Q.; Zhong, H.; Chen, M.; Zhao, C.; Liu, Y.; Xi, F.; Luo, T. Functional nanostructure-loaded three-dimensional graphene foam as a nonenzymatic electrochemical sensor for reagentless glucose detection. *RSC Advances,* **2020**, *10*(56), 33739-33746.
[http://dx.doi.org/10.1039/D0RA05553K]

[60]   Xue, Y.; Tian, B.; Wang, M.; Zhai, T.; Li, R.; Tan, L. Well-dispersed poly (cysteine)-Ni(OH)$_2$ nanocomposites on graphene-modified electrode surface for highly sensitive nonenzymatic glucose detection. *Colloids Surf. A Physicochem. Eng. Asp.,* **2020**, *591*, 124549.
[http://dx.doi.org/10.1016/j.colsurfa.2020.124549]

[61]   Deshmukh, M.A.; Kang, B.C.; Ha, T.J. Nonenzymatic electrochemical glucose sensors based on polyaniline/reduced-graphene-oxide nanocomposites functionalized with silver nanoparticles. *J. Mater. Chem. C Mater. Opt. Electron. Devices,* **2020**, *8*(15), 5112-5123.
[http://dx.doi.org/10.1039/C9TC06836H]

[62]   Wuyun, X.; Guang, Y. Mingqi. X.; Xiangjie, B. A universal laser-assistant growth of transition metal nanoparticle on flexible graphene electrode for nonenzymatic glucose sensor. *New J. Chem.,* **2020**, *44*, 17954-17960.
[http://dx.doi.org/10.1039/D0NJ04200E]

[63]   Guan, J.F.; Huang, Z.N.; Zou, J.; Jiang, X.Y.; Peng, D.M.; Yu, J.G. A sensitive non-enzymatic electrochemical sensor based on acicular manganese dioxide modified graphene nanosheets composite for hydrogen peroxide detection. *Ecotoxicol. Environ. Saf.,* **2020**, *190*, 110123.
[http://dx.doi.org/10.1016/j.ecoenv.2019.110123] [PMID: 31891837]

[64]   Nonaka, L.H.; Almeida, T.S.; Aquino, C.B.; Domingues, S.H.V.; Salvatierra, R.; Souza, V.H. Crumpled graphene decorated with manganese ferrite nanoparticles for hydrogen peroxide sensing and electrochemical supercapacitors. *ACS Appl. Nano Mater.,* **2020**, *3*, 4859-4869.

[http://dx.doi.org/10.1021/acsanm.0c01012]

[65]   Jiang, L.; Zhao, Y.; Zhao, P.; Zhou, S.; Ji, Z.; Huo, D.; Zhong, D.; Hou, C. Electrochemical sensor based on reduced graphene oxide supported dumbbell-shaped $CuCo_2O_4$ for real-time monitoring of H_2O_2 released from cells. *Microchem. J.,* **2021**, *160*, 105521.
       [http://dx.doi.org/10.1016/j.microc.2020.105521]

[66]   Chen, X.; Gao, J.; Zhao, G.; Wu, C. *In situ* growth of FeOOH nanoparticles on physically-exfoliated graphene nanosheets as high performance H_2O_2 electrochemical sensor. *Sens. Actuators B Chem.,* **2020**, *313*, 128038.
       [http://dx.doi.org/10.1016/j.snb.2020.128038]

[67]   Golsheikh, A.M.; Yeap, G.Y.; Yam, F.K.; San Lim, H. Facile fabrication and enhanced properties of copper-based metal organic framework incorporated with graphene for nonenzymatic detection of hydrogen peroxide. *Synth. Met.,* **2020**, *260*, 116272.
       [http://dx.doi.org/10.1016/j.synthmet.2019.116272]

[68]   Xu, S.; Liu, Y.; Zhao, W.; Wu, Q.; Chen, Y.; Huang, X.; Sun, Z.; Zhu, Y.; Liu, X. Hierarchical 0D-2D bio-composite film based on enzyme-loaded polymeric nanoparticles decorating graphene nanosheets as a high-performance bio-sensing platform. *Biosens. Bioelectron.,* **2020**, *156*, 112134.
       [http://dx.doi.org/10.1016/j.bios.2020.112134] [PMID: 32275578]

[69]   Lu, Z.; Wu, L.; Dai, X.; Wang, Y.; Sun, M.; Zhou, C.; Du, H.; Rao, H. Novel flexible bifunctional amperometric biosensor based on laser engraved porous graphene array electrodes: Highly sensitive electrochemical determination of hydrogen peroxide and glucose. *J. Hazard. Mater.,* **2021**, *402*, 123774.
       [http://dx.doi.org/10.1016/j.jhazmat.2020.123774] [PMID: 33254785]

CHAPTER 5

Conducting Polymer Based Nanocomposites for Sensing

D. Navadeepthy[1], **G. Srividhya**[1] and **N. Ponpandian**[1,*]

[1] *Department of Nanoscience and Technology, Bharathiar University, Coimbatore 641 046, India*

Abstract: The enormous development in the industrial and ecological base has led to increasing concern over the advent of new materials with implicit properties. A considerable interest has grown in conducting polymer nanocomposites as they are widely attracted for diverse applications due to the combinatory effect of conducting polymers and electrical and chemical properties of inorganic nanoparticles. The result was incredible with unique functionality, conductivity, structure, reactivity, processability (colloidal stability or mechanical strength), and sensitivity with bio-degradable and bio-compatible nature. The unique features of conducting polymer nanocomposites have gained attention in multiple applications. Recently, conducting polymer based nanocomposites are widely utilized as nanosensors in detecting temperature, stress, toxic gases and bio-elements. The present chapter deeply discusses the types of conducting polymer nanocomposites, as novel hybrid materials for sensor applications.

Keywords: Conducting polymers, Hybrid materials, Nanocomposites, Oxidation, Polymerization, Reduction, Sensors.

INTRODUCTION

The advancement in the field of material science has led to the development of new materials with unique properties and revolutionary combinations. One such novel invention is conducting polymers. Polymers are large compounds that are made of smaller subunits (monomers) in various ways or patterns to form a chain. Basically, polymers are insulators commonly used as coatings on wires and cables for electrical applications. A new era has begun with the invention of conducting polymers in the year 1977 by Hideki Shirakawa, Alan G. MacDiarmid, and Alan J Heeger. The discovery won the polymers arises from the π- conjugated polymer network.

---

* **Corresponding author N. Ponpandian**: Department of Nanoscience and Technology, Bharathiar University, Coimbatore 641 046, India; Email: ponpandian@buc.edu.in

Manorama Singh, Vijai K Rai and Ankita Rai (Eds.)

The Conducting polymers, for the past two decades, have been highly employed in a wide range of applications such as sensors, catalysis, energy storage, shielding electromagnetic radiations, corrosion resistance, medicinal device constructions, and electronic devices (displays, memory cells, transistors) [3]. Conducting polymers possess inimitable electrical, mechanical, optical, and conducting properties with high tunable capability, low cost, less weight, easy preparation methods and resistance to corrosion [4]. With the improvement in researches on nanoparticles and new functional materials, nanocomposites have evolved. Nanocomposites are the combination of two or more nanoparticles or nanoclusters in such a way that the properties of the parent constituents are not affected. By successful choice of combinations, highly potential nanocomposites applicable for a wide range of applications can be obtained [5].

Conducting polymer nanocomposites is the combination of conducting polymers (Polyaniline, polypyrrole, polyacetelene, polythiophene and PEDOT) and other inorganic nanoparticles or metal oxides to attain new characteristics. Among the various fields utilizing such improved physical and chemical properties, the conducting polymer nanocomposites are used to design sensors to detect various types of analytes such as solvent, vapor, strain/stress, pressure, and temperature [6, 7].

In this regard, the present chapter summarizes the introduction and types of conducting polymers and its nanocomposites, their preparation methods and applications in general. More specifically, the employment of conducting polymer nanocomposites for sensor applications is further discussed in detail.

CONDUCTING POLYMERS - AN INTRODUCTION

Conducting Polymers (CP) are novel materials with a combination of properties of organic polymers and semiconductors which are technically called as smart polymers. The electrical and optical properties of CPs are similar to a metal, whereas the mechanical properties of polymers are retained. For more than three decades, conducting polymers are dominating all fields of applications due to their intrinsic properties and flexibility. The high surface area of conducting polymers is highly advantageous for various types of applications, since the natural polymers generally do not possess these properties and do not come under the category of nanoparticles which are the most wanted and explored component of researchers. Conducting polymers are semi-conducting in nature in their pristine state, which can be tuned to become highly conducting by doping [8,9]. Conducting Polymers exert charge mobility along their p-electron back-bones, which are responsible for their unusual electronic properties. Conductivity of Cps thus exclusively depends on the delocalized π- electrons. On doping the oxidation and reduction states of the π-conjugated chains are modified to form p- or n-doped conducting polymers. Based on the methods and chemical components used for doping, the oxidized or reduced forms of conducting polymers are generated. Mainly polymers exit as either cation or anion radical or di-cation, or di-anion species [2,9]. The conductivity of these doped polymers can be tuned by changing the nature of dopant, degree of doping, manipulating the polymer chain backbone or by blending with other polymers.The conducting polymers are polyacetylene (PA), polyaniline (PANI), polypyrrole(PPy),polythiophene (PTh), poly(para-phenylene) (PPP), poly(phenylenevinylene) (PPV), polyfuran (PF), *etc.* Among them, PANI, PPy, PEDOT and PThpossess some unmatchable properties which are highly scrutinized for various applications [10].

CLASSIFICATIONS OF CONDUCTING POLYMERS

Conducting polymers can be classified based on their mode or nature of conduction. Initially the conducting polymers are broadly classified into two types

 i. Intrinsically conducting polymers
 ii. Extrinsically conducting polymers

In intrinsically conducting polymers, the conductivity depends either on the π-conjugated chains of the polymers or the dopants added to the polymers. They are highly advantageous, since they are easily tunable by ion exchanges and undergo oxidation or reduction easily. They have a high ability to store charges and are also optically efficient [11]. The classification of polymers is clearly shown in the chart in Fig. (**1**). In the latter type of polymers, conduction depends on the fillers in the polymers. The conductive fillers such as carbon black or fibers are added to the network to make them conductive. Extrinsically conducting polymers can also be formed by blending the insulators with conducting polymers. Such polymers possess bulk conductivity and are highly stable. They are cheap and easy processable. There is a third category of conducting polymers; they are inorganic conducting polymers which has charge transfer complexes [5]. The well-known CPs are polyaniline (PANI), polypyrrole (PPy), polythiophene (PTh), polyvinyl pyrrolidone (PVP), poly(3,4-ethylene dioxythiophene) (PEDOT), poly(m-phenylene diamine) (PMPD), polynaphthylamine (PNA), poly(p- phenylene sulfide) (PPS), poly(p-phenylene vinylene) (PPV), polyacetylene (PAC), polyfluorene, poly- phenylene and polypyrene [12]. Most of them are intrinsically conducting through their π- conjugated network, and their conductivity falls in the range 10^{-16} and 10^{-5} S/cm. Conducting polymers are either made directly by electro- or oxidative-polymerization or polymerized and then oxidized chemically or electrochemically. In the next section conducting polymers and their synthesis, structure, properties, and applications are briefly discussed.

polymers and their synthesis, structure, properties, and applications are briefly discussed.

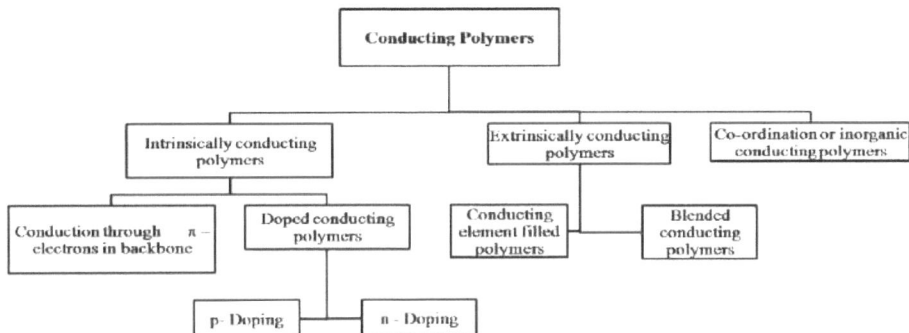

Fig. (1). Classification of Conducting polymers based on their conducting nature or mode.

CONDUCTING POLYMERS; SYNTHESIS, STRUCTURE, PROPERTIES AND APPLICATIONS

Among the diverse fields of nanoscience and its applications, CPs are the most widely used materials. Their intrinsic conductivity and exquisite optoelectronic properties lead them to the next generation of novel materials. Based on the chain structure the conducting polymers are classified and shown in the flow chart in Fig. (**2**) [13]. An organic polymer generally possess a conjugated structure, to conduct charge carriers they need a overlapping set of molecular orbitals [11]. Though the CPs shows good conductivity they fall under the category of semiconductors in their un-doped state. Thus, doping the CPs are the only way to increase their conductivity close to metals. Doping a CP can be done by various methods upon which they can be classified as below

  i. Chemical doping by charge transfer
 ii. Acid-base doping
iii. Electrochemical doping
 iv. Photo-doping
  v. Charge injection at polymer interface

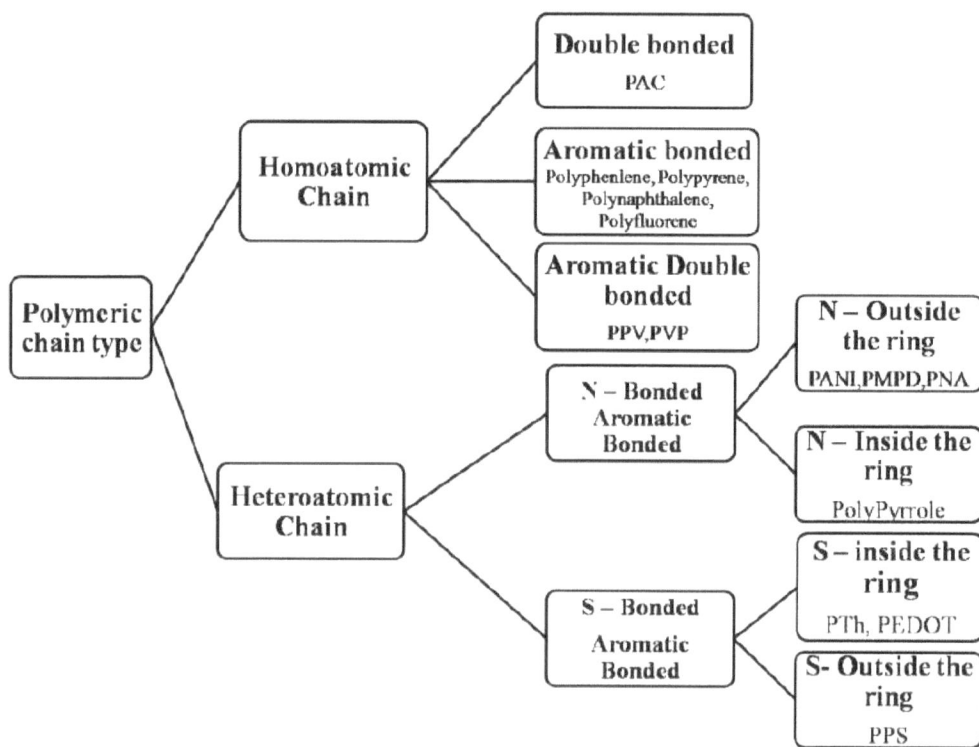

Fig. (2). Conducting polymer classification based on the chain structure.

Conducting polymers can be synthesized by either direct polymerization or by indirect polymerization. In the direct polymerizations the monomers are polymerized by addition or condensation directly to form the final product. Whereas in the indirect route, the addition or condensation of monomers lead to the formation of precursors which then undergoes conversion to form conjugated polymers. The conversion of precursor polymers can happen through isomerization or the addition or elimination of components in the precursor molecule [14, 15]. Based on this, the Polymerization can be carried out by following ways

- *Chemical synthesis:* the oxidation of reduction monomers chemically done and polymerization occurs under chemical reactions. The Monomers should be of low molecular weight to be easily soluble.
- *Electrochemical synthesis:* the anodic oxidation or cathodic reduction of function monomers of our choice. It is a most employed method of synthesis because of its simplicity and cost of process.
- *Photochemical synthesis:* Light irradiation to polymerise the monomers. It is

also simple, less destructive and cost effective route.

- *Metathesis synthesis:* the chemical reaction between two different monomer compounds which results in the swaping of any one part of each of the compound to form two different compounds.
- *Concentrated emulsion method:* the radical polymerization takes place in the emulsion form. The water, monomer and the surfactants are present in the emulsion and the polymerization takes place latex/colloidal particles. The monomer droplets are covered by the surfactants and form the chain structure. Mostly this process helps preparing high molecular weight polymers. Since water-soluble polymers cannot form chain structure on the latex form with surfactants.
- *Inclusion method:* the composite structures are made with this method. Useful to create low dimensional molecular level composites.
- *Solid state method:* the polymer chain lengths are increased by heat application. Usually carried under vacuum conditions to push away the by-products of the reaction process.
- *Plasma polymerization:* the process of forming thin films of organic polymers under plasma conditions. Polymerization is highly stable and adhesive which can be used for various novel applications.
- *Pyrolysis method:* the chemical decay of organic molecules under applied high temperature. Mostly employed in plastic and rubber production.

Every polymer structure is different and can be synthesized by any of the above methods based on their structure, molecular weight and solubility [13, 16]. The list of conducting polymers, their structure and properties are tabulated in Table **1**.

Table 1. Chemical formula, structure, synthesis method opted and conductivity of conducting polymers [15].

Conducting Polymer	Formula	Structure	Synthesis Method	Conductivity
PANI	$(C_6H_5NH_2)_n$		Chemical, electrochemical method, metathesis	10^{-4}
PPy	$(C_4H_5N)_n$		Chemical, electrochemical method, photochemical	$10^{-10} - 10^{-5}$

(Table 1) cont.....

Conducting Polymer	Formula	Structure	Synthesis Method	Conductivity
PTh	$(C_4H_4S)_n$		Chemical, electrochemical method	10^{-4}–10^{-3}
PVP	$(C_6H_9NO)_n$		Chemical polymerization, Radical polymerization, solid state method	7.42×10^{-8}
PEDOT	$(C_{12}H_8S_2O_4)_n$		Emulsion polymerization, chemical polymerization	23.8
PMPD	$(C_6H_4(NH_2))_n$		Chemical oxidative polymerization	-
PNA	$(C_{10}H_6NH)_n$		Electrochemical method	5.4×10^{-9}
PPS	$(C_6H_4S)_n$		Chemical polymerization	10^{-20}

Conducting Polymer	Formula	Structure	Synthesis Method	Conductivity
PPV	$(C_8H_6)_n$		electrochemical	10^{-10}
PAC	$(C_2H_2)_n$		metathesis	10^{-9}
Polyfluorene	$(C_{13}H_8)_n$		Oxidative, cross-coupling polymerization, pyrolysis	10^{-4}
Polyphenyelene	$(C_6H_4S)_n$		Chemical polymerization	10^{-14}

CONDUCTING POLYMER NANOCOMPOSITES

Nanocomposites are a novel kind of functional materials obtained by combining two different materials resulting in a formation of a unique product with extraordinary properties. The incorporation of metals, metal oxides, or carbon based material into the polymer matrix modifies the structural, functional, and physical properties of a conducting polymer in a large amount [5 - 17]. Thus, making it a novel material applicable for a wide variety of applications. The conducting polymer nanocomposites are being used extensively in the following areas (i) Supercapacitors (ii) Light Emitting Diodes (iii) Solar cells (iv) Sensors(v) Transistors and data storage (vi) Corrosion inhibitors (vii) Lasers used in flat televisions (viii) Anti-static substances for photographic film (ix) Electromagnetic shielding for computers 'smart windows' (x) Displays in mobile telephones and mini-format television screens (xi) Batteries (xii) Electrochromic device (xiii) Compact capacitors (xiv) Anti-static coating. The preparation of nanocomposites of conducting polymers can be done based on the type of incorporation. Mostly the composites are prepared by chemical polymerization, sol gel polymerization, emulsion polymerization or electrochemical polymerization [18]. The element to be composited with the polymer is incorporated during the polymerization process along with the monomers. Among the various types of nanocomposites and their diverse applications, sensors are the most widely explored field [8 - 20]. The next section deals with the conducting polymer nanocomposites for various sensors.

CONDUCTIVE POLYMERS NANOCOMPOSITES FOR SENSING APPLICATIONS

A sensor is a device that can detect changes in the environment by measuring a physical quantity and can furnish the sensing data in a user understandable output form. There are many types of sensors like chemi-resistive sensors, electrochemical sensors, optical sensors, piezoelectric sensors and so on, classified based on the type of stimulation and sensor response. Generally, the sensors produce a response signal which directly corresponds to the quantity of the target analyte by any of the transduction methods. Sensors for practical applications are expected to be selective, highly sensitive, and reversible. Polymer nanocomposites are considered as one of most promising materials for electroanalytical sensors owing to their exceptional characteristics like tunable conductivity, variety of morphology, ease of synthesis and fabrication. Compositing conducting polymers with metals, metal oxides, metal sulfides and other inorganic materials improve the performance and stability of the sensor by synergistic charge transfer effects. Carbon based materials like, carbon nanotubes (CNTs), graphene, graphene oxide (GO) and reduced graphene oxide (r-GO) improve the surface properties of the polymer composites, thus enhancing the sensing performance. In this section, we will discuss the nanocomposite of conductive polymers, particularly PANI, Ppy, PTh and PEDOTwith other organic materials, metals, metal oxides and chalcogenides to fabricate the most prominent chemi-resistive, electrochemical sensors among other types.

Electrochemical Biosensors

The development of electrochemical sensors has received an extensive interest among the global research community owing to its advantageous features like simplicity of design, low-cost and efficient sensing of a wide range of metal ions, pharmaceutical compounds and biomolecules. Especially, the development and commercialization of electrochemical biosensors for on-time clinical analysis is in the limelight of research, as a result of their efficacious performance and lower detection limits. The important components of electrochemical and biosensors are

  i. The recognition element like suitable bioreceptors
 ii. The sensing element with a suitable electrode material on a substrate
iii. The electronic components for the processing of signals produced by the sensing element

The exchange of electrons in the redox chemical reactions causes an electronic signal that can be assessed by using different electroanalytical techniques such as potentiometry, amperometry, voltammetry and impedance spectroscopy. Conductive polymers and their composites are among the potential contenders as viable electrode materials for the sensing element in sensors based on electrochemical detection, as they provide a conductive matrix for the immobilization of recognition elements (bioreceptors). Many electrochemical biosensors have been fabricated based on conductive polymer electrodes, and their properties of excellent electroactivity, pH sensitivity, and interaction with bio-compounds have been established [21 - 24]. Among the frequently integrated materials with conductive polymers are the noble and transition metals because of the improved catalytic activity and electron transfer properties of the composite. Xian *et al.* fabricated an enzymatic glucose sensor based on Au/PANI nanocomposite in which they found that electron transfer is facilitated by the presence of metal in the polymer matrix [25]. The detection of glucose was done by amperometric sensing of hydrogen peroxide (H_2O_2) released enzymatically during the oxidation reaction [26]. A higher response in terms of anodic current was obtained for the metal/polymer composite than the bare polymer-based electrode. Similarly, Srivastava *et al.* fabricated an enzymatic cholesterol sensor based on PANI/Au/chitosan nanocomposite by immobilizing a cholesterol oxidase (ChO_x) on the polymer matrix [27]. Their results validated the fact that the presence of metal particles improves the kinetics of the reaction for a faster release of enzymes.

Similar to metal nanoparticles, composites of metal oxides and metal chalcogenides also contribute to the enhanced performance of polymer nanocomposites based sensors. Among the frequently composited metal oxide is ZnO due to its wide bandgap and excellent stability. Ghanbari *et al.* fabricated an electrochemical sensor for the detection of ascorbic acid (AA), dopamine (DA) and uric acid (UA) using ZnO/Cu_2O decorated Ppy nanofibers [28]. Glassy carbon electrode (GCE) coated with this novel composite showed better electroctalytic performance with improved peak current when compared to GCE, PPy/GCE, Cu_xO-PPy/GCE, and ZnO-PPy/GCE. Similarly, ZnO nanocomposite with PTh also improved the sensing response of the polymer-based sensor towards hyrdazine, as revealed by the voltametric studies [23]. Conductive polymer nanocomposites with transitional metal dichalcogenides (TMDs) are also equally explored, as TMDs have unique properties of 2-dimensional morphology and high catalytic activity. Li *et al.* incorporated MoS_2 nanosheets into PEDOT polymer and employed the nanocomposite coated GCE for the simultaneous detection AA, Da and UA, which are important bio-compounds of human metabolism [29].

Electrochemical biosensors based on polymer composites with carbon derivatives such as CNTs, graphene, GO, r-GO and graphene quantum dots are expansively studied and explored as well. The structural and mechanical properties of carbon derivatives modify the polymer matrix in a beneficial way to improve the sensing properties of the polymer material. Meshram *et al.* prepared a Lactate biosensor employing MWCNT/PPy composites deposited on a stainless-steel electrode [30]. The sensor was sensitized by immobilizing the lactate oxidase enzyme. Compared to the bare polymer matrix, modification with CNT enabled high loading of the enzyme, increase in the stability and lifetime (2 weeks) and the rapid sensor response (8 s). Ternary composites of polymers with carbon derivatives and metal oxides and sulphides are also studied to stabilize the electroanalytical performance of electrochemical sensors. For example, Ramachandran *et al.* designed a MnO_2/PTh/r-GO nanocomposite modified GCE for the electrochemical sensing of methyl parathion (MP) [31]. The synergic combination of the three materials resulted in a low charge transfer resistance of 93 Ω as determined from the impedance spectroscopy. This is two times lower than the value of the bare electrode. Ye *et al.* proposed a DNA sensor using zinc sulphur coated PEDOT/r-GO nanocomposite thin films with sensitized with thioacetamide (TAA) reducing agent [32]. Liu *et al.* designed an immunosensor to detect interleukin-6which is both a pro-inflammatory cytokine and an anti-inflammatory myokine using a hybrid ternary GO/PANI/CdSe composite [33]. The nanocomposite based immunosensor was highly stable and reproducible with a very low LOD of 0.17 pg mL^{-1}.

Chemi-resistive Gas and Vapor Sensors

A gas or vapor sensor is used to detect the presence and quantity of harmful gases and volatile vapor in a working atmosphere. The most popular technique of detecting toxic gases and vapors is through chemi-resistive sensing. Chemi-resistive sensing is a simple and effective method that mainly involves a sensing element coated onto the electrodes on a substrate by using any of the coating methods like spin coating, dip coating, sputtering, atomic layer deposition, *etc.* The response of the sensing element is observed as a change in its resistance that is measured using a source meter unit. The response of the sensing element is due to the adsorption of the target gas on its surface, resulting in heterogeneous charge transfer. Sensors based on conducting polymers and their composites with other polymers have been developed for sensing different gases and volatile organic compounds like ammonia and ethanol [34].

The unique electronic and catalytic properties of metal nanoparticles are very useful in enhancing the sensing properties of the polymers by acting as active centers for effective adsorption and desorption of gas molecules. Generally noble metals like platinum (Pt), gold (Au), silver (Ag) and palladium (Pd) are the chosen candidates for compositing with polymers. The sensing characteristics such as enhanced sensor response, short response and recovery time were observed for Au/Ppy and Ag/Ppy based ammonia sensors [35]. Sensors that can selectively detect ammonia vapor at room temperature were developed by Patil *et al.* by incorporating Cu nanoparticles with PANI [36]. The as developed thin-film sensor could efficiently detect ammonia vapor as low as 1 ppm concentration with excellent reversibility. Earlier, Sharma *et al.* developed Cu/PANI chemi-resistive sensor that could detect chloroform vapors at room temperature. The team investigated the sensing mechanism by studying the FT-IR spectra of the nanocomposite in the presence of chloroform vapor which revealed that the metal incorporated polymers are effective in sensing applications, and the performance of the sensor increases with increasing concentration of the metal clusters [37].

Metal oxide semiconductors (MOS) are low-cost alternatives for precious noble metals. MOS based sensors show high stability and optimum absorbency due to the variety of oxygen stoichiometry. MOS like zinc oxide (ZnO), titanium oxide (TiO_2) and tin oxide (SnO_2) are widely explored for their excellent sensing behavior of various gases and volatile vapors with improved characteristics like good sensor response, low limit of detection and short response and recovery time [38 - 40]. Suri *et al.* developed a gas and humidity sensor based on iron oxide/Ppy nanocomposite and found out that the sensing performance of iron oxide increases by the inclusion of PPy and the sensing response is directly proportional to the concentration of Ppy [41]. Similarly, PTh/SnO_2 composite and PTh/ zirconium oxide (ZrO_2) nanocomposite were developed for ethanol and ethene gas sensing [42, 43] In both the studies, improved conductivity and higher surface area played the major role. Integration of metal oxides with polymers affects the electrical conductivity and the morphology of the composite, both of which influence the sensor response.

Polymer nanocomposites with carbon-based materials like graphene, CNTs, graphene oxide can be easily prepared by attaching the polymeric moieties to the above mentioned lower dimensional carbon materials by covalent bonding [44, 45]. Based on this, Hwang *et al.* fabricated a stretchable ammonia sensor based on Ppy/SWCNT/Ag nanowire [46]. SWCNT/Ag network forms a high conductive network with a low resistivity value of 30 Ω.

Apart from chemi-resistive and electrochemical sensors, conductive polymer nanocomposites are also used to fabricate strain sensors, temperature and humidity sensors as well. The combinational effect of conductive polymer with suitable composite material improves the mechanical and electrical properties which makes them useful in wearable piezoelectric sensors to detect human motion [47 - 50]. The list of recently developed conducting polymer based sensors with different methods of detection is given in Table **2**.

Table 2. Different conducting polymer based nanosensors.

S.No.	Conductive Polymer	Composite Material	Detection Technique	Analyte	Ref
1	PANI/Ppy	Au nanoparticles	Electrochemical	Glucose	[51]
2	PANI	MWCNT & MWCNT/hydroxyl-terminated polybutadiene polyurethane	Chemi-resistive	Trichloromethane	[52]
3	PANI	Ag	Electrochemical	H_2O_2 and DA	[53]
4	PANI	r-GOand Ag functionalization	Electrochemical	Glucose	[54]
5	PANI	TiO_2	Electrochemical	1,2-diaminobenzene	[55]
6	PANI	ZnO	Photoluminescence	Acetic acid vapor	[56]
7	PANI	$SrGe_4O_7$	Chemi-resistive	Ammonia	[57]
8	PANI	MWCNT	Fluorescence	Ammonia	[58]
9	Ppy	Phenothiazine/Ag doped ZnO	Chemi-resistive	Ammonia	[59]
10	Ppy	NiO	Chemi-resistive	NO_2	[60]
11	Ppy	TPU/Fe_2O_3	Chemi-resistive	Ethanol and other VOCs	[61]
12	Ppy	ZnO	Photoelectrochemical	Bisphenol S	[62]
13	Ppy	CuO	Electrochemical	Tramadol	[63]
14	PTh	$Pt/g-C_3N_4$	Electrochemical	Hg^{2+}	[64]
15	PTh	graphene/zinc tungstate	Chemi-resistive	Cigarette Smoke	[65]
16	PTh	SWCNT	Chemi-resistive	NO_2	[66]
17	PEDOT	PSS@CNT-poly(glycerol sebacate urethane)	Piezoelectric (Pressure sensor)	Swelling of pathological tissue	[67]
18	PEDOT	PSS@ammonium vanadate	Chemi-resistive	Ammonia	[68]
19	PEDOT	$g-C_3N_4$	Electrochemical	Paracetamol	[69]
20	PEDOT	PSS/r-GO	OFET	Hg^{2+}	[70]

CONCLUDING REMARKS

The conducting polymers, otherwise called synthetic metals, are new inventions of the past two decades with improved functional, electrical and optical properties. The conducting polymers with π- conjugated system on its backbone make them highly conductive in nature. These polymers are highly investigated for their unique properties in diverse applications. The nanocomposite of the conducting polymers depicts combined qualities of both the polymer and the combined metal or metal oxide or carbon-based fillers. Because of the enhanced functional applicability and processability, they are used as detectors in various sensing applications. As a nanocomposite sensor, conducting polymer nanocomposites are one of the challenging inventions based on future perspective.

CONSENT FOR PUBLICATION

Not applicable.

CONFLICT OF INTEREST

The author declares no conflict of interest, financial or otherwise.

ACKNOWLEDGEMENTS

The authors thank the DST-FIST; DST-PURSE and UGC-SAP, Government of India, for financial support to build the instrumental facilities.

REFERENCES

[1]   Kumar, A.B.V.K.; Mishra, D.K.; Vijayakumari, S.C.; Kim, D. Conducting polymer nanocomposite based temperature sensors: A review Conducting polymer nanocomposite based temperature sensors: A review. *Inorg. Chem. Commun.,* **2018**, *98*, 11-28.
       [http://dx.doi.org/10.1016/j.inoche.2018.09.040]

[2]   Lu, X.; Zhang, W.; Wang, C.; Wen, T.; Wei, Y. Progress in Polymer Science One-dimensional conducting polymer nanocomposites : Synthesis, properties and applications. *Prog. Polym. Sci.,* **2011**, *36*(5), 671-712.
       [http://dx.doi.org/10.1016/j.progpolymsci.2010.07.010]

[3]   By, P. World ' s largest Science, Technology & Medicine Open Access book publisher Oxidative Stress : Cause and Consequence of Diseases.

[4]   Naveen, M.H.; Gurudatt, N.G.; Shim, Y. Applications of conducting polymer composites to electrochemical sensors : A review. *Appl. Mater. Today,* **2017**, *9*, 419-433.
       [http://dx.doi.org/10.1016/j.apmt.2017.09.001]

[5]   Gangopadhyay, R.; De, A. *Conducting Polymer Nanocomposites : A Brief Overview,* **2000**, 608-622.

[6]   Gerard, M.; Chaubey, A.; Malhotra, B.D. Application of conducting polymers to biosensors. *Biosens. Bioelectron.,* **2002**, *17*(5), 345-359.
       [http://dx.doi.org/10.1016/S0956-5663(01)00312-8] [PMID: 11888724]

[7]   Gerard, M.; Gerard, M.; Chaubey, A.; Malhotra, B.D. Application of conducting polymers to biosensors. Biosens Bioelectron Application of conducting polymers to biosensors. *Biosens. Bioelectron.,* **2002**, *2002*(17), 345-359.
       [http://dx.doi.org/10.1016/S0956-5663(01)00312-8]

[8]   Kaur, G.; Kaur, A.; Kaur, H. Review on nanomaterials / conducting polymer based nanocomposites for the development of biosensors and electrochemical sensors, Polym. Technol. Mater., **2020**, 1-8.

[http://dx.doi.org/10.1080/25740881.2020.1844233]

[9]   Alan, G. *Introduction of Conducting Polymers*; , **2000**, pp. 1-15.

[10]   Bakhshi, A.K.; Bhalla, G. Electrically conducting polymers: Materials of the twentyfirst century. *J. Sci. Ind. Res. (India),* **2004**, *63*, 715-728.

[11]   Mishra, A.K. Conducting Polymers. *Concepts and Applications,* **2018**, *5*(2), 159-193.
       [http://dx.doi.org/10.26713/jamcnp.v5i2.842]

[12]   Taylor, P.; Das, T.K.; Prusty, S. *Polymer-Plastics Technology and Engineering Review on Conducting Polymers and Their Applications,* **2012**, *37–41*, 2012.
       [http://dx.doi.org/10.1080/03602559.2012.710697]

[13]   Meer, S.; Kausar, A.; Iqbal, T. Trends in Conducting Polymer and Hybrids of Conducting Polymer / Carbon Nanotube : A Review Trends in Conducting Polymer and Hybrids of Conducting Polymer / Carbon, **2016**, 2559.
       [http://dx.doi.org/10.1080/03602559.2016.1163601]

[14]   Kambour, R.P. Polymers with Metal-Like Conductivity- A Review of their Synthesis. *Structure and Properties,* **1981**, *1680*, 361-381.

[15]   Kumar, R.; Singh, S.; Yadav, B.C. Conducting Polymers : Synthesis. *Properties and Applications,* **2016**, (July)
       [http://dx.doi.org/10.17148/IARJSET.2015.21123]

[16]   Kaner, R.B. Synthesis and applications of conducting polymer nanofibers. *MRS Bull.,* **2016**, *2016*(41), 785-790.
       [http://dx.doi.org/10.1557/mrs.2016.213]

[17]   Fu, S.; Sun, Z.; Huang, P.; Li, Y.; Hu, N. Nano Materials Science Some basic aspects of polymer nanocomposites : A critical review. *Nano Mater. Sci.,* **2019**, *1*(1), 2-30.
       [http://dx.doi.org/10.1016/j.nanoms.2019.02.006]

[18]   Iqbal, S.; Ahmad, S. Journal of Industrial and Engineering Chemistry Recent development in hybrid conducting polymers : Synthesis, applications and future prospects. *J. Ind. Eng. Chem.,* **2018**, *60*, 53-84.
       [http://dx.doi.org/10.1016/j.jiec.2017.09.038]

[19]   Ates, M. A review study of (bio)sensor systems based on conducting polymers. *Mater. Sci. Eng. C,* **2013**, *33*(4), 1853-1859.
       [http://dx.doi.org/10.1016/j.msec.2013.01.035] [PMID: 23498205]

[20]   Naseri, M.; Fotouhi, L.; Ehsani, A. *Recent Progress in the Development of Conducting Polymer-Based Nanocomposites for Electrochemical Biosensors Applications : A Mini-Review,* **2018**, 1-21.
       [http://dx.doi.org/10.1002/tcr.201700101]

[21]   Wei, D.; Ivaska, A. Electrochemical biosensors based on polyaniline. *Chem. Anal. (Pol.),* **2006**, *51*, 839-852.

[22]   Nezhadali, A.; Rouki, Z.; Nezhadali, M. Electrochemical preparation of a molecularly imprinted polypyrrole modified pencil graphite electrode for the determination of phenothiazine in model and real biological samples. *Talanta,* **2015**, *144*, 456-465.
       [http://dx.doi.org/10.1016/j.talanta.2015.06.082] [PMID: 26452848]

[23]   Faisal, M.; Harraz, F.A.; Al-Salami, A.E.; Al-Sayari, S.A.; Al-Hajry, A.; Al-Assiri, M.S. F. M, H., AE, A.-S., S. A, A. S., A., A. H. A, Al-Assiri, M., Polythiophene/ZnO nanocomposite-modified glassy carbon electrode as efficient electrochemical hydrazine sensor. *Mater. Chem. Phys.,* **2018**, *214*, 126-134.
       [http://dx.doi.org/10.1016/j.matchemphys.2018.04.085]

[24]   Park, C. S.; Lee, C.; Kwon, O. S. Conducting polymer based nanobiosensors, Polymers (Basel). **2016**, 8-249.

[25]   Xian, Y.; Hu, Y.; Liu, F.; Xian, Y.; Wang, H.; Jin, L. Glucose biosensor based on Au nanoparticles-conductive polyaniline nanocomposite. *Biosens. Bioelectron.,* **2006**, *21*(10), 1996-2000.
        [http://dx.doi.org/10.1016/j.bios.2005.09.014] [PMID: 16275055]

[26]   Thangapandian, M.; Viswanathan, C.; Ponpandian, N. *Nanoscale Advances electrocatalyst toward the sensitive colorimetric and electrochemical sensing of ascorbic acid,* **2020**, 3481-3493.
        [http://dx.doi.org/10.1039/D0NA00283F]

[27]   Srivastava, M.; Srivastava, S.K.; Nirala, N.R.; Prakash, R. A chitosan-based polyaniline–Au nanocomposite biosensor for determination of cholesterol. *Anal. Methods,* **2014**, *6*(3), 817-824.
        [http://dx.doi.org/10.1039/C3AY41812J]

[28]   Ghanbari, Kh.; Babaei, Z. Fabrication and characterization of non-enzymatic glucose sensor based on ternary NiO/CuO/polyaniline nanocomposite. *Anal. Biochem.,* **2016**, *498*, 37-46.
        [http://dx.doi.org/10.1016/j.ab.2016.01.006] [PMID: 26778527]

[29]   Li, Y.; Lin, H.; Peng, H.; Qi, R.; Luo, C. A glassy carbon electrode modified with MoS 2 nanosheets and poly (3, 4-ethylenedioxythiophene) for simultaneous electrochemical detection of ascorbic acid, dopamine and uric acid. *Mikrochim. Acta,* **2016**, *183*(9), 2517-2523.
        [http://dx.doi.org/10.1007/s00604-016-1897-1]

[30]   Meshram, B. H.; Mahore, R. P.; Virutkar, P. D.; Kondawar, S. B. Polypyrrole / carbon nanotubes / lactate oxidase nanobiocomposite film based modified stainless steel electrode lactate biosensor, Procedia Mater. Sci., **2015**, *10* Cont, 176–185

[31]   Dhayabaran, T.R. V., Utilization of a MnO 2 / polythiophene / rGO nanocomposite modified glassy carbon electrode as an electrochemical sensor for methyl parathion. *J. Mater. Sci. Mater. Electron.,* **2019**, *30*(13), 12315-12327.
        [http://dx.doi.org/10.1007/s10854-019-01590-9]

[32]   Ye, X.; Du, Y.; Duan, K.; Lu, D.; Wang, C.; Shi, X. Fabrication of nano-ZnS coated PEDOT-reduced graphene oxide hybrids modified glassy carbon-rotating disk electrode and its application for simultaneous determination of adenine, guanine, and thymine. *Sens. Actuators B Chem.,* **2014**, *203*, 271-281.
        [http://dx.doi.org/10.1016/j.snb.2014.06.135]

[33]   Liu, Q.; Liu, X-P.; Wei, Y-P.; Mao, C-J.; Niu, H-L.; Song, J-M.; Jin, B-K.; Zhang, S-Y. Electrochemiluminescence immunoassay for the carcinoembryonic antigen using CdSe:Eu nanocrystals. *Mikrochim. Acta,* **2017**, *184*(5), 1353-1360.
        [http://dx.doi.org/10.1007/s00604-017-2114-6]

[34]   Lv, D.; Shen, W.; Chen, W.; Tan, R.; Xu, L.; Song, W. PSS-PANI/PVDF composite based flexible NH3 sensors with sub-ppm detection at room temperature. *Sens. Actuators B Chem.,* **2020**, *128*, 129085.

[35]   Zhang, J.; Liu, X.; Wu, S.; Xu, H.; Cao, B. One-pot Fabrication of Uniform Polypyrrole/Au Nanocomposites and Investigation for Gas Sensing. *Sens. Actuators B Chem.,* **2013**, *168*, 695-700.
        [http://dx.doi.org/10.1016/j.snb.2013.06.063]

[36]   Patil, U.V.; Ramgir, N.S.; Karmakar, N.; Bhogale, A.; Debnath, A.K.; Aswal, D.K.; Gupta, S.K.; Kothari, D.C. Room temperature ammonia sensor based on copper nanoparticle intercalated polyaniline nanocomposite thin films. *Appl. Surf. Sci.,* **2015**, *339*, 69-74.
        [http://dx.doi.org/10.1016/j.apsusc.2015.02.164]

[37]   Athawale, A.A.; Pethkar, S.; Sharma, S.; Nirkhe, C. Chloroform vapour sensor based on copper/polyaniline nanocomposite. *Sens. Actuators B Chem.,* **2002**, *85*(1-2), 131-136.
        [http://dx.doi.org/10.1016/S0925-4005(02)00064-3]

[38]   Das, M.; Sarkar, D. One-pot synthesis of zinc oxide - polyaniline nanocomposite for fabrication of e fficient room temperature ammonia gas sensor, Ceram. *Int,* **2017**.

[39]   Moghaddam, H.M.; Malkeshi, H. Self-assembly synthesis and ammonia gas-sensing properties of

ZnO/Polythiophene nanofibers. *J. Mater. Sci. Mater. Electron.,* **2016**, *27*(8), 8807-8815.
[http://dx.doi.org/10.1007/s10854-016-4906-6]

[40]   Ganesan, S.; Muruganandham, A.; Mounasamy, V.; Kannan, V.P.; Madanagurusamy, S. Highly Selective Dimethylamine Sensing Performance of TiO_2 Thin Films at Room Temperature. *J. Nanosci. Nanotechnol.,* **2020**, *20*(5), 3131-3139.
[http://dx.doi.org/10.1166/jnn.2020.17199] [PMID: 31635657]

[41]   Suri, K.; Annapoorni, S.; Sarkar, A.K.; Tandon, R.P. Gas and humidity sensors based on iron oxide-polypyrrole nanocomposites. *Sens. Actuators B Chem.,* **2002**, *81*(2-3), 277-282.
[http://dx.doi.org/10.1016/S0925-4005(01)00966-2]

[42]   Husain, A.; Ahmad, S.; Mohammad, F. Electrical conductivity and alcohol sensing studies on polythiophene / tin oxide nanocomposites. *J. Sci. Adv. Mater. Devices,* **2020**, *5*(1), 84-94.
[http://dx.doi.org/10.1016/j.jsamd.2020.01.002]

[43]   Husain, A.; Ahmad, S.; Mohammad, F. Thermally stable and highly sensitive ethene gas sensor based on polythiophene/zirconium oxide nanocomposites. *Mater. Today Commun.,* **2019**, *20*, 100574.
[http://dx.doi.org/10.1016/j.mtcomm.2019.100574]

[44]   Wu, Z.; Chen, X.; Zhu, S.; Zhou, Z.; Yao, Y.; Quan, W.; Liu, B. Enhanced sensitivity of ammonia sensor using graphene / polyaniline nanocomposite. *Sens. Actuators B Chem.,* **2013**, *178*, 485-493.
[http://dx.doi.org/10.1016/j.snb.2013.01.014]

[45]   Guo, Y.; Wang, T.; Chen, F.; Sun, X.; Li, X.; Yu, Z.; Wan, P.; Chen, X. Hierarchical graphene-polyaniline nanocomposite films for high-performance flexible electronic gas sensors. *Nanoscale,* **2016**, *8*(23), 12073-12080.
[http://dx.doi.org/10.1039/C6NR02540D] [PMID: 27249547]

[46]   Hwang, B.Y.; Du, W.X.; Lee, H.J.; Kang, S.; Takada, M.; Kim, J.Y. Stretchable and High-performance Sensor films Based on Nanocomposite of Polypyrrole/SWCNT/Silver Nanowire. *Nanomaterials (Basel),* **2020**, *10*(4), E696.
[http://dx.doi.org/10.3390/nano10040696] [PMID: 32272556]

[47]   Zhang, F. Multi-modal strain and temperature sensor by hybridizing reduced graphene oxide and PEDOT : PSS, Compos. Sci. Technol., **2020**, *187*, 9-107959.

[48]   Shen, G. Transparent and Stretchable Strain Sensors with Improved Sensitivity and Reliability Based on Ag NWs and PEDOT. *PSS Patterned Microstructures,* **2020**, *1901360*, 1-8.

[49]   Huang, Y.; Gao, L. Highly flexible fabric strain sensor based on graphene nanoplatelet – polyaniline nanocomposites for human gesture recognition, **2017**, *45340*, 1-8.

[50]   Hanif, Z.; Shin, D.; Choi, D.; Jea, S. Development of a vapor phase polymerization method using a wet-on-wet process to coat polypyrrole on never-dried nanocellulose crystals for fabrication of compression strain sensor. *Chem. Eng. J.,* **2019**, *381*, 122700.
[http://dx.doi.org/10.1016/j.cej.2019.122700]

[51]   German, N.; Ramanaviciene, A.; Ramanavicius, A. Formation and Electrochemical Evaluation of Polyaniline and Polypyrrole Nanocomposites Based on Glucose Oxidase and Gold Nanostructures. *Polymers (Basel),* **2020**, *12*(12), 12.
[http://dx.doi.org/10.3390/polym12123026] [PMID: 33348805]

[52]   Tian, X.; Zhang, S.; Ma, Y.Q.; Luo, Y.L.; Xu, F.; Chen, Y.S. Preparation and vapor-sensitive properties of hydroxyl-terminated polybutadiene polyurethane conductive polymer nanocomposites based on polyaniline-coated multiwalled carbon nanotubes. *Nanotechnology,* **2020**, *31*(19), 195504.
[http://dx.doi.org/10.1088/1361-6528/ab704c] [PMID: 31986500]

[53]   Paulraj, P.; Umar, A.; Rajendran, K.; Manikandan, A.; Kumar, R.; Manikandan, E.; Pandian, K.; Mahnashi, M.H.; Alsaiari, M.A.; Ibrahim, A.A.; Bouropoulos, N.; Baskoutas, S. Solid-state synthesis of Ag-doped PANI nanocomposites for their end-use as an electrochemical sensor for hydrogen peroxide and dopamine. *Electrochim. Acta,* **2020**, *363*, 137158.

[http://dx.doi.org/10.1016/j.electacta.2020.137158]

[54] Deshmukh, M.A.; Kang, B.C.; Ha, T.J. Non-enzymatic electrochemical glucose sensors based on polyaniline/reduced-graphene-oxide nanocomposites functionalized with silver nanoparticles. *J. Mater. Chem. C Mater. Opt. Electron. Devices,* **2020**, *8*(15), 5112-5123.
[http://dx.doi.org/10.1039/C9TC06836H]

[55] Karim, M.R.; Alam, M.M.; Aijaz, M.O.; Asiri, A.M.; Almubaddel, F.S.; Rahman, M.M. The fabrication of a chemical sensor with PANI-TiO$_2$ nanocomposites. *RSC Advances,* **2020**, *10*(21), 12224-12233.
[http://dx.doi.org/10.1039/C9RA09315J]

[56] Turemis, M.; Zappi, D.; Giardi, M.T.; Basile, G.; Ramanaviciene, A.; Kapralovs, A.; Ramanavicius, A.; Viter, R. ZnO/polyaniline composite based photoluminescence sensor for the determination of acetic acid vapor. *Talanta,* **2020**, *211*, 120658.
[http://dx.doi.org/10.1016/j.talanta.2019.120658] [PMID: 32070567]

[57] Zhang, Y.; Zhang, J.; Jiang, Y.; Duan, Z.; Liu, B.; Zhao, Q.; Wang, S.; Yuan, Z.; Tai, H. Ultrasensitive flexible NH3 gas sensor based on polyaniline/SrGe4O9 nanocomposite with ppt-level detection ability at room temperature. *Sens. Actuators B Chem.,* **2020**, *319*, 128293.
[http://dx.doi.org/10.1016/j.snb.2020.128293]

[58] Maity, D.; Manoharan, M.; Rajendra Kumar, R.T. Development of the PANI/MWCNT Nanocomposite-Based Fluorescent Sensor for Selective Detection of Aqueous Ammonia. *ACS Omega,* **2020**, *5*(15), 8414-8422.
[http://dx.doi.org/10.1021/acsomega.9b02885] [PMID: 32337403]

[59] Rameshan, M.T.; Greeshma, K.P.; Parvathi, K.; Anilkumar, T. Structural, electrical, thermal, and gas sensing properties of new conductive blend nanocomposites based on polypyrrole/phenothiazine/silver-doped zinc oxide. *J. Vinyl Addit. Technol.,* **2020**, *26*(2), 187-195.
[http://dx.doi.org/10.1002/vnl.21732]

[60] Zhang, J.; Wu, C.; Li, T.; Xie, C.; Zeng, D. Highly sensitive and ultralow detection limit of room-temperature NO2 sensors using *in-situ* growth of PPy on mesoporous NiO nanosheets. *Org. Electron.,* **2020**, *77*, 105504.
[http://dx.doi.org/10.1016/j.orgel.2019.105504]

[61] Beniwal, A.; Sunny, A. Novel TPU/Fe$_2$O$_3$ and TPU/Fe$_2$O$_3$/PPy nanocomposites synthesized using electrospun nanofibers investigated for analyte sensing applications at room temperature. *Sens. Actuators B Chem.,* **2020**, *304*, 127384.
[http://dx.doi.org/10.1016/j.snb.2019.127384]

[62] Viter, R.; Kunene, K.; Genys, P.; Jevdokimovs, D.; Erts, D.; Sutka, A.; Bisetty, K.; Viksna, A.; Ramanaviciene, A.; Ramanavicius, A. Photoelectrochemical Bisphenol S Sensor Based on ZnO-Nanoroads Modified by Molecularly Imprinted Polypyrrole. *Macromol. Chem. Phys.,* **2020**, *221*(2), 1900232.
[http://dx.doi.org/10.1002/macp.201900232]

[63] Arabali, V.; Malekmohammadi, S.; Karimi, F. Surface amplification of pencil graphite electrode using CuO nanoparticle/polypyrrole nanocomposite; a powerful electrochemical strategy for determination of tramadol. *Microchem. J.,* **2020**, *158*, 105179.
[http://dx.doi.org/10.1016/j.microc.2020.105179]

[64] Mahmoudian, M.R.; Basirun, W.J.; Alias, Y.; Meng, P.; Woi, P. Investigating the effectiveness of g-C3N4 on Pt /g-C3N4/ polythiophene nanocomposites performance as an electrochemical sensor for Hg2+ detection. *J. Environ. Chem. Eng.,* **2020**, *8*(5), 104204.
[http://dx.doi.org/10.1016/j.jece.2020.104204]

[65] Husain, A.; Ahmad, S.; Mohammad, F. Polythiophene/graphene/zinc tungstate nanocomposite: Synthesis, characterization, DC electrical conductivity and cigarette smoke sensing application, Polym. *Polym. Compos,* **2020**, 0967391120929079.

[http://dx.doi.org/10.1177/0967391120929079]

[66]   Ramadhan, A.A.; Ibrahim, F.T.; Nasir, E.M. (PTh/SWCNT) for NO2 Gas Sensing. *J. Phys. Conf. Ser.,* **2020**, *1660*(1), 12093.
        [http://dx.doi.org/10.1088/1742-6596/1660/1/012093]

[67]   Tadayyon, G.; Krukiewicz, K.; Britton, J.; Larrañaga, A.; Vallejo-Giraldo, C.; Fernandez-Yague, M.; Guo, Y.; Orpella-Aceret, G.; Li, L.; Poudel, A.; Biggs, M.J.P. *In vitro* analysis of a physiological strain sensor formulated from a PEDOT:PSS functionalized carbon nanotube-poly(glycerol sebacate urethane) composite. *Mater. Sci. Eng. C,* **2021**, *121*, 111857.
        [http://dx.doi.org/10.1016/j.msec.2020.111857] [PMID: 33579489]

[68]   Lee, S.H.; Bang, J.H.; Kim, J.; Park, C.; Choi, M.S.; Mirzaei, A.; Im, S.S.; Ahn, H.; Kim, H.W. Sonochemical synthesis of PEDOT:PSS intercalated ammonium vanadate nanofiber composite for room-temperature NH3 sensing. *Sens. Actuators B Chem.,* **2021**, *327*, 128924.
        [http://dx.doi.org/10.1016/j.snb.2020.128924]

[69]   Yan, Y.; Jamal, R.; Yu, Z.; Zhang, R.; Zhang, W.; Ge, Y.; Liu, Y.; Abdiryim, T. Composites of thiol-grafted PEDOT with N-doped graphene or graphitic carbon nitride as an electrochemical sensor for the detection of paracetamol. *J. Mater. Sci.,* **2020**, *55*(13), 5571-5586.
        [http://dx.doi.org/10.1007/s10853-020-04351-w]

[70]   Sayyad, P.W.; Ingle, N.N.; Al-Gahouari, T.; Mahadik, M.M.; Bodkhe, G.A.; Shirsat, S.M.; Shirsat, M.D. Selective Hg2+ sensor: rGO-blended PEDOT:PSS conducting polymer OFET. *Appl. Phys., A Mater. Sci. Process.,* **2021**, *127*(3), 167.
        [http://dx.doi.org/10.1007/s00339-021-04314-1]

Nanostructured Molecularly Imprinted Polymers in Electrochemical Sensing

Sajini T[1] and **Beena Mathew**[2,*]

[1] *Research & Post Graduate Department of Chemistry, St. Berchmans College (Autonomous), Affiliated to Mahatma Gandhi University, Changanassery-686101, Kerala, India*

[2] *School of Chemical Sciences, Mahatma Gandhi University, Priyadarsini Hills P O, Kottayam - 686560, Kerala, India*

Abstract: Molecular imprinted polymers (MIP) are one of the promising method in various research area in which artificial receptor sites of targeted molecule were fabricated on a polymer matrix. These polymers are analogues to naturally occuring antigen-antibody system. Due to its high recognition capabilty and structural specificity towards the target molecule, these kind of polymers exhibits wide variety of applications in various fileds. Among the temendous applications, MIPs in electrochemical sensing got much attention in recent years. Innovative developments in nanochemistry again improve its applications in electrochemical sensing. In this chapter, we detailed the significance of nanostructured MIP focusing on multiwalled carbon nanotubes as supporting material in elecrochemical sensing applications. It presents recent progresses associated to molecularly imprinted electrochemical sensors based multiwalled carbon nanotubes.

Keywords: Applications, Electrochemical sensor, Molecularly imprinted polymer, MWCNTs, Nanostructured MIP.

INTRODUCTION

Molecular imprinting technology is the most developing area of research in which artificial binding sites of target or template molecule were created on a polymer network. A typical molecular imprinting process involves the fabrication of a pre-organized complex of chosen template molecule and its complementary functional monomer followed by a crosslinking polymerization in the addition of a suitable initiator [1]. Subsequently, the extraction of a target molecule using suitable eluents, complementary cavities of template molecule remains as such in the poly-

---

* **Corresponding author Beena Mathew:** School of Chemical Sciences, Mahatma Gandhi University, Priyadarsini Hills P O, Kottayam -686560, Kerala, India; Email: beenamscs@gmail.com

Manorama Singh, Vijai K Rai and Ankita Rai (Eds.)

mer matrix and the polymer composite formed were designated as molecularly imprinted polymer (MIP).

(Fig. **1**) depicted the schematic representation of the formation of conventional MIP.

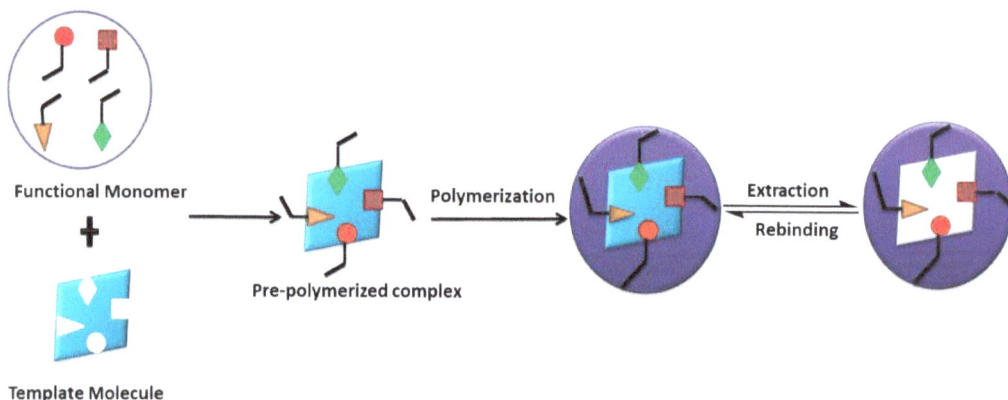

Functional Monomer

Pre-polymerized complex

Polymerization

Extraction

Rebinding

Template Molecule

Fig. (1). Schematic representation of molecular imprinting process.

Electrochemical sensing mechanism involves the transformation of the interaction of the template molecule with a receptor on an electrode surface, which acts as a transducer, into a useful analytical signal. Mosbach and Haupt were the first to discover the electrochemical sensor with MIPs, which they used to create a MIPs-coated electrode known as a molecularly imprinted electrochemical sensor (MIECS) in 1999 [2]. The electrochemical sensor generated from MIPs possesses both the properties of detection and transduction. The advantage of MIECS includes high detection characteristics, low cost and ease of fabrication and simple automation. These MIECS have got a lot of coverage for their ability to track a wide range of molecules, including organic compounds, emerging pollutants, biomolecules, heavy metals and more.

Although biosensor technology has progressed dramatically [3], there are still some problems with biological materials used in biosensing, such as (i) they are less stable, (ii) expensive, and (iii) lack of receptor sites that are capable of differentiating certain template molecules. Due to its high stability, MIP can be ideal substitute for the biological receptors in sensors [4]. The common mechanism of MIP based electrochemical sensors comprises the utilization of MIP electropolymerized or immobilized onto the electrode surface in order to accomplish the different electrochemical analysis [5]. MIP has been established for sensing sugars, herbicides, derivatives of nucleic and amino acids, toxins, drugs and solvents [6]. MIP is used in other emerging fields such as mass sensitive sensing devices, microbalance-based and fluorescence-based sensors, and in electrochemical methods includes, chemical luminance, surface plasmon resonance spectroscopy, and conductivity sensing devices [7 - 9].

In order to improve the high sensitivity and specific selectivity of the MIP based chemo-biosensors, nanomaterials, which have remarkable physicochemical properties and characteristics, were comprehensively introduced in recent years. Nanomaterials have got much attention as the fundamental materials during the MIP fabrication and in order to improve and strengthen the electrochemical signal, it acts as an electrode surface modifier. Nanostructured materials have the extremely large surface area and high electrical and thermal conductivity and hence exhibit an electrocatalytic property. The majority of literature reports that the MIP based electrochemical sensors are fabricated in association with nanomaterials, including graphene, magnetic nanoparticles, MWCNT, and gold nanoparticles which are illustrated in Fig. (**2**).

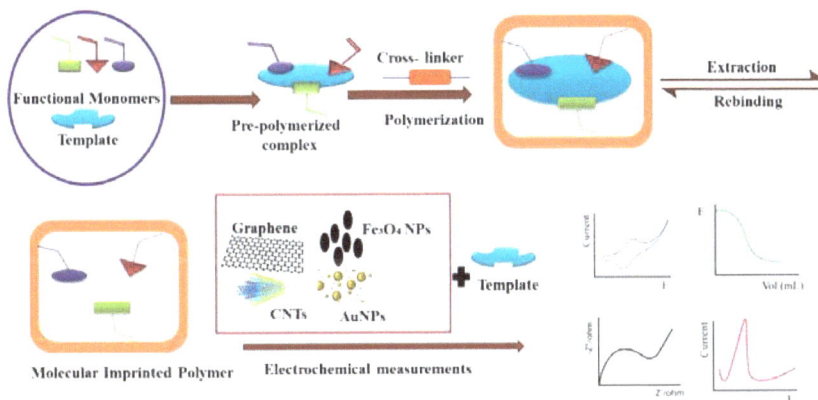

Fig. (2). Schematic representation of nanostructured MIP fabrication as electrochemical sensor.

NANOSTRUCTURED MOLECULARLY IMPRINTED POLYMER (MIP)

The process of molecular imprinting can generate complementary recognition sites with high selectivity and specificity of template molecule in a crosslinked polymeric matrix. The efficacy of MIP technique is directly dependent upon the bonding nature of template and functional monomer pre-polymerized complex [10 - 13], nature of the imprinted materials [14 - 17] and the flexibility of the polymeric network [18, 19]. Though the traditional imprinting protocol is effective and simple, it possesses major difficulties like (i) highly crosslinked bulky imprinted cavities are formed which exhibit uneven size and shape, (ii) the removal of target molecules from the bulk part of the polymeric network is somewhat challenging which diminishes the capacity of rebinding analytes [19, 20], (iii) rate of the binding reaction reduces [21, 22], (iv) high rigid polymeric network diminishes the conformational choice of target detection [23, 24] (v) most of the bulk imprinted cavities are unavailable for further rebinding due to the heterogeneous binding site formation [25 - 27], and (vi) template binding at conventional MIP materials lacks output signal due to the poor assembly at the electrode surface which restraining its usage in chemo-bio sensor applications [28]. These limitations and heterogeneous binding sites of non-covalent MIP can be endured in some analytical applications; some novel methods have been adopted to enhance the formation of homogeneous binding cavities of MIP [29 - 31].

The emergence of novel nanomaterials in MIP methodology have got significant research interest due to its extensive applications in chemosensors and bioassays [14,-16, 32, 33]. Nanomaterials possess extremely high electrical, mechanical, catalytic, thermal, optical and magnetic characteristics [34, 35] which are different from bulk materials synthesized using similar chemical composition. The good dispersion capacity and the high surface-to-volume ratio delivers large homogeneous adsorption sites which are complementary to template molecules [14 - 16]. Specifically, the chemical immobilization of biological receptors like enzymes and antibodies or carbon-based functional moieties at the surface of nanomaterials should promote its specificity and selectivity of recognition and detection [34, 36] Functionalized inorganic or organic nanostructured based chemo-bio sensors have been discovered for the sensing of many templates, like quantum dots for the fluorescent detections of nitroaromatic explosives [37], enzyme activity [38] and DNA [39], and functionalized carbon nanotubes for the electrochemical recognition of various biological/organic templates molecule [40, 38]. The nanotechnology based MIP simply nano-imprinting is anticipated to improve the interaction and affinity of imprinted materials and hence offer various applications in biological receptors [41]. Nano-imprinted materials possess a large surface-to-volume ratio, and thus, many target molecules are located on the surface. The imprinted cavities on the nanostructures exhibit regular shapes and sizes, enhanced dispersion in template solutions and hence diminish the mass transfer resistance, displaying high binding kinetics [14, 15,42].

Numerous nano-imprinted materials have already been established, including imprinted nanoparticles (Imp-NPs) [43], imprinted hybrid materials [44] and imprinted nanocomposites (Imp-NCs) [45], for diverse applications, such as in separation science [46], chemical sensing [47] and molecular recognition [48]. Among these, imprinted nanomaterials have got more attention due to their high surface area, high chemical, physical and thermal stabilities, formation of an excellent number of complementary binding cavities, less expensive production methods, comparative simplicity, and diverse application to a different nature of analytes [49, 50].

MULTI-WALLED CARBON NANOTUBE-BASED MIP

Among various nanomaterials, multiwalled carbon nanotube (MWCNT) has gained extensive consideration due to its high electrical, mechanical and thermal conduction characteristics [51 - 53]. MWCNT, with an exceedingly high surface-to-volume ratio, is an outstanding candidate as the support material in imprinting technology. When MIP was fabricated on the exterior surface of MWCNT, it results the enhancement of number of homogeneous binding cavities [54 - 56]. Subsequently formed binding cavities present in the exterior part of the composite would develop the template accessibility and hence diminish the binding time [57]. Some problems associated with the usage of carbon nanotubes are its low dispersion and aggregation of nanotube tubules which affects its chemical reactivity. One method adopted to overcome these limitations is surface functionalization of nanotubes *via* chemical modification. For the fabrication of a homogeneous MIP layer on the surface of MWCNT, initially MWCNT is converted to carboxyl functionalized MWCNT *via* simple, strong acid treatment [58]. Subsequent functionalization to vinyl group incorporated MWCNT followed by polymerization in the presence of an initiator, cross-linker and functional monomer would help the formation of a homogeneous MIP on the outer layer surface of MWCNT.

MWCNT-MIP COMPOSITES IN ELECTROCHEMICAL SENSING

Among the various carbon nanomaterial, MWCNT based electrochemical sensors possess significant features such as a) enhanced electroactive surface area distribution, b) low electron transfer resistance, c) and support for the adsorption of template molecule [59]. Hence these materials are ideal for numerous applications in chemo and biosensor. A general schematic sketch for the fabrication of an electrochemical sensor by MWCNT- imprinted polymer is depicted in Fig. (**3**).

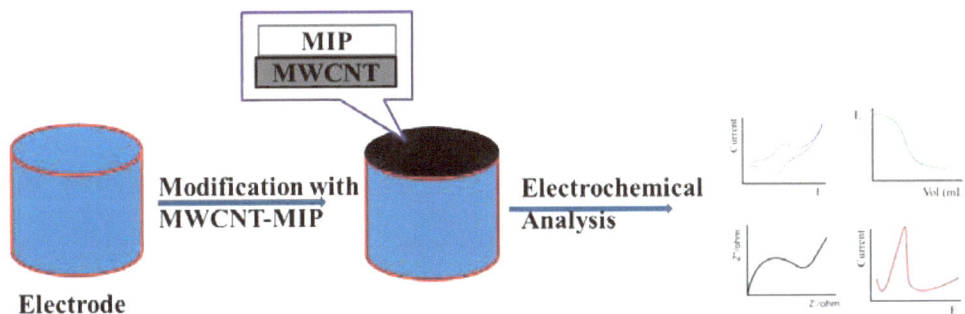

Fig. (3). General schematic sketch of formation of MWCNT- MIP as electrochemical sensor.

Functionalised surfaces of MWCNT were used for the specific detection and recognition of cimetidine [60], aspirin [61], tyramine [62], ribavirin [57] and S-ibuprofen [63] from its structurally related compounds. Electrochemical sensors for the detection of environmentally hazardous metal ions, including lead [64], cadmium [65], chromium [66] and cobalt [67] ions were also discovered where MIP was fabricated on the surface layer of vinyl functionalized MWCNT *via* imprinting approach. Another electrochemical sensor for the detection of bisphenol A was developed through the combination of a MIP approach with a MWCNT paste electrode [68]. Chen *et al.* investigated that potentiometric sensor based MWCNT-MIP was a good support material for the effective sensing of γ-hexachlorocyclohexane (γ-HCCH) molecules [69].

A molecularly imprinted film on a glassy carbon electrode (GCE) for the selective and specific recognition for 2-nonylphenol (2-NP) was discovered by Zhang *et al.*, which is established on the differential pulse voltammetry [70]. Here, the MIP film was immobilised on the glassy carbon electrode by electrodeposition of imprinted sol-gel involving MWCNT-nafion composite. Kan *et al.* investigated composites of MWCNT and MIP for the recognition of dopamine [71]. For the detection of dopamine in real samples, Anirudhan's group fabricated a surface modified MWCNT based MIP using the potentiometric method [72].

The synthesis of polymer-nano composites based on molecularly imprinted poly(methacrylic acid) attached on an iniferter-modified carbon nanotube using living-radical polymerization for the adsorption of 17β-estradiol is developed by Prete's group [73]. MIP based on the magnetic MWCNT for specific separation and spectrophotometric detection of melamine in milk samples was fabricated by Hashemi's group [74]. An electrochemical sensor for warfarin detection was also reported, which developed *via* molecular imprinting methodology using electropolymerization of o-phenylenediamine (o-PD) on a GCE by cyclic voltammetry. To improve the sensitivity of electrode surface and electronic transmission in this system, carboxyl functionalized MWCNT were introduced on the surface of GCE [75].

CNTs have established abundant applications for the improvement of electro-chemical sensors based MIP [75 - 79]. Zhang *et al.* reported a nanocomposite of MIP wrapped on the exterior part of MWCNT and fabricated the respective sensor *via* potentiodynamic electropolymerization of poly(3-aminophenyl boronic acid) (PAPBA) and it acts as an active electrochemical sensor for the detection of epinephrine [77]. The synthesized PAPBA (MIP)/MWCNT modified electrode exhibited a lesser charge transfer resistance and displayed a very good electrochemical performance for epinephrine recognition compared with electrodes modified with non-imprinted polymers layered on MWCNT. Sensor for the recognition of environmentally hazardous metal ions such as lead [64], cadmium [65], chromium [66] and cobalt [67] ion were prepared on vinyl functionalized MWCNT *via* molecularly imprinting approach. Sensors for bisphenol A [68] and tripiramine [78] were also fabricated on MWCNT paste electrode. Chen *et al.* investigated that potentiometric sensor based MWCNT-MIP is a good material for the effective detection of γ-hexachlorocyclohexane (γ-HCCH) molecules [69]. Rezaei *et al.* developed a MIP composite of Au nanoparticle and MWCNT for the sensing of warfarine [75].

Table 1. MWCNT based MIP on electrochemical sensing applications.

MIP-MWCNT Composites	Target Molecule	LOD (mol/L)	Reference
MIP-MWCNT/PMAA	Uric acid	22×10^{-11}	[80]
MIP-MWCNTs	Rutin	5.0×10^{-5}	[81]
MIP-polypyrrole sulfonated graphene/ hyaluronicacid-MWNTs/paraaminobenzoic acid	Tryptamine	7.4×10^{-8}	[82]
MWCNTs-MIP/glassy carbon electrode	Nitrocellulose	3.45×10^{-10}	[83]
MIP/MWCNT/GCE	Ochratoxin A	1.7×10^{-6}	[84]
MIP/acrylamide-MWCNT/GCE	Nonylphenol	0.02×10^{-6}	[85]
MIP-f-MWCNTs	Parathion-methyl	6.7×10^{-8}	[86]
MIP/MWCNT/SPCE	Naloxone	0.20×10^{-6}	[87]

Recently Francois *et al.* developed a gold nanoparticle (AuNP)-coated MWCNT wrapped in polymeric chitosan (CS) network and the fabricated nano-chitosan composite used to create a MIP in the presence of catechol on a boron-doped diamond (BDD) electrode [88]. The electrochemical response of the sensor has analyzed by CV measurements where outstanding reproducibility and specificity to catechol detection in the range of 0 to 1 mM was found, with a limit of detection value 3.7×10^{-5} M.

MIP based electrochemical sensor for the specific detection of dopamine (DA) was fabricated with MWCNT spaced graphene aerogels (MWCNT/GAs) as the sensing substrate and polypyrrole (PPy) as the functional monomer in MIP by Limin's group [89]. Their work again proves the conductive nature of MWCNTs, which greatly enhanced the electrical conductivity and hence an increased electrochemical performance of the composite aerogel. A voltammetric sensor based on molecularly imprinted poly(acrylic acid)-MWCNT nanocomposite drop-coated onto GCE was fabricated and applied to tramadol (TR) determination in pharmaceutical samples by Ricardo's group [90].

In comparison with other classical approaches of electrochemical sensing analysis, MWCNT supported electrodes can efficiently enhance the recognition signal of the electrochemical sensors. The hydrophobic nature of MWCNTs limits its range of recognition in other biomedical fields. MWCNTs possess the least stability and adhesion, cytotoxic effects, and they are biologically incompatible, which limit the emergence of the electrochemical sensors. Different surface functionalized MWCNTs were used to synthesize the functional groups on the nanotube surface, and these groups can inhibit the agglomeration of nanotube, but the performances of the MIP membrane are also affected [59]. Hence some more works have to be done on the functionalized nanotubes in MIPs.

CONCLUDING REMARKS

The present chapter has attempted to summarize various features of MWCNT nanostructured molecular imprinting technologies for electrochemical applications. MIP possesses fabulous advantages like less expense, stability, durability, excellent reusability, and biomimetic properties in comparison with natural receptors like antibodies and enzymes. Recently, there have been trends in the growth of nano-sized MIP for various applications especially in biomimetic sensors. The advantage of catalytic and high electrical conductivity characteristics delivered by the nano counterpart and the greater selectivity provided by the MIP made them an effective tool for electrochemical sensing of different kinds of templates. MIP is an auspicious constituent that could be attached to different kinds of nanostructured materials to develop different sensitive, specific and selective electrochemical devices. The outstanding performances of MIP make these kinds of polymers the most favourable artificial materials for template recognition and detection in various fields of science and technology. Novel alternatives for MIP synthesis strategies to enhance the properties of MIP particles are needed and it should promote future developments in sensing applications.

CONSENT FOR PUBLICATION

Not applicable.

CONFLICT OF INTEREST

The author declares no conflict of interest, financial or otherwise.

ACKNOWLEDGEMENTS

Declared none.

REFERENCES

[1]   Sajini, T.; Gigimol, M.G.; Mathew, B. A Brief Overview of Molecularly Imprinted Polymers Supported on Titanium Dioxide Matrices. *Mater. Today Chem.,* **2019**, *11*, 283-295.
[http://dx.doi.org/10.1016/j.mtchem.2018.11.010]

[2]   Cui, B.; Liu, P.; Liu, X.; Liu, S.; Zhang, Z. Molecularly Imprinted Polymers for Electrochemical Detection and Analysis: Progress and Perspectives. *J. Mater. Res. Technol.,* **2020**, *9*(6), 12568-12584.
[http://dx.doi.org/10.1016/j.jmrt.2020.08.052]

[3]   Cormack, P.A.G.; Mosbach, K. Molecular Imprinting: Recent Developments and the Road Ahead. *React. Funct. Polym.,* **1999**, *41*(1–3), 115-124.
[http://dx.doi.org/10.1016/S1381-5148(99)00024-3]

[4]   Domb, A.J.; Israel, E. *Preparation of Biologically Active Molecules by Molecular Imprinting. Patent ,* **1997**. 5630978.

[5]   Lahcen, A.A.; Amine, A. Recent Advances in Electrochemical Sensors Based on Molecularly Imprinted Polymers and Nanomaterials. *Electroanalysis,* **2019**, *31*(2), 188-201.
[http://dx.doi.org/10.1002/elan.201800623]

[6]   Hillberg, A.L.; Brain, K.R.; Allender, C.J. Molecular imprinted polymer sensors: implications for therapeutics. *Adv. Drug Deliv. Rev.,* **2005**, *57*(12), 1875-1889.
[http://dx.doi.org/10.1016/j.addr.2005.07.016] [PMID: 16257082]

[7]   Haupt, K.; Mosbach, K. Molecularly imprinted polymers in chemical and biological sensing. *Biochem. Soc. Trans.,* **1999**, *27*(2), 344-350.
[http://dx.doi.org/10.1042/bst0270344] [PMID: 10093761]

[8]   Sing, M.N.; Narayanaswamy, R. Molecularly Imprinted β-Cyclodextrin Polymer as Potential Optical Receptor for the Detection of Organic Compound. *Sens. Actuators B Chem.,* **2009**, *139*(1), 156-165.
[http://dx.doi.org/10.1016/j.snb.2008.10.035]

[9]   Öpik, A.; Menaker, A.; Reut, J.; Syritski, V. Molecularly Imprinted Polymers: A New Approach to the Preparation of Functional Materials. *Proc. Est. Acad. Sci.,* **2009**, *58*(1), 3-11.
[http://dx.doi.org/10.3176/proc.2009.1.01]

[10]   Andersson, L.I.; Mosbach, K. Enantiomeric resolution on molecularly imprinted polymers prepared with only non-covalent and non-ionic interactions. *J. Chromatogr. A,* **1990**, *516*(2), 313-322.
[http://dx.doi.org/10.1016/S0021-9673(01)89273-6] [PMID: 2079492]

[11]   Andersson, L.I.; Miyabayashi, A.; O'Shannessy, D.J.; Mosbach, K. Enantiomeric resolution of amino acid derivatives on molecularly imprinted polymers as monitored by potentiometric measurements. *J. Chromatogr. A,* **1990**, *516*(2), 323-331.
[http://dx.doi.org/10.1016/S0021-9673(01)89274-8] [PMID: 2079493]

[12]   Ramström, O.; Ye, L.; Mosbach, K. Artificial antibodies to corticosteroids prepared by molecular imprinting. *Chem. Biol.,* **1996**, *3*(6), 471-477.
[http://dx.doi.org/10.1016/S1074-5521(96)90095-2] [PMID: 8807877]

[13]   Spivak, D.A. Optimization, evaluation, and characterization of molecularly imprinted polymers. *Adv. Drug Deliv. Rev.,* **2005**, *57*(12), 1779-1794.
[http://dx.doi.org/10.1016/j.addr.2005.07.012] [PMID: 16260064]

[14]   Xie, C.; Zhang, Z.; Wang, D.; Guan, G.; Gao, D.; Liu, J. Surface molecular self-assembly strategy for TNT imprinting of polymer nanowire/nanotube arrays. *Anal. Chem.,* **2006**, *78*(24), 8339-8346.
[http://dx.doi.org/10.1021/ac0615044] [PMID: 17165825]

[15]   Xie, C.; Liu, B.; Wang, Z.; Gao, D.; Guan, G.; Zhang, Z. Molecular imprinting at walls of silica nanotubes for TNT recognition. *Anal. Chem.,* **2008**, *80*(2), 437-443.
[http://dx.doi.org/10.1021/ac701767h] [PMID: 18088103]

[16]   Gao, D.; Zhang, Z.; Wu, M.; Xie, C.; Guan, G.; Wang, D. A surface functional monomer-directing

strategy for highly dense imprinting of TNT at surface of silica nanoparticles. *J. Am. Chem. Soc.,* **2007**, *129*(25), 7859-7866.
[http://dx.doi.org/10.1021/ja070975k] [PMID: 17550249]

[17]   Guan, G.; Zhang, Z.; Wang, Z.; Liu, B.; Gao, D.; Xie, C. Single-Hole Hollow Polymer Microspheres toward Specific High-Capacity Uptake of Target Species. *Adv. Mater.,* **2007**, *19*(17), 2370-2374.
[http://dx.doi.org/10.1002/adma.200700984]

[18]   Ki, C.D.; Oh, C.; Oh, S.G.; Chang, J.Y. The use of a thermally reversible bond for molecular imprinting of silica spheres. *J. Am. Chem. Soc.,* **2002**, *124*(50), 14838-14839.
[http://dx.doi.org/10.1021/ja0277881] [PMID: 12475312]

[19]   Michael, M.; Paul, K.; Gang, D.; Paul, S.; Jonathan, D.; Douglas, C.; Gaber, B.P. Catalytic Silica Particles *via* Template-Directed Molecular Imprinting. *Langmuir,* **2000**, *16*(4), 1759-1765.
[http://dx.doi.org/10.1021/la990809t]

[20]   Mukti, S. R.; Dave, B. C. Selective Intake and Release of Proteins by Organically-Modified Silica Sol–gels. *J. Am. Chem. Soc.,* **1998**, *120*(50), 13270-13271.
[http://dx.doi.org/10.1021/ja982468w]

[21]   Carter, S.R.; Rimmer, S. Surface Molecularly Imprinted Polymer Core–shell Particles. *Adv. Funct. Mater.,* **2004**, *14*(6), 553-561.
[http://dx.doi.org/10.1002/adfm.200305069]

[22]   Carter, S.R.; Rimmer, S. Molecular Recognition of Caffeine by Shell Molecular Imprinted Core-Shell Polymer Particles in Aqueous Media. *Adv. Mater.,* **2002**, *14*(9), 667-670.
[http://dx.doi.org/10.1002/1521-4095(20020503)14:9<667::AID-ADMA667>3.0.CO;2-3]

[23]   Espinosa-García, B.M.; Argüelles-Monal, W.M.; Hernández, J.; Félix-Valenzuela, L.; Acosta, N.; Goycoolea, F.M. Molecularly imprinted chitosan-genipin hydrogels with recognition capacity toward o-xylene. *Biomacromolecules,* **2007**, *8*(11), 3355-3364.
[http://dx.doi.org/10.1021/bm700458a] [PMID: 17944514]

[24]   Aburto, J.; Le Borgne, S. Selective Adsorption of Dibenzothiophene Sulfone by an Imprinted and Stimuli-Responsive Chitosan Hydrogel. *Macromolecules,* **2004**, *37*(8), 2938-2943.
[http://dx.doi.org/10.1021/ma049961e]

[25]   Rampey, A.M.; Umpleby, R.J., II; Rushton, G.T.; Iseman, J.C.; Shah, R.N.; Shimizu, K.D. Characterization of the imprint effect and the influence of imprinting conditions on affinity, capacity, and heterogeneity in molecularly imprinted polymers using the Freundlich isotherm-affinity distribution analysis. *Anal. Chem.,* **2004**, *76*(4), 1123-1133.
[http://dx.doi.org/10.1021/ac0345345] [PMID: 14961747]

[26]   Katz, A.; Davis, M.E. Molecular imprinting of bulk, microporous silica. *Nature,* **2000**, *403*(6767), 286-289.
[http://dx.doi.org/10.1038/35002032] [PMID: 10659842]

[27]   Bass, J.D.; Katz, A. Thermolytic Synthesis of Imprinted Amines in Bulk Silica. *Chem. Mater.,* **2003**, *15*(14), 57-2763.
[http://dx.doi.org/10.1021/cm021822t]

[28]   Holthoff, E.L.; Bright, F.V. Molecularly templated materials in chemical sensing. *Anal. Chim. Acta,* **2007**, *594*(2), 147-161.
[http://dx.doi.org/10.1016/j.aca.2007.05.044] [PMID: 17586109]

[29]   Umpleby, R.J.; Rushton, G.T.; Shah, R.N.; Rampey, A.M.; Bradshaw, J.C.; Berch, J.K.; Shimizu, K.D. Recognition Directed Site-Selective Chemical Modification of Molecularly Imprinted Polymers. *Macromolecules,* **2001**, *34*(24), 8446-8452.
[http://dx.doi.org/10.1021/ma010903s]

[30]   Kirsch, N.; Alexander, C.; Lübke, M.; Whitcombe, M.J.; Vulfson, E.N. Enhancement of Selectivity of Imprinted Polymers *via* Post-Imprinting Modification of Recognition Sites. *Polymer (Guildf.),* **2000**,

41(15), 5583-5590.
[http://dx.doi.org/10.1016/S0032-3861(99)00782-X]

[31]   Patel, A.; Fouace, S.; Steinke, J.H. Novel Stereoselective Molecularly Imprinted Polymers *via* Ring-Opening Metathesis Polymerisation. *Anal. Chim. Acta,* **2004**, *504*(1), 53-62.
[http://dx.doi.org/10.1016/j.aca.2003.08.069]

[32]   Lu, C.H.; Zhou, W.H.; Han, B.; Yang, H.H.; Chen, X.; Wang, X.R. Surface-imprinted core-shell nanoparticles for sorbent assays. *Anal. Chem.,* **2007**, *79*(14), 5457-5461.
[http://dx.doi.org/10.1021/ac070282m] [PMID: 17563116]

[33]   Vandevelde, F.; Belmont, A.S.; Pantigny, J.; Haupt, K. Hierarchically Nanostructured Polymer Films Based on Molecularly Imprinted Surface-Bound Nanofilaments. *Adv. Mater.,* **2007**, *19*(21), 3717-3720.
[http://dx.doi.org/10.1002/adma.200700600]

[34]   Jianrong, C.; Yuqing, M.; Nongyue, H.; Xiaohua, W.; Sijiao, L. Nanotechnology and biosensors. *Biotechnol. Adv.,* **2004**, *22*(7), 505-518.
[http://dx.doi.org/10.1016/j.biotechadv.2004.03.004] [PMID: 15262314]

[35]   Banholzer, M.J.; Millstone, J.E.; Qin, L.; Mirkin, C.A. Rationally designed nanostructures for surface-enhanced Raman spectroscopy. *Chem. Soc. Rev.,* **2008**, *37*(5), 885-897.
[http://dx.doi.org/10.1039/b710915f] [PMID: 18443674]

[36]   Zhuo, Y.; Yuan, P.X.; Yuan, R.; Chai, Y.Q.; Hong, C.L. Nanostructured conductive material containing ferrocenyl for reagentless amperometric immunosensors. *Biomaterials,* **2008**, *29*(10), 1501-1508.
[http://dx.doi.org/10.1016/j.biomaterials.2007.12.007] [PMID: 18166224]

[37]   Tu, R.; Liu, B.; Wang, Z.; Gao, D.; Wang, F.; Fang, Q.; Zhang, Z. Amine-capped ZnS-Mn2+ nanocrystals for fluorescence detection of trace TNT explosive. *Anal. Chem.,* **2008**, *80*(9), 3458-3465.
[http://dx.doi.org/10.1021/ac800060f] [PMID: 18336012]

[38]   Jiang, H.; Ju, H. Enzyme-quantum dots architecture for highly sensitive electrochemiluminescence biosensing of oxidase substrates. *Chem. Commun. (Camb.),* **2007**, (4), 404-406.
[http://dx.doi.org/10.1039/B616007G] [PMID: 17220985]

[39]   Suzuki, M.; Husimi, Y.; Komatsu, H.; Suzuki, K.; Douglas, K.T. Quantum dot FRET biosensors that respond to pH, to proteolytic or nucleolytic cleavage, to DNA synthesis, or to a multiplexing combination. *J. Am. Chem. Soc.,* **2008**, *130*(17), 5720-5725.
[http://dx.doi.org/10.1021/ja710870e] [PMID: 18393422]

[40]   Huang, F.; Peng, Y.; Jin, G.; Zhang, S.; Kong, J. Sensitive Detection of Haloperidol and Hydroxyzine at Multi-Walled Carbon Nanotubes-Modified Glassy Carbon Electrodes. *Sensors (Basel),* **2008**, *8*(3), 1879-1889.
[http://dx.doi.org/10.3390/s8031879] [PMID: 27879799]

[41]   O'Connor, N.A.; Paisner, D.A.; Huryn, D.; Shea, K.J. Screening of 5-HT1A receptor antagonists using molecularly imprinted polymers. *J. Am. Chem. Soc.,* **2007**, *129*(6), 1680-1689.
[http://dx.doi.org/10.1021/ja067276c] [PMID: 17284006]

[42]   Yoshimatsu, K.; Ye, L.; Stenlund, P.; Chronakis, I.S. A simple method for preparation of molecularly imprinted nanofiber materials with signal transduction ability. *Chem. Commun. (Camb.),* **2008**, (17), 2022-2024.
[http://dx.doi.org/10.1039/b719586a] [PMID: 18536808]

[43]   Poma, A.; Guerreiro, A.; Whitcombe, M.J.; Piletska, E.V.; Turner, A.P.; Piletsky, S.A. Solid-Phase Synthesis of Molecularly Imprinted Polymer Nanoparticles with a Reusable Template - "Plastic Antibodies". *Adv. Funct. Mater.,* **2013**, *23*(22), 2821-2827.
[http://dx.doi.org/10.1002/adfm.201202397] [PMID: 26869870]

[44]   Lakshmi, D.; Bossi, A.; Whitcombe, M.J.; Chianella, I.; Fowler, S.A.; Subrahmanyam, S.; Piletska,

E.V.; Piletsky, S.A. Electrochemical sensor for catechol and dopamine based on a catalytic molecularly imprinted polymer-conducting polymer hybrid recognition element. *Anal. Chem.,* **2009**, *81*(9), 3576-3584.
[http://dx.doi.org/10.1021/ac802536p] [PMID: 19354259]

[45]   Matsui, J.; Takayose, M.; Akamatsu, K.; Nawafune, H.; Tamaki, K.; Sugimoto, N. Molecularly imprinted nanocomposites for highly sensitive SPR detection of a non-aqueous atrazine sample. *Analyst (Lond.),* **2009**, *134*(1), 80-86.
[http://dx.doi.org/10.1039/B803350A] [PMID: 19082178]

[46]   Cheong, W.J.; Yang, S.H.; Ali, F. Molecular imprinted polymers for separation science: a review of reviews. *J. Sep. Sci.,* **2013**, *36*(3), 609-628.
[http://dx.doi.org/10.1002/jssc.201200784] [PMID: 23281278]

[47]   Li, J.; Wei, G.; Zhang, Y. Molecularly Imprinted Polymers as Recognition Elements in Sensors.*Molecularly Imprinted Sensors*; Elsevier, **2012**, pp. 35-55.
[http://dx.doi.org/10.1016/B978-0-444-56331-6.00002-5]

[48]   Guan, G.; Liu, B.; Wang, Z.; Zhang, Z. Imprinting of Molecular Recognition Sites on Nanostructures and Its Applications in Chemosensors. *Sensors (Basel),* **2008**, *8*(12), 8291-8320.
[http://dx.doi.org/10.3390/s8128291] [PMID: 27873989]

[49]   He, C.; Long, Y.; Pan, J.; Li, K.; Liu, F. Application of molecularly imprinted polymers to solid-phase extraction of analytes from real samples. *J. Biochem. Biophys. Methods,* **2007**, *70*(2), 133-150.
[http://dx.doi.org/10.1016/j.jbbm.2006.07.005] [PMID: 17107715]

[50]   Menaker, A.; Syritski, V.; Reut, J.; Öpik, A.; Horváth, V.; Gyurcsányi, R.E. Electrosynthesized Surface-Imprinted Conducting Polymer Microrods for Selective Protein Recognition. *Adv. Mater.,* **2009**, *21*(22), 2271-2275.
[http://dx.doi.org/10.1002/adma.200803597]

[51]   Hone, J. *Carbon Nanotubes : Thermal Properties; Dekker Encyclopedia of Nanoscience and Nanotechnology*; Marcel Dekker: Newyork, **2004**.

[52]   Ciraci, S.; Dag, S.; Yildirim, T.; Gulseren, O.; Senger, R.T. Functionalized Carbon Nanotubes and Device Applications. *J. Phys. Condens. Matter,* **2004**, *16*(29), 901-960.
[http://dx.doi.org/10.1088/0953-8984/16/29/R01]

[53]   Valle, M.; Pumera, M.; Llopis, X.; Pe, B. New Materials for Electrochemical Sensing VI. *Trends Analyt. Chem.,* **2005**, *24*(9), 826-838.
[http://dx.doi.org/10.1016/j.trac.2005.03.019]

[54]   Belin, T.; Epron, F. Characterization Methods of Carbon Nanotubes : A Review. *Mater. Sci. Eng. B,* **2005**, *119*(2), 105-118.
[http://dx.doi.org/10.1016/j.mseb.2005.02.046]

[55]   Seetharamappa, J.; Yellappa, S.; D'Souza, F. Carbon Nanotubes : Next Generation of Electronic Materials. *Electrochem. Soc. Interface,* **2006**, *15*(2), 23-26.
[http://dx.doi.org/10.1149/2.F02062IF]

[56]   Yang, W.; Thordarson, P.; Gooding, J.J.; Ringer, S.P.; Braet, F. Carbon Nanotubes for Biological and Biomedical Applications. *Nanotechnology,* **2007**, *18*(41), 412001.
[http://dx.doi.org/10.1088/0957-4484/18/41/412001]

[57]   Xu, L.; Xu, Z. Molecularly Imprinted Polymer Based on Multiwalled Carbon Nanotubes for Ribavirin Recognition. *J. Polym. Res.,* **2012**, *19*(8), 9942.
[http://dx.doi.org/10.1007/s10965-012-9942-1]

[58]   Rosca, I.D.; Watari, F.; Uo, M.; Akasaka, T. Oxidation of Multiwalled Carbon Nanotubes by Nitric Acid. *Carbon,* **2005**, *43*(15), 3124-3131.
[http://dx.doi.org/10.1016/j.carbon.2005.06.019]

[59]   Zhong, C.; Yang, B.; Jiang, X.; Li, J. Current Progress of Nanomaterials in Molecularly Imprinted

Electrochemical Sensing. *Crit. Rev. Anal. Chem.,* **2018**, *48*(1), 15-32.
[http://dx.doi.org/10.1080/10408347.2017.1360762] [PMID: 28777018]

[60] Sooraj, M.P.; Mathew, B. Histamine H 2 -Receptor Antagonist Imprinted-Poly (Vinylimidazole). *Grafted Multiwalled Carbon Nanotubes. J. Chem. Pharm. Res.,* **2014**, *6*(12), 394-402.

[61] Sooraj, M.P.; Mathew, B. Artificial Recognition Sorbents on Multiwalled Carbon Nanotubes for the Separation of Aspirin from Its Structural Analogues. *Int. J. Sci. Eng. Res.,* **2015**, *6*(9), 723-733.

[62] Ashok, R.; Mathew, B. Tailoring of Sorbents for the Selective Recognition of Tyramine Based on Molecular Imprinting Approach on Multiwalled Carbon Nanotubes. *Res. J. Recent Sci.,* **2014**, *3*, 163-169.

[63] Sooraj, M.P.; Mathew, B. Highly Selective Nanocomposite Sorbents for the Specific Recognition of S-Ibuprofen from Structurally Related Compounds. *Appl. Nanosci.,* **2016**, *6*(5), 659-671.
[http://dx.doi.org/10.1007/s13204-015-0476-9]

[64] Sebastian, M.; Mathew, B. Ion Imprinting Approach for the Fabrication of an Electrochemical Sensor and Sorbent for Lead Ions in Real Samples Using Modified Multiwalled Carbon Nanotubes. *J. Mater. Sci.,* **2018**, *53*(5), 3557-3572.
[http://dx.doi.org/10.1007/s10853-017-1787-x]

[65] Aravind, A.; Mathew, B. Tailoring of Nanostructured Material as an Electrochemical Sensor and Sorbent for Toxic Cd (II) Ions from Various Real Samples. *J. Anal. Sci. Technol.,* **2018**, *9*(1), 22.
[http://dx.doi.org/10.1186/s40543-018-0153-1]

[66] Aravind, A.; Mathew, B. Electrochemical Sensor Based on Nanostructured Ion Imprinted Polymer for the Sensing and Extraction of Cr (III) Ions from Industrial Wastewater. *Polym. Int.,* **2018**, *67*(12), 1595-1604.
[http://dx.doi.org/10.1002/pi.5683]

[67] Sebastian, M.; Mathew, B. Multiwalled Carbon Nanotube Based Ion Imprinted Polymer as Sensor and Sorbent for Environmental Hazardous Cobalt Ion and Sorbent for Environmental Hazardous Cobalt Ion. *J. Macromol. Sci. Part A Pure Appl. Chem.,* **2018**, *55*(6), 455-465.
[http://dx.doi.org/10.1080/10601325.2018.1470463]

[68] Chen, Z.; Tang, C.; Zeng, Y.; Liu, H.; Yin, Z.; Li, L. Determination of Bisphenol a Using an Electrochemical Sensor Based on a Molecularly Imprinted Polymer-Modified Multiwalled Carbon Nanotube Paste Electrode. *Anal. Lett.,* **2014**, *47*(6), 996-1014.
[http://dx.doi.org/10.1080/00032719.2013.862624]

[69] Anirudhan, T.S.; Alexander, S. Design and fabrication of molecularly imprinted polymer-based potentiometric sensor from the surface modified multiwalled carbon nanotube for the determination of lindane (γ-hexachlorocyclohexane), an organochlorine pesticide. *Biosens. Bioelectron.,* **2015**, *64*, 586-593.
[http://dx.doi.org/10.1016/j.bios.2014.09.074] [PMID: 25310493]

[70] Zhang, J.; Niu, Y.; Li, S.; Luo, R.; Wang, C. A Molecularly Imprinted Electrochemical Sensor Based on Sol-Gel Technology and Multiwalled Carbon Nanotubes-Nafion Functional Layer for Determination of 2-Nonylphenol in Environmental Samples. *Sens. Actuators B Chem.,* **2014**, *193*, 844-850.
[http://dx.doi.org/10.1016/j.snb.2013.11.081]

[71] Anirudhan, T.S.; Alexander, S.; Lilly, A. Surface Modified Multiwalled Carbon Nanotube Based Molecularly Imprinted Polymer for the Sensing of Dopamine in Real Samples Using Potentiometric Method. *Polymer (Guildf.),* **2014**, *55*(19), 4820-4831.
[http://dx.doi.org/10.1016/j.polymer.2014.07.057]

[72] Kan, X.; Zhao, Y.; Geng, Z.; Wang, Z.; Zhu, J. Composites of Multiwalled Carbon Nanotubes and Molecularly Imprinted Polymers for Dopamine Recognition. *J. Phys. Chem. C,* **2008**, *112*(13), 4849-4854.
[http://dx.doi.org/10.1021/jp077445v]

[73]   Prete, M.C.; Dos Santos, D.M.; Effting, L.; Tarley, C.R.T. Preparation of Molecularly Imprinted Poly(Methacrylic Acid) Grafted on Iniferter-Modified Multiwalled Carbon Nanotubes by Living-Radical Polymerization for 17β-Estradiol Extraction. *J. Chem. Eng. Data,* **2019,** *64*(5), 1978-1990.
[http://dx.doi.org/10.1021/acs.jced.8b01010]

[74]   Hashemi, M.; Nazari, Z. Preparation of Molecularly Imprinted Polymer Based on the Magnetic Multiwalled Carbon Nanotubes for Selective Separation and Spectrophotometric Determination of Melamine in Milk Samples. *J. Food Compos. Anal.,* **2018,** *69*(February), 98-106.
[http://dx.doi.org/10.1016/j.jfca.2018.02.010]

[75]   Rezaei, B.; Rahmanian, O.; Ensafi, A.A. An Electrochemical Sensor Based on Multiwall Carbon Nanotubes and Molecular Imprinting Strategy for Warfarin Recognition and Determination. *Sens. Actuators B Chem.,* **2014,** *196,* 539-545.
[http://dx.doi.org/10.1016/j.snb.2014.02.037]

[76]   Valentini, F.; Amine, A.; Orlanducci, S.; Terranova, M.L.; Palleschi, G. Carbon nanotube purification: preparation and characterization of carbon nanotube paste electrodes. *Anal. Chem.,* **2003,** *75*(20), 5413-5421.
[http://dx.doi.org/10.1021/ac0300237] [PMID: 14710820]

[77]   Zhang, J.; Guo, X.T.; Zhou, J.P.; Liu, G.Z.; Zhang, S.Y. Electrochemical preparation of surface molecularly imprinted poly(3-aminophenylboronic acid)/MWCNTs nanocomposite for sensitive sensing of epinephrine. *Mater. Sci. Eng. C,* **2018,** *91,* 696-704.
[http://dx.doi.org/10.1016/j.msec.2018.06.011] [PMID: 30033304]

[78]   Akhoundian, M.; Alizadeh, T.; Ganjali, M.R.; Rafiei, F. A new carbon paste electrode modified with MWCNTs and nano-structured molecularly imprinted polymer for ultratrace determination of trimipramine: The crucial effect of electrode components mixing on its performance. *Biosens. Bioelectron.,* **2018,** *111,* 27-33.
[http://dx.doi.org/10.1016/j.bios.2018.03.061] [PMID: 29631160]

[79]   Bagheri, H.; Khoshsafar, H.; Amidi, S.; Hosseinzadeh Ardakani, Y. Fabrication of an Electrochemical Sensor Based on Magnetic Multi-Walled Carbon Nanotubes for the Determination of Ciprofloxacin. *Anal. Methods,* **2016,** *8*(16), 3383-3390.
[http://dx.doi.org/10.1039/C5AY03410H]

[80]   Chen, P.; Nien, P.; Hu, C.; Ho, K. Chemical Detection of Uric Acid Based on Multi-Walled Carbon Nanotubes Polymerized with a Layer of Molecularly Imprinted PMAA. *Sens. Actuators B Chem.,* **2010,** *146*(2), 466-471.
[http://dx.doi.org/10.1016/j.snb.2009.11.035]

[81]   Rezaei, B.; Majidi, N.; Ensafi, A.A.; Karimi-Maleh, H. Molecularly Imprinted-Multiwall Carbon Nanotube Paste Electrode as a Biosensor for Voltammetric Detection of Rutin. *Anal. Methods,* **2011,** *3*(11), 2510.
[http://dx.doi.org/10.1039/c1ay05271c]

[82]   Xing, X.; Liu, S.; Yu, J.; Lian, W.; Huang, J. Electrochemical sensor based on molecularly imprinted film at polypyrrole-sulfonated graphene/hyaluronic acid-multiwalled carbon nanotubes modified electrode for determination of tryptamine. *Biosens. Bioelectron.,* **2012,** *31*(1), 277-283.
[http://dx.doi.org/10.1016/j.bios.2011.10.032] [PMID: 22074810]

[83]   Meng, X.; Xiao, Z.; Scott, S.K. Preparation and Application of Electrochemical Sensor Based on Molecularly Imprinted Polymer Coated Multi-Walled Carbon Nanotubes for Nitrocellulose Detection. *Propellants Explos. Pyrotech.,* **2019,** *44*(10), 1337-1346.
[http://dx.doi.org/10.1002/prep.201900055]

[84]   Pacheco, J.G.; Castro, M.; Machado, S.; Barroso, M.F.; Nouws, H.P.A.; Delerue-Matos, C. Molecularly Imprinted Electrochemical Sensor for Ochratoxin A Detection in Food Samples. *Sens. Actuators B Chem.,* **2015,** *215,* 107-112.
[http://dx.doi.org/10.1016/j.snb.2015.03.046]

[85]   Liu, B.; Yan, J.; Wang, M.; Wu, X. Electrochemical Sensor Based on Molecularly Imprinted Polymer for Determination of Nonylphenol. *Int. J. Electrochem. Sci.,* **2018**, *13*, 11953-11960.
[http://dx.doi.org/10.20964/2018.12.17]

[86]   Zhang, D.; Yu, D.; Zhao, W.; Yang, Q.; Kajiura, H.; Li, Y.; Zhou, T.; Shi, G. A molecularly imprinted polymer based on functionalized multiwalled carbon nanotubes for the electrochemical detection of parathion-methyl. *Analyst (Lond.),* **2012**, *137*(11), 2629-2636.
[http://dx.doi.org/10.1039/c2an35338e] [PMID: 22498757]

[87]   Lopes, F.; Pacheco, J. G.; Rebelo, P.; Delerue-Matos, C. *Molecularly Imprinted Electrochemical Sensor Prepared on a Screen Printed Carbon Electrode for Naloxone Detection,* **2017**.
[http://dx.doi.org/10.1016/j.snb.2016.12.031]

[88]   Salvo-Comino, C.; Rassas, I.; Minot, S.; Bessueille, F.; Arab, M.; Chevallier, V.; Luz Rodriguez-Mendez, M.; Errachid, A.; Jaffrezic-Renault, N. *Metric Sensor Based on Molecularly Imprinted Chitosan-Carbon Nanotubes Decorated with Gold Nanoparticles Nanocomposite Deposited on Boron-Doped Diamond Electrodes for Catechol Detec-Tion. Mater. MDPI,* **2020**, *13* (3).

[89]   Ma, X.; Gao, F.; Dai, R.; Liu, G.; Zhang, Y.; Lu, L.; Yu, Y. Novel Electrochemical Sensing Platform Based on a Molecularly Imprinted Polymer-Decorated 3D-Multi-Walled Carbon Nanotube Intercalated Graphene Aerogel for Selective and Sensitive Detection of Dopamine. *Anal. Methods,* **2020**, *12*(14), 1845-1851.
[http://dx.doi.org/10.1039/D0AY00033G]

[90]   Ricardo Teixeira Tarley, C.; Cássia Mendonça, J.; Rianne da Rocha, L.; Boareto Capelari, T.; Carolyne Prete, M.; Cecílio Fonseca, M.; Midori de Oliveira, F.; César Pereira, A.; Luiz Scheel, G.; Bastos Borges, K.; Gava Segatelli, M. Development of a Molecularly Imprinted Poly(Acrylic Acid)-MWCNT Nanocomposite Electrochemical Sensor for Tramadol Determination in Pharmaceutical Samples. *Electroanalysis,* **2020**, *32*(5), 1130-1137.
[http://dx.doi.org/10.1002/elan.201900148]

CHAPTER 7

Multi-walled Carbon Nanotubes Based Molecular Imprinted Polymers for Sensing

Archana Aravind[1]**, Anu Rose Chacko**[1] **and Beena Mathew**[1,*]

[1] *School of Chemical Sciences, Mahatma Gandhi University, Kottayam, Kerala 686560, India*

Abstract: Molecular imprinting technique (MIT) has been commonly and effectively used to prepare polymers which have some unique features like structure predictability, recognition specificity, low cost, remarkable robustness, physical stability, and application universality compared to other reported recognition systems. The application of molecular imprinting technology to the surface of carbon nanotubes, resulting in MWCNT-MIPs. Surface imprinting polymers have a high mass transfer power, high sensitivity, and a fast response time since the imprinting sites are on or near the surface of the substrates. These materials spread out their applications in many fields such as biomaterials, chemo/biosensors, catalysis, molecular/ionic separation and drug delivery. Even though there are enormous applications of such nanomaterials in various fields, this chapter proceeds with the sensing applications.

Keywords: Multi-walled carbon nanotubes, Molecularly imprinted polymers, Sensor.

INTRODUCTION

The molecular recognition method has an infinite number of applications in nature; it is used in enzymatic catalysis, antibody-antigen recognition, cell communication, and other biological processes [1]. Molecularly imprinted polymers (MIPs) have been shown to be the most promising synthetic materials carrying selective molecular recognition sites, as they mimic the mechanism of biomolecular recognition while having high stability, ease of preparation, and low cost [2]. In the 1970s, Wulff *et al.* used a covalent approach to implement Molecular Imprinting Technology (MIT), which includes the creation of binding sites in a synthetic polymer matrix that are functionally and structurally complementary to the 'substrate' molecule. The analytes (or templates) were first compounded with the monomers, which were then polymerized [3].

---

* **Corresponding Author Beena Mathew:** School of Chemical Sciences, Mahatma Gandhi University, Kottayam, Kerala 686560, India; Email: beenamscs@gmail.com

Manorama Singh, Vijai K Rai and Ankita Rai (Eds.)

The polymers are hollowed out to create cavities that are sterically and chemically complementary to the templates. When the templates are removed, the precise binding cavities will tell the difference between the templates and their analogues [4].

In the molecular imprinting approach, functional and crosslinking monomers are copolymerized in the presence of template molecule (analyte) and a three-dimensional polymer network is created. The functional monomers primarily form a complex with the molecule to be imprinted and polymerized. The functional monomers are assumed to be in position by a highly crosslinked polymeric structure. The elimination of the template molecule discloses binding sites that are complementary in size and shape to the template. So a molecular memory is left into the polymer matrix which is now capable of rebinding the template with high specificity. The schematic representation of the general imprinting procedure is shown in Fig. (**1**).

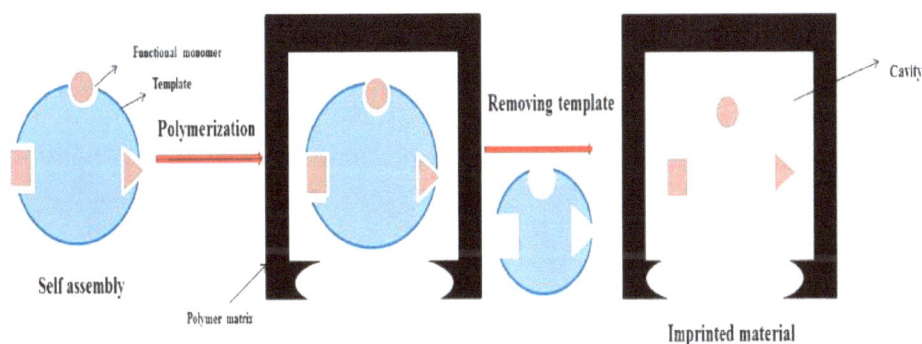

Fig. (1). Schematic representation of general procedure of imprinting.

The molecular imprinting methods can be divided into four categories according to the mechanism of interaction between the monomers and templates: (1) pre-organized approach (covalent bonding), (2) self-assembly approach (non-covalent interactions), (3) hybrid molecularly imprinting method, and (4) metal-chelating method [5]. They were confined to a small range of use due to the restrictive choices of templates. While conventionally prepared bulk MIPs have high selectivity, coarse post-treatment causes some inborn defects. Restricted mass transferability, another major flaw in bulk MIPs, has become a major problem in the macromolecule imprinting or biosensor fields [6, 7].

FACTORS AFFECTING THE IMPRINTING PROCESS

The methodology of molecular imprinting is effortless. The design of the imprinting process is very complicated since it depends upon the number of variables such as templates, functional monomers, crosslinkers, solvents, initiators, temperature and pressure. Brief descriptions of some factors are described below.

Template

The template molecule should possess at least one functional group through which it can connect to the functional monomer as well as a part of the distinct three-dimensional structure. The kind of functional group directs the imprinting approach that can be utilized. Not all templates will form a covalent bond with a functional monomer that is effortlessly cleaved. On the other hand, the number of functional groups has an effect on the affinity of the template for the molecularly imprinted polymer.

As the number of interactions between the template and functional monomer increases, the affinity with which the molecularly imprinted polymer rebinds the template molecule also increases. Conversely, it also raises the non-specific binding of the template to the polymer. The removal of the template after polymerization is essential to reveal the imprinted cavities. If the template molecule remains, these can disclose while performing analysis on the polymers [8]. The compounds such as drugs, amino acids, proteins, carbohydrates, hormones, pesticides, and co-enzymes have been effectively used for the synthesis of the discerning recognition matrices. A few of the templates used in the molecular imprinting process are shown in Fig. (2).

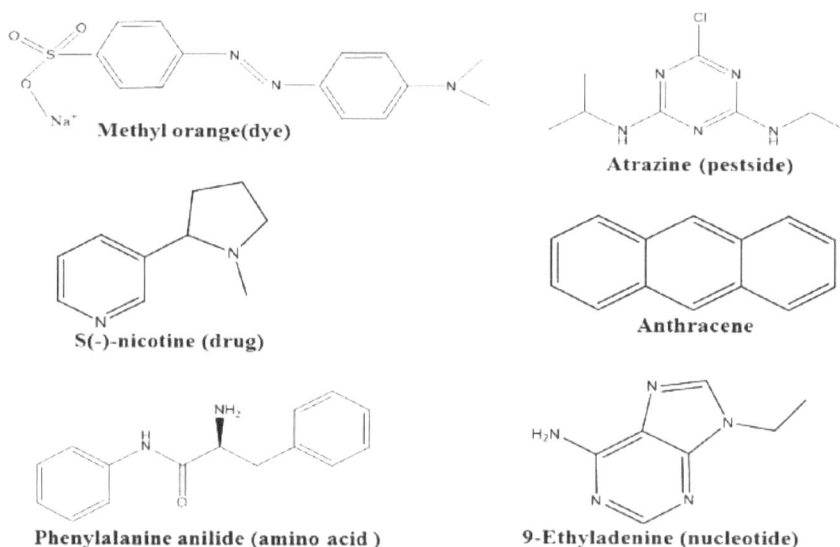

Fig. (2). Commonly used templates in imprinting process.

Functional Monomer

The selectivity of a molecularly imprinted polymer takes place in the interactions between the template molecules and the functional monomers. Functional monomer is the polymerizable unit that interacts with the template molecule and is responsible for its rebinding interactions in the imprinted sites. It is evidently very important that functionalities of the template and functional monomer should be in a matching fashion [9]. For example, an H-bond donor with H-bond acceptor exploits the complex formation and consequently the imprinting effect. The functional monomers can be basic, acidic or neutral (Fig. **3**). In general, methacrylic acid (MAA) is used as the functional monomer for the synthesis of polymers with the molecular imprints of organic compounds containing basic groups such as triazines [10, 11], whereas 4-vinyl pyridine (4-VP) is commonly used in the case of compounds containing acidic groups [12, 13].

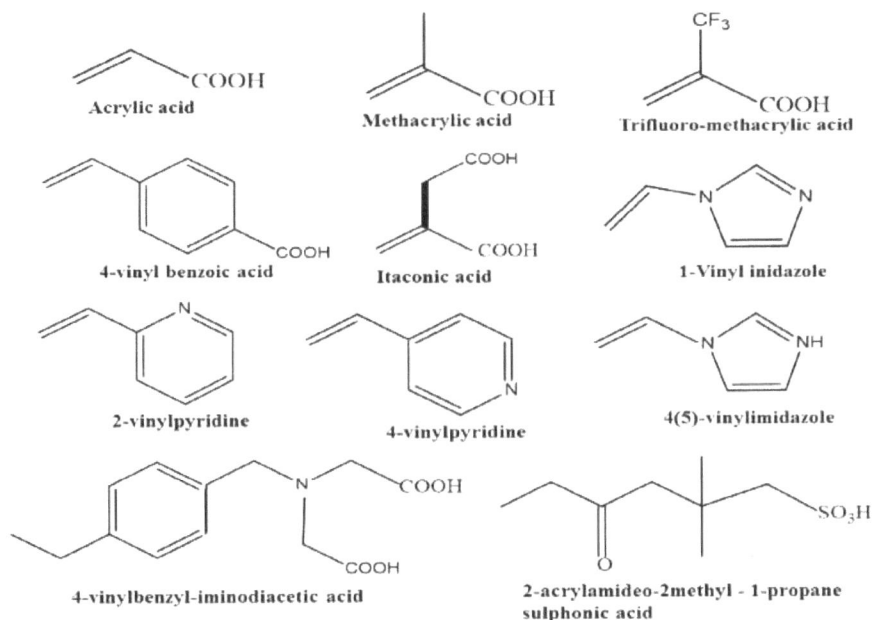

Fig. (3). Commonly used monomers in the imprinting process.

Crosslinker

The main responsibility of the crosslinking monomer is to catch the position of the functional monomers relative to the template to form a three-dimensional structure that stays unchanged after the removal of the template. The type and the amount of crosslinker used in the polymer matrix are very vital as they influence the physicochemical properties and rigidity of molecularly imprinted polymers, which consecutively affect the template selectivity of the polymer matrix [14].

Previously, it has been traditionally thought that the major use of the crosslinking monomer has been as an inert component in the polymer matrix that does not interrelate with the functional groups that bind the template. Ethylene glycol dimethylacrylate (EGDMA) has been verified to be the most efficient crosslinking monomer in the molecular imprinting process [15]. The crosslinking monomer having an optimum length such as TTEGDMA or PEG400 DMA are suitable for the templates like protein in imprinting procedures. The divinylbenzene (DVB) crosslinking imparts rigidity and hydrophobicity to the polymer support, which increases with an increase in crosslinking. By using diverse kinds of crosslinking agents we can direct both the structure of the guest binding sites and the chemical environment around them. The most commonly used crosslinking monomers are reviewed in Fig. (**4**).

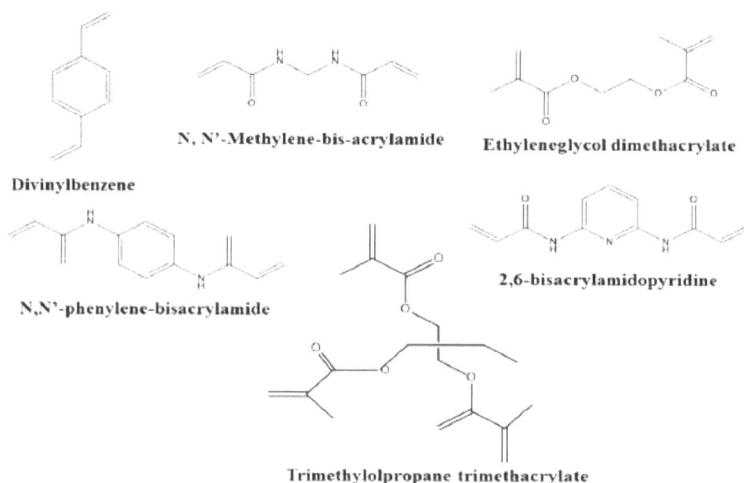

Fig. (4). Commonly used crosslinking agents in the imprinting process.

Initiator

The time of free radical polymerization depends upon the concentration of the initiator used in the imprinting process [16]. 2, 2'-Azobisisobutyronitrile (AIBN) is the most efficient initiator used in the synthesis of molecular imprinting technology. The selection of the initiator directly depends upon the nature of the template used in imprinting. If the template used is thermally or photochemically unstable, then the initiators which can be triggered photochemically or thermally should not be employed in the process. If the complexation occurs by hydrogen bonding, the low temperature polymerization is preferred and under such conditions, photochemical active initiators are well recommended. The chemical structure of some selected polymerization initiators is shown in Fig. (**5**).

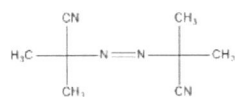

2,2'-azobisisobutyronitrile

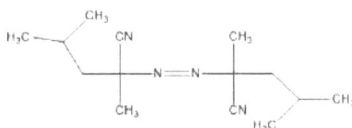

Azo-bis-dimethylvaleronitrile

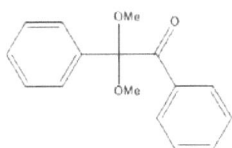

Dimethylacetal benzyl

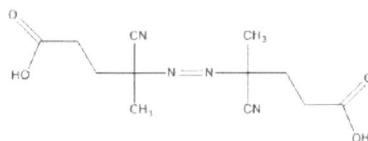

4,4'-azo(4-cyanovaleric acid)

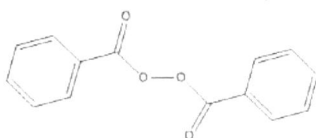

Benzoyl peroxide

Fig. (5). Commonly used initiators in the imprinting process.

Porogen/Solvent

The selectivity and competence of imprinted polymers directly depend upon the nature of the solvent used in the molecular imprinted processes. The main role of the solvent is to dissolve the monomer, template, crosslinking agent and initiator for polymerization and also break up the heat of the reaction produced during polymerization. In imprinting technology, a solvent is used as a porogen as well as a medium for rebinding studies. As the volume of porogens increases, the pore volume also increases. Imprinted polymers synthesized in the absence of the solvents are dense, too firm, and hardly binds guests so that the removal of the template becomes difficult. It is noted that the solvents with low permittivities are suitable for molecular imprinted technique because it results in strong template-monomer non-covalent interactions compared to polar solvents [17, 18].

Temperature

The effect of polymerization temperature on molecular imprinting technique has been the subject of several studies [19]. The stability of the pre-polymerisation complex and the polymerization reaction itself plays a very significant role in identifying the recognition act by imprinted polymers. The nature of the exothermic polymerization reaction and the consequent rise in temperature will intrinsically change the association of template-monomer complexes [20]. If the reaction is carried out at a higher temperature, it will force the equilibrium away from the template-monomer complex towards the dissociated species resulting in a diminished number of strong binding cavities [21]. Generally at lower temperatures, there has to be a strong pre-polymer complex between template and monomer which is eventually accountable for the formation of the specific binding sites within the polymer matrix [22].

NANOMATERIALS IN MOLECULAR IMPRINTING

The conventional molecular imprinting procedure is easy and valuable, but there are numerous critical factors to hinder the functions of molecularly imprinted materials as artificial receptors in analytical chemistry. Initially, most of the imprinted polymers are highly crosslinked polymers with an asymmetrical shape. The removal of original templates located at the core area of bulky materials is fairly difficult because of the presence of high crosslinked three dimensional natures of imprinted materials as well as the agglomerated bundle-like structure, which decreases the rebinding capacity of the template molecule to the polymer matrix [23]. Finally, the irrepressible random polymerization always experiences from the heterogeneity of the imprinted sites in the polymer matrix [24].

The adsorption process was endothermic and spontaneous. N. Jiang *et al.*; have synthesized a new Ni(II)-imprinted amino-functionalized silica gel sorbent with tremendous selectivity for nickel(II) ion by an easy one-step reaction by combining a surface imprinting technique [25]. The developed sensor of Ni(II) ion was also successfully applied to the identification of trace nickel ions in plants and water samples with good results. Li *et al.* fabricated a new electrochemical sensor for bovine hemoglobin using a surface molecular imprinting technique [26]. In current years many researchers are using nanostructured materials such as metal oxide nanoparticles [27], nanowires [28], nanotubes [29], *etc.*, for the design of molecularly imprinted polymers with high efficiency.

Tan *et al.* have successfully synthesized an electrochemical sensor based on molecularly imprinted polymer using reduced graphene oxide and gold nanoparticle modified electrode for the detection of carbofuran [30]. The limit of detection of the fabricated sensor was found to be 2.0×10^{-8} mol/L. Our group (Archana *et al.*) have synthesized different IIPs for sensing various metal cations such as Mn (II), Cr (III), Ni (II) [31 - 33]. Teng and his co-workers synthesized three-dimensional structures comprising polypyrrole nanowires (PPyNWs) and molecularly imprinted polymer by electropolymerization process on the surfaces of a glassy carbon electrode [34]. The sensors exhibited good electrocatalytic capacity and selectivity for dopamine recognition. The limit of detection of the sensor was found to be 33 nM. *Sooraj et al.* reported highly selective nanocomposite sorbents for the specific detection of S-ibuprofen from closely related structural compounds [35].

MULTI-WALLED CARBON NANOTUBE BASED MOLECULAR IMPRINTED POLYMERS

The amazing physical, chemical, and mechanical properties of multi-walled carbon nanotubes (MWCNTs) have inspired widespread investigations since their invention in the early 1990s by Prof. Sumio Iijima [36]. Among the various variety of nanostructures used for the development of molecularly imprinted polymers, multi-walled carbon nanotubes (MWCNT) have been verified as widely accepted support materials due to their properties such as large surface-t--volume ratio, stability under severe conditions, and also due to their mechanical, thermal, and electronic properties. The molecularly imprinted polymers can be connected to the surface of MWCNTs using two main approaches they are, (i) molecular imprinted polymers immobilized on the surface of carboxylic acid functionalized MWCNTs and (ii) immobilization of molecularly imprinted polymer on the surface of vinyl functionalized MWCNTs.

Kan *et al.* synthesized a novel composite of vinyl functionalized multi-walled carbon nanotubes and molecularly imprinted polymers (MWCNTs-MIP) using dopamine (DA) as a template molecule [37]. Hashemi *et al.* prepared multi-walled carbon nanotubes based molecularly imprinted polymer (MIP/MWCNTs) for the separation and preconcentration of L-cysteine (L-Cys) [38]. The synthesized MIP/MWCNT was characterized by FT-IR, X-ray diffraction and scanning electron microscopy. The MIP/MWCNT is revealed to be a feasible sorbent for L-Cys, which is checked by spectrophotometry during the formation of a charge transfer complex with the DDQ reagent. The limit of detection of the sensor for L-Cys was found to be 2.3 ng mL^{-1}. The intra-day and inter-day precision are in the range of 2.4 to 3.6%. The developed sensor was successfully applied for the detection of L-Cys in spiked human serum and water samples. The results obtained exhibited a recovery percentage from 96.6 to 102.4%.

Sebastian and her co-worker found a new technique for sensing and sorption of Zn(II) ion using MWCNT based ion imprinted polymer [39]. MWCNT based ion imprinted polymer was prepared by using acrylamide as a monomer, N, N'-methylene-bis-acrylamide as a crosslinking agent, and zinc as template ion. The electrochemical performance of electrodes modified with MWCNT/IIP was optimized with cyclic voltammetry (CV). Using differential pulse voltammetry (DPV), the limit of detection for Zn (II) ion was found to be 1.32×10^{-4} µM. The sensor was successfully applied for sensing Zn(II) ions in water effluents collected from the paint industry.

SENSING APPLICATIONS OF MULTI-WALLED CARBON NANOTUBE-BASED MOLECULAR IMPRINTED POLYMERS

The literature confirms that research in molecular imprinting is an ever-rising field as they find applications in many areas due to their tailor-made cavity, selectivity, stability and uncomplicated methods of synthesis. The outstanding properties of MWCNT based imprinted polymers have made them a highly interesting implement for diverse application areas, including separations, drug delivery, sensors and solid phase extraction. An important area where MWCNT based imprinting technology plays a key role, is sensor technology. A sensor is characterized by two important ingredients; one is a specific interaction with an analyte, and the second is its ability to convert this interaction into a measuring effect.

Many sensors are already reported in food analysis, environmental monitoring, biomedical and also spread on biomolecules such as antibodies or enzymes as the specific recognition elements [40]. For example, green synthesized silver nanoparticles are used as a good metal sensor [41 - 47]. Due to the reduced physical and chemical stability of bio molecules, the artificial receptors are achieving much interest. In MWCNT based imprinted polymers having the recognition sites are tailor-made with surface area and at the similar time they are incorporated into a solid polymeric support [48]. Due to these advantages, a number of efforts are carried out to create chemical sensors based on these imprinted materials as the recognition elements. A number of MWCNT based MIP sensors have been proposed to sense the template molecules having environmental or biological significance.

MWCNT based ion-imprinted polymer for the highly selective electrochemical sensor for lead ions is reported [49]. The MWCNT-IIP of lead ion imprinted polymer was synthesized using lead ion as a template and NNMBA as crosslinking agent and acrylamide as the monomer. Ion imprinted polymer without MWCNT was also synthesized for comparison; non-imprinted polymers without MWCNT were also synthesized. In all cases, the ion-imprinted polymer showed high specificity towards lead ions. The synthesized polymers were characterized using FT-IR, XRD, and TEM. The electrochemical response of lead ions was studied by using a platinum electrode which was modified with this synthesized MWCNT-IIP. Cyclic voltammetry (CV) and differential pulse voltammetry (DPV) were used for electrochemical studies. The selectivity studies were also carried out in the presence of other metal ions. The limit of detection of the sensor towards Pb(II) ion was found to be 2×10^{-2} M. The sensor was applied to the samples collected from the lake, mining effluent, food sample, and cosmetics. The MWCNT-IIP was also used for the extraction of Pb (II) ions.

Yang and colleagues have identified novel amoxicillin (AMOX) electrochemical sensor made of multi-walled carbon nanotubes@molecularly imprinted polymer (MWCNTs@MIP), single-walled carbon nanotube (SWCNT), and dendritic Pt-Pd nanoparticles (NP) [50]. Initially, an ionic liquid (IL, *i.e.*, 3-propyl-1- vinyl imidazolium bromide) was attached to MWCNT surface to form MWCNTs@IL by using an ion exchange strategy. The ensuing MWCNTs@IL was used as a monomer to prepare MWCNTs@MIP. Meanwhile, dendritic Pt-Pd bimetallic NP was prepared using hexadecylpyridinium chloride and hexamethylenetriamine as synergistic structure-directing agents, and was dispersed into SWCNT suspension. Afterwards, the hybrid suspension was fabricated on a glassy carbon electrode, followed by coating with MWCNTs@MIP. The fabricated sensor showed a linear response to amoxicillin in the ranges of 1.0×10^{-9}-1.0×10^{-6} mol L^{-1} and 1.0×10^{-6}-6.0×10^{-6}, respectively, and the limit of detection was found to be 8.9×10^{-10} mol L^{-1}. This sensor was also applied to sense amoxicillin in real samples with satisfactory results.

CONCLUDING REMARKS

The benefits of using MWCNT -MIPs, technology has seen rapid growth, not least because it can be manufactured at a low cost and using a variety of synthetic approaches. The capacity to construct molecularly imprinted polymers allows for the creation of artificial structures that can substitute natural biological structures like DNA aptamers or biological recognition displaying proteins (including antibodies and receptors). It's interesting to note that some MIP-based sensors are durable, operate at room temperature, and have good selectivity and sensitivity. For the formation of conducting polymer-based layers, a number of polymerization processes can be used; nevertheless, the most unique and controllable polymerization methods provide the most efficient control and modification of created sensing layers. In conclusion, MWCNT -MIPs retain the affinity and selectivity of antibodies, enzymes, and biological receptors while also providing many advantages such as high mechanical strength, large specific surface area, and chemical stability. We also make predictions about their future research and growth.

CONSENT FOR PUBLICATION

Not applicable.

CONFLICT OF INTEREST

The author declares no conflict of interest, financial or otherwise.

ACKNOWLEDGEMENTS

Declared none.

REFERENCES

[1]   Hinterdorfer, P.; Dufrêne, Y.F. Detection and localization of single molecular recognition events using atomic force microscopy. *Nat. Methods,* **2006**, *3*(5), 347-355.
        [http://dx.doi.org/10.1038/nmeth871] [PMID: 16628204]

[2]   Ye, L.; Mosbach, K. Molecular Imprinting: Synthetic Materials as Substitutes for Biological Antibodies and Receptors. *Chem. Mater.,* **2008**, *20*(3), 859-868.
        [http://dx.doi.org/10.1021/cm703190w]

[3]   Wulff,   G.;   Sarhan,   A.   Über   die   Anwendung   von enzymanaloggebautenPolymerenzurRacemattrennung. *Angew. Chem.,* **1972**, *84*(8), 364.
        [http://dx.doi.org/10.1002/ange.19720840838]

[4]   Nicholls, I.A.; Rosengren, J.P. Molecular imprinting of surfaces. *Bioseparation,* **2001**, *10*(6), 301-305.
        [http://dx.doi.org/10.1023/A:1021541631063] [PMID: 12549873]

[5]   Wulff, G. Enzyme-like catalysis by molecularly imprinted polymers. *Chem. Rev.,* **2002**, *102*(1), 1-27.
        [http://dx.doi.org/10.1021/cr980039a] [PMID: 11782127]

[6]   Tan, C.J.; Tong, Y.W. Molecularly imprinted beads by surface imprinting. *Anal. Bioanal. Chem.,* **2007**, *389*(2), 369-376.
        [http://dx.doi.org/10.1007/s00216-007-1362-4] [PMID: 17563884]

[7]   Ge, Y.; Turner, A.P. Too large to fit? Recent developments in macromolecular imprinting. *Trends Biotechnol.,* **2008**, *26*(4), 218-224.
        [http://dx.doi.org/10.1016/j.tibtech.2008.01.001] [PMID: 18295919]

[8]   Piletska, E.V.; Guerreiro, A.R.; Whitcombe, M.J.; Piletsky, S.A. Influence of the Polymerization Conditions on the Performance of Molecularly Imprinted Polymers. *Macromolecules,* **2009**, *42*(14), 4921-4928.
        [http://dx.doi.org/10.1021/ma900432z]

[9]   Mosbach, K. Molecular imprinting. *Trends Biochem. Sci.,* **1994**, *19*(1), 9-14.
        [http://dx.doi.org/10.1016/0968-0004(94)90166-X] [PMID: 8140624]

[10]   Muldoon, M.T.; Stanker, L.H. Molecularly imprinted solid phase extraction of atrazine from beef liver extracts. *Anal. Chem.,* **1997**, *69*(5), 803-808.
        [http://dx.doi.org/10.1021/ac9604649] [PMID: 9068268]

[11]   Dauwe, C.; Sellergren, B. Influence of Template Basicity and Hydrophobicity on the Molecular Recognition Properties of Molecularly Imprinted Polymers. *J. Chromatogr. A,* **1996**, *753*(2), 191-200.
        [http://dx.doi.org/10.1016/S0021-9673(96)00564-X]

[12]   Masqué, N.; Marcé, R.M.; Borrull, F.; Cormack, P.A.G.; Sherrington, D.C. Synthesis and evaluation of a molecularly imprinted polymer for selective on-line solid-phase extraction of 4-nitrophenol from environmental water. *Anal. Chem.,* **2000**, *72*(17), 4122-4126.
        [http://dx.doi.org/10.1021/ac0000628] [PMID: 10994973]

[13]   Zhu, L.; Chen, L.; Xu, X. Application of a molecularly imprinted polymer for the effective recognition of different anti-epidermal growth factor receptor inhibitors. *Anal. Chem.,* **2003**, *75*(23), 6381-6387.
        [http://dx.doi.org/10.1021/ac026371a] [PMID: 14640704]

[14]   Zhu, Q.Z.; Haupt, K.; Knopp, D.; Niessner, R. Molecularly Imprinted Polymer for Metsulfuron-Methyl and Its Binding Characteristics for Sulfonylurea Herbicides. *Anal. Chim. Acta,* **2002**, *468*(2), 217-227.
        [http://dx.doi.org/10.1016/S0003-2670(01)01437-4]

[15]   Kempe, M.; Mosbach, K. Receptor Binding Mimetics: A Novel Molecularly Imprinted Polymer. *Tetrahedron Lett.,* **1995**, *36*(20), 3563-3566.
[http://dx.doi.org/10.1016/0040-4039(95)00559-U]

[16]   Haupt, K.; Dzgoev, A.; Mosbach, K. Assay system for the herbicide 2,4-dichlorophenoxyacetic Acid using a molecularly imprinted polymer as an artificial recognition element. *Anal. Chem.,* **1998**, *70*(3), 628-631.
[http://dx.doi.org/10.1021/ac9711549] [PMID: 21644761]

[17]   Martin-Esteban, A. Molecularly imprinted polymers: new molecular recognition materials for selective solid-phase extraction of organic compounds. *Fresenius J. Anal. Chem.,* **2001**, *370*(7), 795-802.
[http://dx.doi.org/10.1007/s002160100854] [PMID: 11569855]

[18]   Lanza, F.; Hall, A.J.; Sellergren, B.; Bereczki, A.; Horvai, G.; Bayoudh, S.; Cormack, P.A.G.; Sherrington, D.C. Development of a Semiautomated Procedure for the Synthesis and Evaluation of Molecularly Imprinted Polymers Applied to the Search for Functional Monomers for Phenytoin and Nifedipine. *Anal. Chim. Acta,* **2001**, *435*(1), 91-106.
[http://dx.doi.org/10.1016/S0003-2670(01)00905-9]

[19]   Jacob, R.; Tate, M.; Banti, Y.; Rix, C.; Mainwaring, D.E. Synthesis, characterization, and ab initio theoretical study of a molecularly imprinted polymer selective for biosensor materials. *J. Phys. Chem. A,* **2008**, *112*(2), 322-331.
[http://dx.doi.org/10.1021/jp074405i] [PMID: 18095662]

[20]   Piletsky, S.A.; Piletska, E.V.; Karim, K.; Freebairn, K.W.; Legge, C.H.; Turner, A.P.F. Polymer Cookery: Influence of Polymerization Conditions on the Performance of Molecularly Imprinted Polymers. *Macromolecules,* **2002**, *35*(19), 7499-7504.
[http://dx.doi.org/10.1021/ma0205562]

[21]   O'Shannessy, D.J.; Ekberg, B.; Andersson, L.I.; Mosbach, K. Recent advances in the preparation and use of molecularly imprinted polymers for enantiomeric resolution of amino acid derivatives. *J. Chromatogr. A,* **1989**, *470*(2), 391-399.
[http://dx.doi.org/10.1016/S0021-9673(01)83567-6]

[22]   Spivak, D.; Gilmore, M.A.; Shea, K.J. Evaluation of Binding and Origins of Specificity of 9-Ethyladenine Imprinted Polymers. *J. Am. Chem. Soc.,* **1997**, *119*(19), 4388-4393.
[http://dx.doi.org/10.1021/ja963510v]

[23]   Pagona, G.; Tagmatarchis, N. Carbon nanotubes: materials for medicinal chemistry and biotechnological applications. *Curr. Med. Chem.,* **2006**, *13*(15), 1789-1798.
[http://dx.doi.org/10.2174/092986706777452524] [PMID: 16787221]

[24]   Geng, H.Z.; Rosen, R.; Zheng, B.; Shimoda, H.; Fleming, L.; Liu, J.; Zhou, O. Fabrication and Properties of Composites of Poly (Ethylene Oxide) and Functionalized Carbon Nanotubes. *Adv. Mater.,* **2002**, *14*(19), 1387-1390.
[http://dx.doi.org/10.1002/1521-4095(20021002)14:19<1387::AID-ADMA1387>3.0.CO;2-Q]

[25]   Jiang, N.; Chang, X.; Zheng, H.; He, Q.; Hu, Z. Selective solid-phase extraction of nickel(II) using a surface-imprinted silica gel sorbent. *Anal. Chim. Acta,* **2006**, *577*(2), 225-231.
[http://dx.doi.org/10.1016/j.aca.2006.06.049] [PMID: 17723676]

[26]   Li, L.; Yang, L.; Xing, Z.; Lu, X.; Kan, X. Surface molecularly imprinted polymers-based electrochemical sensor for bovine hemoglobin recognition. *Analyst (Lond.),* **2013**, *138*(22), 6962-6968.
[http://dx.doi.org/10.1039/c3an01435e] [PMID: 24089215]

[27]   Zengin, A.; Yildirim, E.; Tamer, U.; Caykara, T. Molecularly imprinted superparamagnetic iron oxide nanoparticles for rapid enrichment and separation of cholesterol. *Analyst (Lond.),* **2013**, *138*(23), 7238-7245.
[http://dx.doi.org/10.1039/c3an01458d] [PMID: 24133677]

[28]   Chen, T.; Shao, M.; Xu, H.; Zhuo, S.; Liu, S.; Lee, S.T. Molecularly Imprinted Polymer-Coated Silicon Nanowires for Protein-Specific Recognition and Fast Separation. *J. Mater. Chem.,* **2012**, *22*(9), 3990.
[http://dx.doi.org/10.1039/c2jm14329a]

[29]   Pan, J.; Yao, H.; Xu, L.; Ou, H.; Huo, P.; Li, X.; Yan, Y. Selective Recognition of 2,4,6-Trichlorophenol by Molecularly Imprinted Polymers Based on Magnetic Halloysite Nanotubes Composites. *J. Phys. Chem. C,* **2011**, *115*(13), 5440-5449.
[http://dx.doi.org/10.1021/jp111120x]

[30]   Tan, X.; Hu, Q.; Wu, J.; Li, X.; Li, P.; Yu, H.; Li, X.; Lei, F. Electrochemical Sensor Based on Molecularly Imprinted Polymer Reduced Graphene Oxide and Gold Nanoparticles Modified Electrode for Detection of Carbofuran. *Sens. Actuators B Chem.,* **2015**, *220*, 216-221.
[http://dx.doi.org/10.1016/j.snb.2015.05.048]

[31]   Aravind, A.; Mathew, B. Nano Layered Ion Imprinted Polymer Based Electrochemical Sensor and Sorbent for Mn (II) Ions from Real Samples. *J. Macromol. Sci. Part A Pure Appl. Chem.,* **2020**, *57*(4), 256-265.
[http://dx.doi.org/10.1080/10601325.2019.1691451]

[32]   Aravind, A.; Mathew, B. Electrochemical Sensor Based on Nanostructured Ion Imprinted Polymer for the Sensing and Extraction of Cr(III) Ions from Industrial Wastewater. *Polym. Int.,* **2018**, *67*(12), 1595-1604.
[http://dx.doi.org/10.1002/pi.5683]

[33]   Aravind, A.; Mathew, B. . An Electrochemical Sensor and Sorbent Based on Mutiwalled Carbon Nanotube Supported Ion Imprinting Technique for Ni(II) Ion from Electroplating and Steel Industries. In: *S.N. Appl. Sci;* , **2019**; p. 1.

[34]   Teng, Y.; Liu, F.; Kan, X. Voltammetric Dopamine Sensor Based on Three-Dimensional Electrosynthesized Molecularly Imprinted Polymers and Polypyrrole Nanowires. *Mikrochim. Acta,* **2017**, *184*(8), 2515-2522.
[http://dx.doi.org/10.1007/s00604-017-2243-y]

[35]   Sooraj, M.P.; Mathew, B. Highly Selective Nanocomposite Sorbents for the Specific Recognition of S-Ibuprofen from Structurally Related Compounds. *Appl. Nanosci.,* **2016**, *6*(5), 659-671.
[http://dx.doi.org/10.1007/s13204-015-0476-9]

[36]   Iijima, S. Helical Microtubules of Graphitic Carbon. *Nature,* **1991**, *354*(6348), 56-58.
[http://dx.doi.org/10.1038/354056a0]

[37]   Kan, X.; Zhao, Y.; Geng, Z.; Wang, Z.; Zhu, J.J. Composites of Multi-walled Carbon Nanotubes and Molecularly Imprinted Polymers for Dopamine Recognition. *J. Phys. Chem. C,* **2008**, *112*(13), 4849-4854.
[http://dx.doi.org/10.1021/jp077445v]

[38]   Hashemi, M.; Nazari, Z.; Bigdelifam, D. A Molecularly Imprinted Polymer Based on Multi-walled Carbon Nanotubes for Separation and Spectrophotometric Determination of L-Cysteine. *Mikrochim. Acta,* **2017**, *184*(8), 2523-2532.
[http://dx.doi.org/10.1007/s00604-017-2236-x]

[39]   Sebastian, M.; Mathew, B. Carbon Nanotube-Based Ion Imprinted Polymer as Electrochemical Sensor and Sorbent for Zn(II) Ion from Paint Industry Wastewater. *Int. J. Polym. Anal. Char,* **2017**, *1*

[40]   Andrea, P.; Miroslav, S.; Silvia, S.; Stanislav, M. A Solid Binding Matrix/Molecularly Imprinted Polymer-Based Sensor System for the Determination of Clenbuterol in Bovine Liver Using Differential-Pulse Voltammetry. *Sens. Actuators B Chem.,* **2001**, *76*(1-3), 286-294.
[http://dx.doi.org/10.1016/S0925-4005(01)00586-X]

[41]   Sebastian, M.; Aravind, A.; Mathew, B. Green silver-nanoparticle-based dual sensor for toxic Hg(II) ions. *Nanotechnology,* **2018**, *29*(35), 355502.

[http://dx.doi.org/10.1088/1361-6528/aacb9a] [PMID: 29889047]

[42]   Sebastian, M.; Aravind, A.; Mathew, B. Simple Unmodified Green Silver Nanoparticles as Fluorescent Sensor for Hg(II). *Mater. Res. Express,* **2018**, *5*(8), 085015.
       [http://dx.doi.org/10.1088/2053-1591/aad317]

[43]   Sebastian, M.; Aravind, A.; Mathew, B. Green Silver Nanoparticles Based Multi-Technique Sensor for Environmental Hazardous Cu(II) Ion. *Bionanoscience,* **2019**, *9*(2), 373-385.
       [http://dx.doi.org/10.1007/s12668-019-0608-x]

[44]   Aravind, A.; Sebastian, M.; Mathew, B. Green Silver Nanoparticles as a Multifunctional Sensor for Toxic Cd(ii). *New J. Chem.,* **2018**, *42*(18), 15022-15031.
       [http://dx.doi.org/10.1039/C8NJ03696A]

[45]   Aravind, A.; Sebastian, M.; Mathew, B. Green Synthesized Unmodified Silver Nanoparticles as a Multi-Sensor for Cr(iii). *Environ. Sci. Water Res. Technol.,* **2018**, *4*(10), 1531-1542.
       [http://dx.doi.org/10.1039/C8EW00374B]

[46]   Aravind, A.; Sebastian, M.; Mathew, B. Unmodified Silver Nanoparticles Based Multisensor for Ni (II) Ions in Real Samples. *Int. J. Environ. Anal. Chem.,* **2019**, *99*(4), 380-395.
       [http://dx.doi.org/10.1080/03067319.2019.1599874]

[47]   Aravind, A.; Mathew, B. Tailoring of Nanostructured Material as an Electrochemical Sensor and Sorbent for Toxic Cd(II) Ions from Various Real Samples. *J. Anal. Sci. Technol.,* **2018**, *9*(1), 9.
       [http://dx.doi.org/10.1186/s40543-018-0153-1]

[48]   Soleimani, M.; Afshar, M.G.; Shafaat, A.; Crespo, G.A. High-Selective Tramadol Sensor Based on Modified Molecularly Imprinted Polymer Carbon Paste Electrode with Multi-walled Carbon Nanotubes. *Electroanalysis,* **2013**, *25*(5), 1159-1168.
       [http://dx.doi.org/10.1002/elan.201200601]

[49]   Sebastian, M.; Mathew, B. Ion Imprinting Approach for the Fabrication of an Electrochemical Sensor and Sorbent for Lead Ions in Real Samples Using Modified Multi-walled Carbon Nanotubes. *J. Mater. Sci.,* **2018**, *53*(5), 3557-3572.
       [http://dx.doi.org/10.1007/s10853-017-1787-x]

[50]   Yang, G.; Zhao, F. Molecularly Imprinted Polymer Grown on Multi-walled Carbon Nanotube Surface for the Sensitive Electrochemical Determination of Amoxicillin. *Electrochim. Acta,* **2015**, *174*, 33-40.
       [http://dx.doi.org/10.1016/j.electacta.2015.05.156]

CHAPTER 8

Molecularly Imprinted Polymer (MIP) Nanocomposites–based Sensors

Juhi Srivastava[1] and **Meenakshi Singh**[1,*]

[1] *Department of Chemistry, MMV, Banaras Hindu University, Varanasi, U.P., India*

Abstract: Molecular recognition in biological systems drives and controls all the activities related to 'Life.' The accuracy, specificity, and selectivity of biological elements led to their use as biosensors for 'sensing'. An ideal molecular recognition agent must comprise a stable, reproducible, reusable, robust, specific and preferably nonbiological material. Molecular imprinting has almost all attributes that qualify it to be an 'ideal' recognition agent. As a surrogate to biological receptors, synthetic MIPs have shown aspiring futuristic tools. Next-generation sensors could be visualized by a collaboration of synthetic polymers (MIPs) with innovative technologies replacing biosensors. Over the period of the last three decades, the introduction of specific binding sites within synthetic polymers by utilizing target-directed cross linking of functional monomers has attracted substantial consideration for the sake of the formation of molecularly imprinted polymer (MIP) based sensors. MIP seems like a reasonable tool for the creation of various sensors with broad practical relevance.

This chapter outlines the sensors prepared on nanocomposite as an imprinting matrix. Strategic planning in synthesizing these novel matrices is praiseworthy. Hopefully, such measures would bring down the economic burden by devising cheaper sensing tools, especially diagnostic kits in such pandemic times.

Keywords: Chitosan nanoparticle, Graphene, Molecularly imprinted polymer, Nanocomposite, Sensor, Starch Nanoparticle.

INTRODUCTION

In the last decade, engineered nanoparticles with diverse functionality and purpose received utmost attention from almost all affiliations of science. As the scientists worked upon them and came with novel feats and achievements regarding their design, innovative ideas popped up while working with them. Conventionally, metallic nanoparticles are favourites, followed by bimetallic, magnetic, organic, polymeric and biopolymeric particles, which have been

* **Corresponding author Meenakshi Singh**: Department of Chemistry, MMV, Banaras Hindu University, Varanasi, U.P., India; Email: meenakshi@bhu.ac.in

Manorama Singh, Vijai K Rai and Ankita Rai (Eds.)

synthesized and explored well. Their exploration fetched varieties of characteristics, some of them unique to their class only. To employ them optimally for diverse needs, hybrid nanocomposites were also tested successfully. The typical definition of a composite is "a material which is produced from two or more constituent materials having dissimilar chemical or physical properties, merged to create a material with properties, unlike the individual elements." So, for visualizing these 'new' characteristics, many attempts are made in almost all sections of science to prepare nanocomposites. They offer new applications with these hybrid materials expecting synergy between them.

In this chapter, an attempt is made to summarize the scope of molecularly imprinted polymers (MIP) – composites in sensing devices, especially to serve the healthcare sector of our society. To achieve this purpose, accurate detection and quantification of analytes are required in clinical analysis, biological and chemical security, environmental protection and food safety. These analyses involve a huge economic burden on our society [1 - 5]. Efforts are being made to reduce this burden by designing and fabricating cost-effective arrays of sensors, sensing devices and also smart devices. Tailoring such devices needs a combined effort of interdisciplinary research and their application in 'real' samples.

Sensors or sensing devices, whether chemosensors or biosensors, comprise of two main units; recognition unit and transduction unit. The recognition unit predicts selectivity *via* chemical interactions, whereas the transduction unit transduces these interactions into analytical signals [6, 7]. Among the approaches adopted for chemosensors, molecular imprinting is one of the most aspiring and realistic approaches for fabricating synthetic receptors, which can be used as a recognition unit in chemosensors [8 - 12]. Generally, MIPs in thin-film form with suitable transducers are used for the fabrication of such sensing devices for different analytes [8, 10, 13]. Commonly used transducers for the fabrication of such chemosensors are voltammetric, amperometric, peizoeletrogravimetric, electrochemical impedance spectroscopy (EIS) and surface plasmon resonance (SPR) spectroscopic techniques.

MIPs are commonly used 'recognition sites' for chemosensing devices. Molecular imprinting has come up as an almost foolproof method for generating artificial receptors or, in other words, as chemical sensors competing with the biological analogues - biosensors. The molecular imprinting approach is able to form cavities in polymer matrices, which are the exact mirror images of analytes; thus, it is able to induce the movement of analyte molecules only towards these imprinted cavities, rather than their analogues or other structurally similar molecules. These polymers have an affinity for the original target molecules and have been used for various purposes, such as chemical separation, molecular sensors, biosensors, bioseparation and drug delivery, *etc.* [14, 15]. MIPs are known for their stability under ambient conditions, whether mechanical and/or chemical stability. Their purposive characteristics of selectivity, stability under ambient conditions, sturdiness and steadiness, reproducibility, reusability and specificity are instrumental for their wide applications in fields of biosensors, bioseparation, chromatography, molecular receptor and drug delivery, *etc.*

MIP synthesis includes the arrangement of a complex with template and useful monomers during self-organization, either by covalent or non-covalent bonds pursued by polymerization with cross linkers in an appropriate solvent (porogen). Upon complete extraction of template molecules, the specific imprints are made in the polymeric grid corresponding to format fit as a fiddle and contain properly arranged recognizing elements valuable in rebinding and consequently, after extraction of the analyte, the subsequent polymer can rebind analyte with high inclination and specificity. Rebinding of the template with imprinted polymer is made conceivable by the formation of shape-corresponding cavities inside the polymeric system. Fundamentally, an atomic memory is imprinted in polymer, or, in other words, the template is rebound specifically. A pictorial presentation of the development of MIP is shown in Fig. (**1**).

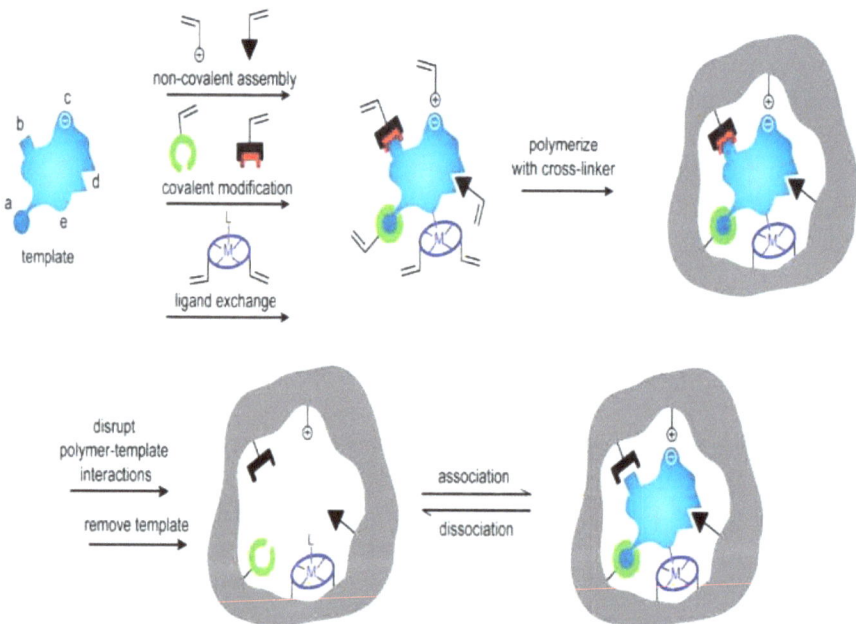

Fig. (1). Schematic representation for the synthesis of MIP [12].

MIP offers the synthesis of various artificial recognition materials and they are proving themselves best suited for the development of novel synthetic receptors and biorecognition elements for environmental pollutants, toxic substances and many more analytes [16 - 19]. The scope of molecular imprinting has been realized in recent years, as attested by the scientific literature boom covering this area. Molecular imprinting has emerged from sensing simple isomers to complicated biomolecules, such as proteins, viruses, *etc.*, but still, many fundamental intricacies of imprinting dynamics need to be explored [20]. In this journey, imprinting technology has effectuated many successful isolations, separations, extractions and sensing of a wide range of analytes from almost all scientific and industrial fields with exemplary operational efficacy and stability. According to Whitcombe *et al.,* molecular imprinting is defined as "The development of ligand-specific recognition locales in synthetic polymers where a template (particle, particle, atom, complex or an atomic, ionic or macromolecular get together, including smaller scale life forms) is utilized to encourage recognition site development during the covalent get together of the mass stage by a polymerization or on the other hand polycondensation process, with resulting expulsion of a few or the majority of the analyte being essential for recognition to happen in the spaces emptied by the templating species" [12].

While the advancement of this field runs swiftly, adding to the exemplary selectivity and specificity, a matching sensitive transducer has to be hyphenated to fabricate instruments to go with the molecular imprinting impact and their ramifications for the nano-level, smaller-scale level and full-scale level structure and capacity of imprinted materials. The foundation of an ability to anticipate material structure and capacity is conceivably the single factor well on the way to rush the advancement of molecular imprinting. Over a few years, the scope of hypothetical and trial techniques has been conveyed to address basic constituents of MIPs at nano-level, miniaturized scale level and large scale level. These endeavors, outlined in the succeeding sections, have yielded new experiences, novel equipment and new designs, having a significant effect on the advancement of MIP application territories.

The intricacies behind the designing of such a MIP demand basic knowledge regarding its skeletal constitution and function attributable under working conditions. The intriguing thing is that, regardless of how the monomer-analyte adduct is formed for forming the MIPs, the exact mechanism and dynamics of imprinting affect performance and recognition characteristics and thus they must be explicitly visualised. The intricacies of binding/rebinding/sensing events need to be discerned completely. Current approaches and future advents are expected to assist in the contrivance of flawless manmade molecularly imprinted receptors and aid in accumulating the acumen for a better comprehension of molecular interactions involved in the molecular recognition of biological systems. Polymeric format for imprinting shall be chosen after giving due considerations to water compatibility and/or biocompatibility of the existing library being used by the imprinting fraternity.

The significance of new techniques for MIP concoction and polymer morphology is a basic factor in the improvement of materials for new applications. MIPs have the basic segment of natural receptors, *i.e.*, the ability to see and bind specific target (bio) molecules. MIPs could be utilized to accumulate subjective and quantitative data for analytes from numerous classes, including biomolecules (proteins, amino acids, and chemicals), drugs from pharmaceutical dose frames, or natural examples, alongside pesticides, which can likewise be followed in their sustenance [21].

Besides the discussion as given above, many other new works establish a more prominent understanding about MIP, MIP nanoparticles (NPs) and their application in sensing technology, including electrochemical MIP sensor, optical sensing, *etc.,* by using various nanocarbon materials like graphene, carbon dots (CDs), multiwalled carbon nanotubes (MWCNTs), *etc.* [22]. The new trends developed in this field also exploit the vast application of MIPs in various fields like gas sensing, quartz crystal microbalance sensor, liquid/ solution sensing, biomarkers, pharmaceutical application, environmental sensing, food analyses and explosive detection, *etc.* [23].

DEVELOPMENTS IN MOLECULAR IMPRINTING TECHNOLOGY

MIPs usually specified as 'artificial enzymes' and/or 'artificial antibodies' are one of the aspiring techniques in creating a synthetic alternate to biosensing elements. However, for achieving this endeavour, and that also for the application of MIPs in 'real' matrices, imprinting technology has to be transferred from 'organic' to 'aqueous' media, *i.e.,* water-incompatibility of MIPs has to be prevailed over as highlighted by many authors [20, 24]. The discordance of sensing elements of MIPs in an aqueous system can be chiefly ascribed to: (1) The electrostatic forces and other non-covalent forces, especially H-bonding responsible for creating specific imprinted cavities within polymeric matrix wane in water as a solvent in comparison to organic solvent, affecting the molecular recognition adversely [25, 26]; (2) Hydrophobic interactions are strengthened in water leading to non-specificity, especially in comparison to organic solvents [27 - 29]. Hence, water incompatibility looks like one of the major apprehensions of imprinting fraternity offering hurdles in 'real' analysis. Lemieux pointed out the' role of water in Molecular Recognition' about 30 years back, highlighting the role of aqueous medium: 'water plays a central role in all biological associations because its interactions with molecular surfaces lead to perturbation by either ordering (nonpolar surface) or strain (polyamphiphilic surface). This bottled energy is always on tap to help drive specific molecular events. Like a chaperon, water accompanies the reactants in their search for each other' [30].

Although the imprinting fraternity has achieved remarkable milestones like discriminating geometrical isomers, molecular recognition, selective catalysis, and many more, still the potential of the MIP technique was not fully exploited through conventional bulk imprinting. Customary planning strategies had not accomplished the maximum capacity of MIPs with the confinements unsolved, for example, poor site availability and low, restricting capacity [31]. Some endeavours have been made to address these issues proposed, for example, surface/film imprinting, surface unite imprinting, what's more, surface centre shell nanoparticles (NPs), *etc.* [32 - 36].

Common substrates utilized for imprinting on the surface layer are silica gel, ferric oxide nanoparticles and graphene, *etc.*, [37 - 39]. Nevertheless, the binding and sensing compatibility between these inorganic moieties and imprinted layers is not up to the mark. As of late, natural materials are connected to the readiness of MIPs and can possibly substitute for conventional imprinting networks. For example, Romana Schirhagl's group has explored a two-fold imprinting methodology utilizing the normal antibodies as a format to produce biomimetic receptors and utilize them for the recognition of extensive biomolecules [40, 41]. In another investigation, the yeast was chosen as the grid material to perceive ciprofloxacin (CIP) from aqueous arrangement, for its favourable circumstances of minimal effort, effortlessly accessible source, and generous practical accumulation, which makes it more helpful to frame molecularly imprinted polymer layer on the yeast surface [42]. Nonetheless, conventional procedures (*e.g.,* mass blend) utilized for MIP manufacture present a few disadvantages, for example, spillage or inadequate expulsion of the format, arbitrary dispersion of restricting locales, moderate mass exchange, and sporadic morphology [43]. Distinctive imprinting techniques, for example, *in situ* polymerization or centre shell MIP union, were connected with the end goal to limit these deficiencies.

Moreover, conventionally MIPs were customized in an organic framework, thus showing poor recognition ability for target analytes in aqueous environments. Attempts 'to transfer the imprinting process from organic medium to aqueous medium' are being continually made [24, 44 - 55]. This seriously affects applications in aqueous media and presents complications in 'real' analysis.

MIP SENSORS BASED ON METAL NANOCOMPOSITE/CARBON NANOTUBES

A piezo-electrogravimmetric MIP sensor for L-ser was prepared [56]. Electropolymerization of AuNPs derivatized with 4-amino thiophenol (4-ATP) on EQCM electrode functionalized with a SAM of 4-ATP. High specificity was obtained for L-serine and the limit of detection (LOD) was 0.41 μM.

A core–shell structure of gold nano/micro-islands (NMIs) was designed for the estimation of heart-fatty acid-binding protein (H-FABP). These gold NMIs were used as a platform for electropolymerization of imprinted ortho-phenylenediamine [57]. The contours of NMIs were attuned with regulated electrodeposition of gold on RGO thin film. The imprinted sensor was applied to serum samples successfully with lower detection time and LOD. Another MIP sensor for tetracycline was attempted with the composite ($ZnO@NH_2$-UiO-66, ZUM) [58]. This composite showed the ability of selective adsorption and good photocatalytic degradation efficiency to tetracycline.

A disposable electrochemical sensor based on an imprinted composite of graphene and gold nanoparticles grafted on screen-printed electrodes was conceptualized and tested for the estimation of rutin [59]. Better electrochemical efficiency was observed with reusability, stability and reproducibility also.

A magnetic molecularly imprinted polymer (Fe_3O_4/GO/DMIPs) capable of detecting multiple analytes was prepared by molecular imprinting using isoprenaline as the dummy target and graphene oxide as the carrier [60]. This nanocomposite having magnetic characteristics was used in the MSPE column for applications in selective extraction, separation, and estimation of trace analytes in 'real' biological samples.

A molecularly imprinted photoelectrochemical (MIP-PEC) sensor utilizing metal-organic frameworks (MOFs) and TiO_2 (NH_2-MIL-125(Ti)–TiO_2) is fabricated by solvothermal method for the analytical estimation of oxytetracycline [61]. Quenching of photocurrent was observed on the binding of the analyte oxytetracycline as they occupy the imprinted cavities and in turn, block the electron transfer pathways. Molecular imprinting improved the selectivity on a larger scale and sensitivity also.

A MIP-sensor for protein amyloid-β ($A\beta_{42}$) was constructed on delaminated titanium carbide MXene (d-$Ti_3C_2T_X$ MXene)-MWCNTs composite [62]. $A\beta_{42}$ imprinted electrode produced a linear range with LOD of 0.3 fg mL^{-1}. MIP sensor on polyoxometalate/carbon nitride nanotubes composite was fabricated for γ-LND detection [63]. The stability, repeatability, reproducibility and reusability of this sensor were found to be impressive, and this imprinted sensor was successfully applied to orange juice samples showing high recovery. Another sensor based on graphitic carbon nitride/N-doped carbon dots composite with imprinted polymer was prepared for the determination of epinephrine using pyrrole as monomer and epinephrine as a template through cyclic voltammetry [64]. Epinephrine imprinted electrode showed a high sensitivity for epinephrine in urine samples also.

A nanocomposite formed by the condensation of formaldehyde, resorcinol and melamine was imprinted for ametryn on the surface of MWNTs@Fe$_3$O$_4$[65]. This blend of monomers formed a composite with superhydrophilic characteristics. Further testing and performance checks showed specific recognition of triazines in water.

A QCM sensor utilizing gold nanoparticles (AuNPs) impregnated with MIP layer and covalent organic frameworks composite was fabricated for the estimation of aflatoxin B1 [66]. This MIP layer uses crosslinking of *o*-aminothiophenol functionalized AuNPs (*o*-ATP@AuNPs) in the presence of templates on the electrode surface. The molecular recognition sites created by extracting the templates from the polymeric layer of poly(*o*-ATP) accompanying AuNPs created the specific three-dimensional imprinted contours casted for the target molecules.

Electropolymerization of *p*-aminobenzoic acid was chosen for imprinting of quercetin on GCE grafted with Pd nanoparticles-porous graphene-carbon nanotubes composite (Pd/pGN-CNTs) made by hydrothermal method and showed good conductivity, large surface as well as good electrocatalysis [67]. Hydrogen bonding between *p*-aminobenzoic acid and quercetin, predicted by density functional theory, is accountable for projecting high specificity and selectivity to the imprinted cavities. The sensor was applied to 'real' samples of medicine and food samples.

Another imprinted nanoparticle composite sensor for ciprofloxacin was formed by blending the fluorescent optical fibre with imprinted nanoparticles composite polyethyleneglycol diacrylate hydrogel (MINs@PEGDA) [68]. The sensor was attached with a laser light source having an optical fibre spectrometer. This kit is proposed as a futuristic portable sensing kit for multiple targets.

A carbon paste electrode coated with MWCNTs and boron-embedded MIP composite membranes was prepared for imprinting tinidazole (Fig. **2**) [69]. The collaborative effort of MWCNTs and boron-embedded MIP composite membranes was observed by a shift of the reduction peak potential of tinidazole at the modified electrode by ~200 mV and multifold amplification of the peak current, compared with a bare electrode.

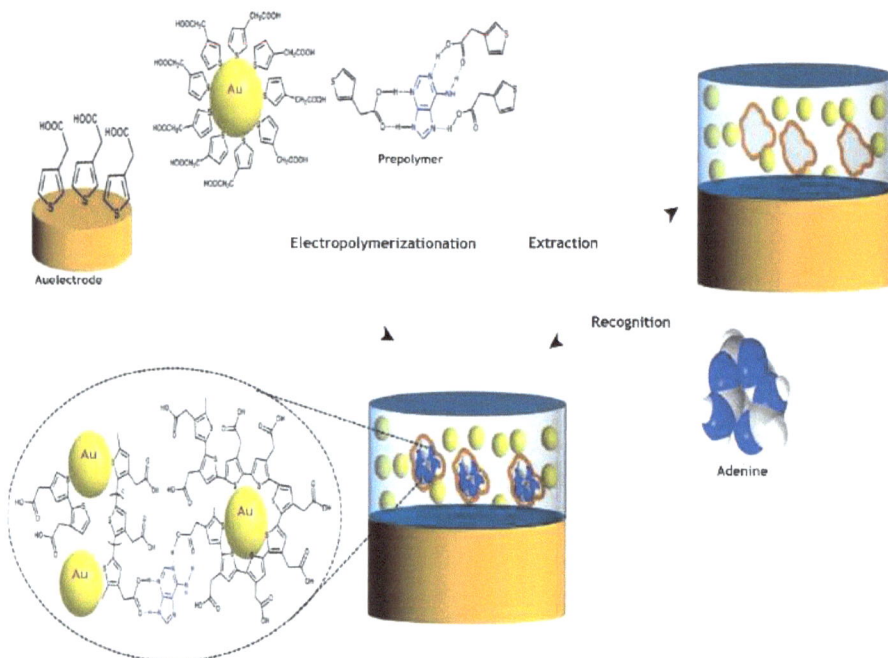

Fig. (2). Fabrication of a MIP-based voltammetric sensor for adenine based on conductive 3D network composite [69].

3-Nitrotyrosine is found to be a biomarker for oxidative stress. A sensor for this through imprinting was formed by polypyrrole doped with AuNPs on GCE coated with multiwalled carbon nanotube @ graphene oxide nanoribbons [70]. The incorporation of AuNPs was found to be able to effectively generate an electrical response by promoting the charge transfer procedure. This electrochemical MIP sensor was effectively able to detect 3-nitrotyrosine in human serum and urine samples.

Au-Pd nanoparticles–ionic liquid derivatized graphene–carbon nanotubes nanocomposite modified GCE was used for electropolymerization and imprinting to fabricate a sensing platform for paracetamol [71]. These nanoparticles were formed *via the* hydrothermal method in the ionic liquid (1- hydroxyethyl-3-methyl imidazolium bis[(trifluoromethyl) sulfonyl] imide), which facilitated the formation of small AuPd alloy nanoparticles, and provided a "spacer" to prevent the π-π stacking and conglomeration of graphene sheets and carbon nanotubes.

MIPs/hemin-graphene nanosheets composite was prepared using thyroglobulin and 3,3′,5,5′-tetramethylbenzidine (probe) as double templates [72]. Here, imprinting of both analyte as well as of a probe molecule was carried out. This double-pronged strategy facilitated the recognition as well as detection *via* monitoring the probe molecule.

A GCE modified with Prussian blue porous carbon CNTs hybrids was used for the electropolymerization of pyrrole and imprinted with cysteine. For the elution of templates from the polymeric matrix, overoxidation/dedoping was found to be appropriate.

Using green-emitting carbon dots as a transducer and dopamine as a functional monomer, 3-nitrotyrosine was imprinted for selective recognition of it [74]. Quenching of fluorescence intensity was used to bind the analyte in imprinted cavities (Fig. **3**)).

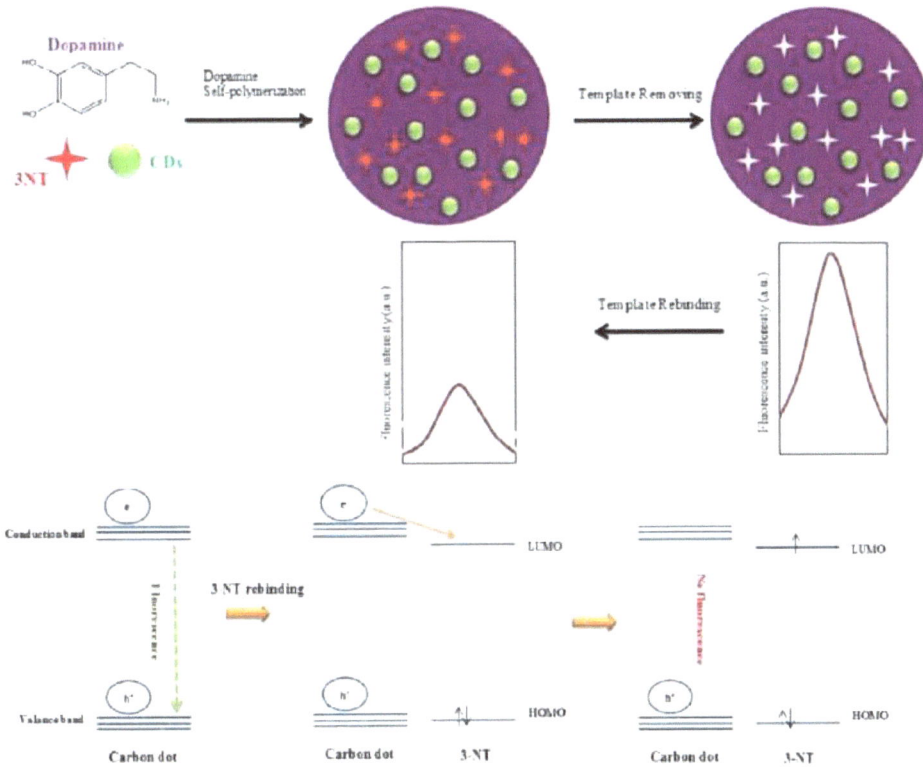

Fig. (3). Schematic illustration of BMIP@CDs sensor and fluorescence quenching mechanism [74].

MIP SENSORS BASED ON GRAPHENE NANOCOMPOSITE

Graphene (GR) is considered as two-dimensional carbon materials with a one-molecule thick planar sheet of sp^2 fortified carbon particles that are thickly pressed in a honeycomb precious stone cross section. The exceptional properties of GR have been widely applied [75 - 78]. The GR derivatives, such as graphene oxide (GO) and reduced graphene oxide (RGO), offer a wide scope of conceivable outcomes to practically sensible materials for diversified applications [79, 80]. RGO can be crafted in substantial amounts through microwave irradiation treatment of GO. It has a higher conductivity and holds unparalleled credibility for electrochemical applications, especially detection [81]. As of late, RGO-based hybrid materials, including the composites of RGO and polymers, have been broadly investigated for diversified applications [79 - 83]. Tailoring these composites, RGO plays as the fortifying component in a polymer grid. These blends of RGO and distinctive polymers show novel properties, *e.g.*, great dispersibility and high conductivity [84 - 86]. MIP, which is customized for an objective analyte, is considered a greater possibility to mix with RGO. The composite of RGO and MIP has an appropriate conductivity, just as a high adsorption limit and great explicitness for the targeted analyte. A standout amongst the most interesting and intriguing fields for the use of this material is the sensing field [87, 88]. GR has essentially been used for the creation of a few sensors for the identification of natural gases like H_2, NH_3, CH_4, NO_2, CO_2, natural vapours and formaldehyde [89 - 94].

Imidacloprid, one of the neonicotinoid pesticides, is generally utilized in horticulture because of its high insecticidal movement and low poisonous quality [95, 96]. An electrochemical sensor dependent on imprinted poly(o-phenylenediamine) (PoPD) films at RGO adjusted terminal for particular and delicate detection of imidacloprid was manufactured. RGO was dropped outside the smooth carbon terminal (SCT) to intensify the electrocatalytic movement. The imprinted PoPD layers were then coordinated onto the RGO surface by electropolymerization utilizing oPD as a monomer and imidacloprid as a template (Fig. **4**). The imprinted electrochemical sensor was effectively utilized for the specific detection of imidacloprid in real samples [97]. 4-Nitrophenol (4-NP) is one of the nitrophenols referred to in the contaminations of the U.S.A. Ecological Protection Agency (EPA) due to its harmfulness and continuances. A composite of RGO and MIP (RGO-MIP) was fabricated utilizing free radical polymerization and further utilized as a recognition component in the development of an electrochemical sensor for 4-NP [98].

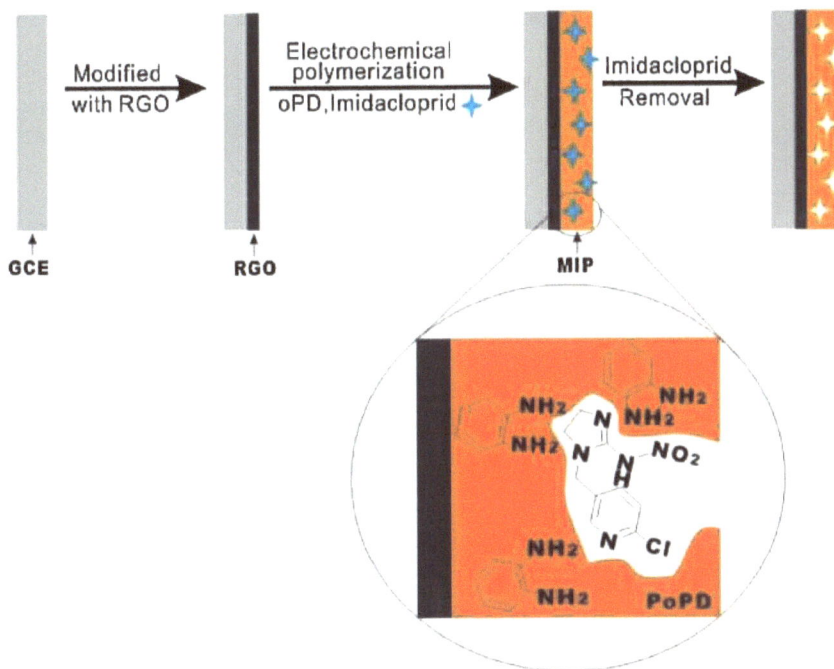

Fig. (4). Schematic representation for the development of imprinted PoPD-RGO electrode [97].

Carbofuran (CBF) is one of the carbamate species that is broadly utilized as a pesticide in farming. It can minimize the harvest development period and increment crop yields. A molecularly imprinted electrochemical sensor was constructed dependent on smooth carbon electrode brightened by RGO and gold NPs (RGO@Au) for the identification of CBF. Electrodes were modified with MIPs using CBF as the analyte, methyl acrylic acid as the imprinting monomer, and ethylene glycol maleic rosinate acrylate (EGMRA) as a cross linker (Fig. **5**). The sensor displayed a high adsorption limit and great selectivity for CBF, and it was effectively applied for the identification of CBF in genuine vegetable examples [99].

Vincristine, a bisindole alkaloids antibiotic, was separated from a plant called vinca. It is applied for the clinical therapy of cancer. An electrochemical sensor fabrication for vincristine through the development of an RGO-Au crossover film utilizing one stage co-electrodeposition of GO and chloroauric acid on a glassy carbon electrode substrate was carried out in a study [100].

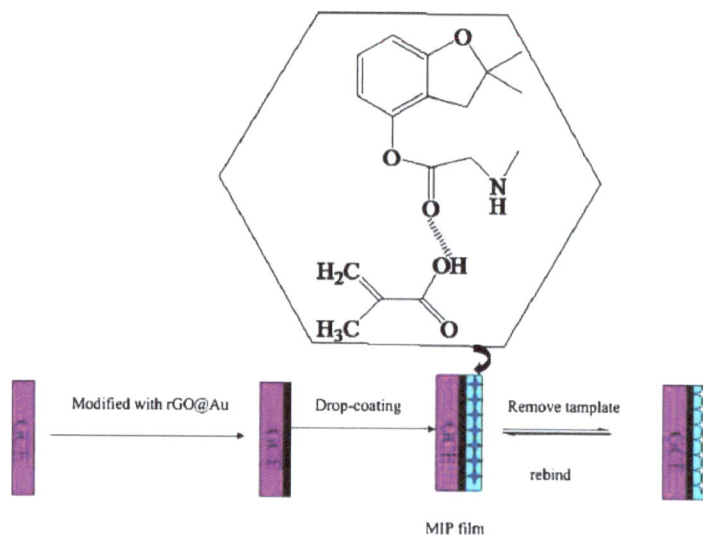

Fig. (5). Scheme of the process for the preparation of the imprinted RGO@Au electrode [100].

An imprinted sensor (chemiresistor) for nitrobenzene vapours was synthesized on a nanocomposite of GR/imprinted polymer, NPs/poly (methyl methacrylate) [101]. The nanosized MIP holding the cavities complementary were blended using methacrylic acid and vinyl benzene as the monomers and divinylbenzene as a cross linker. The copolymerization response was satisfied in acetonitrile by means of precipitation strategy. The MIP NPs were blended with GR to create a chemiresistor gas sensor. The expansion of poly(methyl methacrylate) in the nanocomposite brought about physical sturdiness, improved selectivity and higher opposition against dampness impact [94].

A unique electrochemical sensor for the discovery of thrombin (THR) was created dependent on a polished carbon terminal by the adjustment of MIP and nanocomposites of GR and gold NPs. A MIP polythionine film was shaped by cyclic voltammetric electropolymerization and played the job of the redox test to investigate the sensor execution. Nanocomposites to enhance the sensitivity were characterized with SEM and energy dispersive spectroscopy [102]. An electrochemical sensor for the determination of flutamide was fabricated using RGO-AuNPs and imprinted poly 2-aminophenol [103]. Poly 2-aminophenol enabled selectivity toward flutamide, recognizing it from other structural analogues. A multi-analyte sensing platform was formed using mesoporous silica modified magnetic graphene oxide (MGO@MS@MIP) by surface imprinting through the sol-gel process and was successfully used for the selective and simultaneous recognition of tetracyclines (TCs), even doxycycline (DC), tetracycline (TC), chlorotetracycline (CTC) and oxytetracycline (OTC) in water [104]. The imprinted composite showed good adsorption quantities, high selectivity and quick kinetics for four tetracyclines.

GCE modified with thiol terminated aptamer and gold nanoparticles were utilized with pyrrole as a functional monomer to imprint dopamine [105]. The modified electrode showed an augmented oxidation current for dopamine due to the enhanced conductivity facilitated by the good electron transfer ability of AuNPs/rGO. This combination of aptamer and MIP sensor for electroactive target assay is envisaged as a futuristic way for extending it to other sensors. Another MIP sensor for domperidone was fabricated utilizing the electrochemical method [106]. The performance of this sensor gave a wide concentration range of detection.

A composite of graphene quantum dots (GQD) and poly(indolylboronic acid) was used for imprinting dopamine [107]. This composite on the addition of dopamine drives to an aggregation resulting in quenching of fluorescence. It was discovered that boronic acid groups have a high selectivity for dopamine and that GQDs have a fluorescence advantage; this sensing platform was successful in dopamine sensing with commendable sensitivity and selectivity.

An electrochemical sensor for diclofenac (DCF) was prepared on a modified carbon paste electrode. Polyaniline, RGO and triphenylamine (cross linker) were used as an imprinting matrix for DCF [108]. The selectivity of this sensing matrix was examined in binary solutions of DCF with glucose, urea and ascorbic acid. The sensor was applied to 'real' samples of pharmaceutical and urine. In another instance, ionic liquids were used as the functional monomer (1-vinyl-3-butylimidazolium tetrafluoroborate) and crosslinker (1,4-butanediyl-3,3'-b-s-1-vinylimidazolium dibromide). Graphene composite with these ionic liquids was utilized for preparing a sensing matrix of 6-benzylaminopurine (6-BAP). The ionic liquid used as a crosslinker was found to augment electrocatalytic activity and also adsorption capability for 6-BAP of IL-GR-MIP [109].

An attempt to create a sensor for melamine was made by hyphenating liquid chromatography-tandem mass spectrometry with an imprinting matrix made up of a magnetic composite of graphene and carbon nanotube [110]. Another attempt to fabricate a miniaturized analytical tool was made on a MIP-graphene oxide composite as adsorbent for miniaturized tip solid-phase extraction (MTSPE) for estimating naphthalene-derived plant growth regulators (PGRs) in apples [111]. This method was hyphenated with high-performance liquid chromatography-fluorescence detection. It exhibited good linearity, low detection limits, high accuracy, and precision. A MIP/RGO composite for simultaneous determination of uric acid and tyrosine was reported by Zheng *et al.* [112]. Electropolymerization of the imprinted layer of 2-amino-5-mercapto-1, 3, 4-thiadiazole on RGO modified electrode was carried out in the presence of dual templates, uric acid and tyrosine. The RGO-MIP integration created a regular nanostructure with a large surface area of the sensing platform; the target was first recognized, followed by catalysing the oxidation reactions on the composite. An electrochemical sensor for folic acid (FA) was formed on graphite/restricted access molecularly imprinted poly(methacrylic acid)/SiO$_2$ composite [113]. The cathodic peak current for FA was enhanced. Yet another composite of surface imprinted MIP with graphene oxide (GO) was used for imprinting 2,4-dinitrophenol (DNP) [114]. A voltammetric sensor for bovine hemoglobin (BHb) was attempted on a graphene-MIP composite (GR-MIP) matrix [115]. The GR-MIP composite was prepared by the polymerization of dopamine.

MIP SENSORS BASED ON CHITOSAN NANOPARTICLE/NANOCOMPOSITE

Chitosan (CS) has proven itself as a useful polymer network for imprinting; compared to customary MIP, it also displays delicate properties, efficiently moulding and removing the analyte [116, 117]. The concentration of urea in blood serum can assess the level of renal disappointment, and it is additionally a clinical variable that gives valuable data on heart failure [118]. A urea electrochemical sensor was developed dependent on CS MIP films, which were arranged by potentiostatic electrodeposition of CS within sight of urea. The urea MIP electrochemical sensor demonstrated great selectivity to urea among the structurally similar analogues, high sensitivity to urea in the range from 1.0 x 10^{-8} to 4.0 x 10^{-5} M with a discovery breaking point of 5.0 x 10^{-9} M. Moreover, the recovery was 96.3 to 103.3% and thus offered great potential for clinical finding applications [119].

Trichlorphon [O,O-Dimethyl-2,2,2-trichloro-1-hydroxy-ethyl-phosphonate ester] (tri), a generally utilized organophosphorus (OP) pesticide, was a target analyte in anionic form, which easily bonded with protonated CS by electrostatic communication, one of the weak interactions between particles (Fig. **6**). MIPs were directly blended on GCE in view of electrodeposition of protonated CS as a useful polymer framework together with the targeted anion is trichlorphon [119, 120].

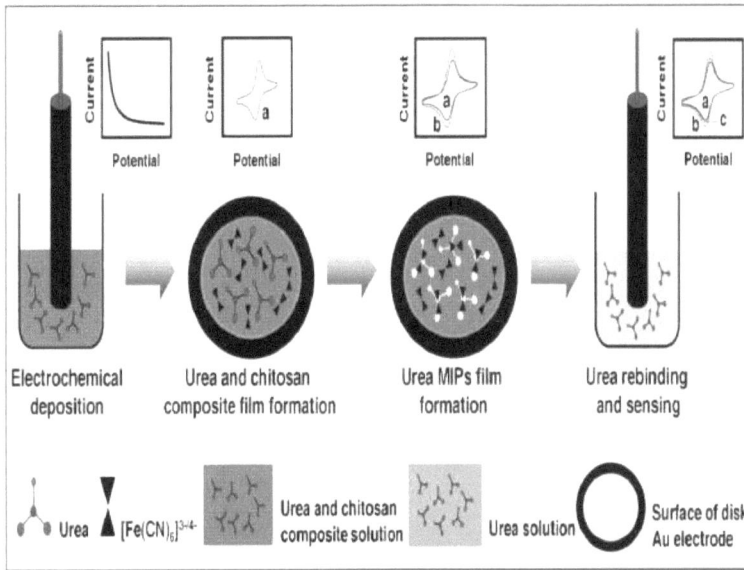

Fig. (6). Schematic illustration of the urea MIP electrochemical sensor fabrication based on CS [118].

Quinoxaline-2-carboxylicacid (QCA) is the residue of carbadox (CBX) left out in tissues. QCA was reported to be present at trace levels in samples of pork, chicken and fish. A fast, sensitive, and selective MIP electrochemical sensor for the estimation of QCA was fabricated using GCE and DPV for signal detection [121]. The GCE was grafted with multi-walled carbon nanotubes (MWNTs) - CS composite and an imprinted film on the surface [122].

A modest and straightforward technique was provided to plan and set up a highly specific electrochemical sensor for glucose at an exceptionally low level. Glucose imprinted matrix was fabricated on nickel oxide electrode through electrodepositing glucose–CS composite film on the electrochemically treated nickel. The created sensor was able to recognise glucose in the presence of existing interfering substances such as oxalic, uric, and ascorbic acid [123].

Nickel is a particularly harmful metal that poses a serious threat to the fauna and flora of bodies of water where wastewater is released. A Ni(II)-imprinted magnetic chitosan/poly (vinyl alcohol) composite (IMCP) was set up for the adsorption for Ni(II) particle. When synthesising Ni(II)- IMCP, using an imprinting system could improve selectivity and adsorption limit with respect to Ni(II) particles [124]. As a significant monomer in the compound industry, bisphenol A(2,2-bis(4-hydroxyphenyl) propane, BPA) is generally utilized in the creation of polycarbonate plastics and epoxy tars. It can be transferred into nature through different ways, such as wastewater, nourishing jugs, and even existing sustenance wrapping packs [125]. Numerous confirmations have demonstrated that BPA has harmful properties, initiating estrogenic endocrine interruption and advancement of tumorigenic movement [126]. An electrochemical sensor based on GCE grafted with imprinted chitosan film impregnated with acetylene black was used for the estimation of BPA (Fig. **7**). For 'real' sample analysis, BPA was estimated in drinking water samples [127].

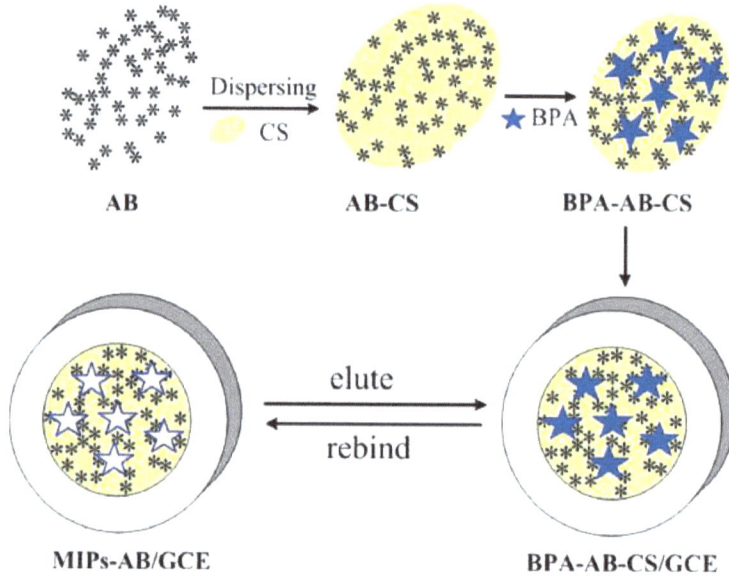

Fig. (7). Preparation of MIPs-AB/GCE and its recognition for BPA [127].

Morphine is used as a pain reliever in the treatment of moderate to serious pain, yet animal studies have shown a short pain-relieving impact of about 80 – 120 min [128 - 132]. A sufficient dimension of morphine focus for an all-inclusive timeframe is fundamental to furnish viable treatment with the medication. CS-based MIP nanogel is set up in close contact with morphine format, completely described, and utilized as another vehicle to broaden the length of morphine pain-relieving impact in Naval Medical Research Institute mice. CS-MIP nanogel can be a conceivable system as a morphine transporter for controlled discharge and augmentation of its pain-relieving efficacy [133]. Rose Bengal is a xanthene dye that is generally utilized in textile and photochemical enterprises. It has a few risky effects on human well being particularly on corneal epithelium [134]. Although this dye was utilized in the treatment of dermatitis and psoriasis, at a higher focus, it caused skin tingling, disturbance, blushing and rankling. A new MIP dependent on CS-TiO$_2$ nanocomposite (CTNC) was set up for the specific removal of rose Bengal from mechanical wastewaters utilizing a minimal effort course. The proficiency of the MIP for binding Rose Bengal colour from wastewater was checked spectrophotometrically (Fig. **8**) [135].

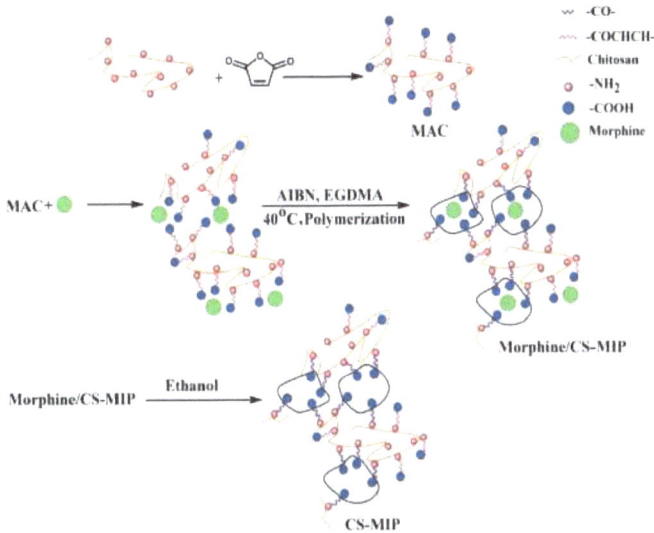

Fig. (8). Diagrammatic presentation of synthesis procedure of CS-MIP [135].

As a biocide, the use of triclosan (TCS) is widespread throughout the world as an additive, disinfectant, cleanser, conditioner, and so on; it is also used to disinfect cultivates [136, 137]. Additionally, TCS has natural poisonous quality, *e.g.,* organic hindrance and environmental improvement. It also has properties like bioaccumulation and biodegradability, so TCS has been identified in ecological media most of the time. The TCS imprinted material was acquired by utilizing chitosan as the functional monomer (having an amine and hydroxyl group), Fe^0 as the attractive framework, and glutaraldehyde as the cross-connecting operator. The TCS imprinted nanomaterial has the advantages of good strength, condition cordial, and quick detachment [138]. Aptamers are oligonucleotides with high restricting affinity and explicitness for target particles, so they can be utilized as a suitable platform for molecular diagnostics [139]. Herein, cross breed MIP receptor utilizing an electrochemical sensor focusing on the quantitative analysis of the HCV centre antigen was reported. A new strategy by hyphenating MIP receptors with apt sensing utilizing MWCNTs-CS nanocomposite was attempted for the successful immobilization of aptamer (Fig. **9**). The high strength/power and simplicity of the nanocomposite-based stage joined with excellent sensitivity and selectivity of the hybrid receptors, additionally advantaging from the two-fold explicit molecular recognition property of MIPs [140].

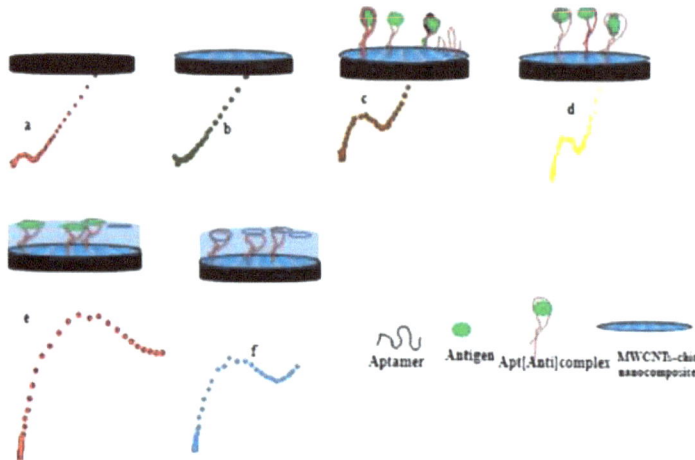

Fig. (9). The schematic presentation for the preparation of the Apt-MIP nanohybrid for HCV core antigen detection [140].

Perfluorooctane sulfonate has been made worldwide since the 1950s and is utilized as a crude material for the production of materials, makeup, and firefighting froths [141 - 143]. Because of appealing solid carbon-fluorine bonds and its trouble to degrade, it is perceived as a powerful contaminating agent in the environment. This sulfonate has been found in serum tests because of its presence in humans [144]. For quantitative estimation of PFOS, an imprinted sensor was prepared with CS doped with fluorescent carbon quantum dots [145].

MIP SENSOR BASED ON CHITOSAN NANOPARTICLE-GRAPHENE NANOCOMPOSITE

Erythromycin, extensively used in the treatment of bacterial infection for over 50 years, is mostly used to protect animals and farm crops from bacterial infections. But the residues of erythromycin may be left behind, leading to spurious effects on end users [149]. Hence, a robust method is needed for their quantitative estimation. Hence, an imprinted electrochemical sensor was prepared established on a gold electrode coated by CS–platinum NP (CS–PtNP) together with GR–gold NP (GR-AuNP) nanocomposites for specific and selective evaluation of erythromycin [150]. Neomycin, an aminoglycoside antibiotic produced by streptomyces fradiae, has excellent activity against gram-negative microscopic organisms and partial activity against gram-positive bacteria [151, 152]. Neomycin, often used to treat diverse infections, is reported to cause side effects also. A MIP sensor for neomycin was prepared to scan clinical treatments as per the need. The electrochemical sensor based on CS-silver NP (CS-AgNP)/G--MWCNTs composites grafted gold electrode using pyrrole as monomer was prepared for the recognition of neomycin. The sensor was successfully applied for the estimation of neomycin in milk, nectar samples and showed good reproducibility and stability [155].

Bisphenol A (2,2-bis (4-hydroxyphenyl) propane, BPA) is a natural compound that is broadly utilized in the plastic business as a monomer for delivering epoxy gums and polycarbonate. BPA is prevalent in our lives since it is unintentionally released into nature to pollute streams and groundwater during the manufacturing of these items. BPA can also find its way into food, for example, by drinking water from a wide range of food contact surfaces made of polycarbonates and epoxy gums, such as newborn child bolstering bottles, flatware, storage compartments, and food can linings [156]. Subsequently, people may routinely inhale trace quantities of BPA. Most vitally, BPA was found to conceivably cause malignancy [157]. In this manner, the dependable, quick and genuine auspicious assurance of follow dimension of BPA is critical for wellbeing insurance and security. The purpose of this study is to investigate if graphene drops doped CS film may be used as an imprinted detecting nanocomposite on an acetylene dark glue terminal (ABPE) for BPA detection [158]. A new chitosan/ionic liquid–graphene (CS/IL–GR) arrangement was tailored on imprinted GCE (MIPs/CS/IL– GR/GCE) for the purpose of specific and selective estimation of BSA. Polymer chosen for imprinting was polypyrrole (PPy), electropolymerized on the surface of CS/IL–GR/GCE (Fig. **10**). The manouvered MIPs/CS/ IL–GR/GCE displayed high sensitivity, great selectivity and strength, which demonstrate potential clinical application [159].

A unique Zn(II) imprinted polymer was integrated by means of a co-precipitation technique utilizing GO/attractive CS nanocomposite as supporting material and connected as a sorbent for particular attractive strong stage extraction of zinc pursued by its assurance by fire nuclear retention spectrometry. The created methodology was effectively connected to the particular extraction and assurance of zinc in different examples, including drinking water, dark tea, rice, and milk [160]. A voltammetric MIP sensor based on ion-imprinted CS GR nanocomposites (IIP-S) was fabricated for the detection of Cr(VI) (Fig. **11**). The analyte imprinted polymers were built by one-step electrodeposition. The IIP-S showed good stability and reproducibility. Additionally, it was effectively combined for the determination of Cr(VI) in tap water and stream water [161].

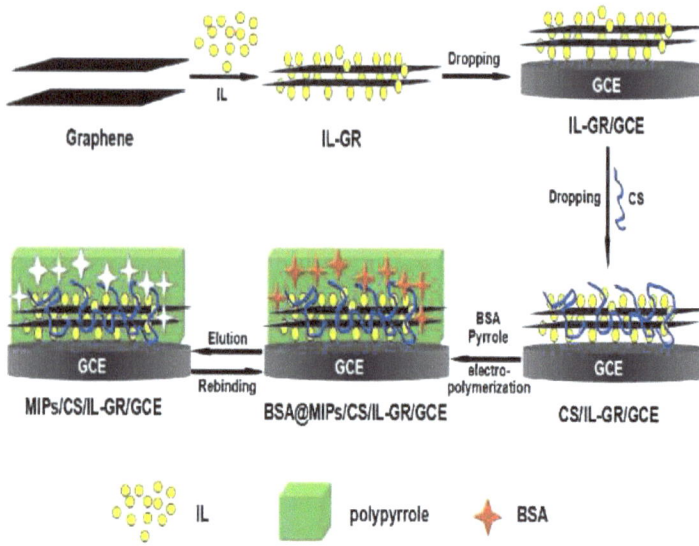

Fig. (10). Schematic representation of fabrication of molecularly imprinted electrochemical sensor based on CS/IL/G [159].

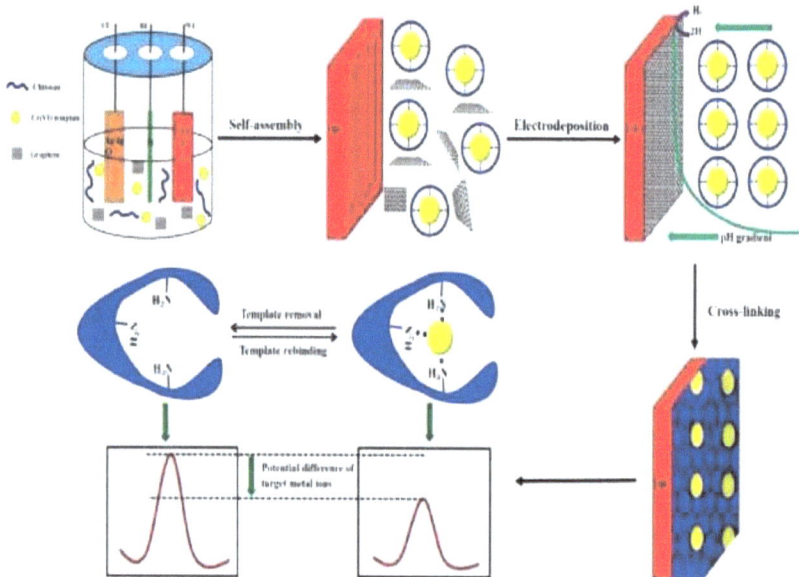

Fig. (11). Schematic representation of the development of the IIP-S for Cr(VI) detection [161].

Aspartame imprinted MIP-EQCM sensing matrix was designed by electrodepositing the aspartame-imprinted polymeric matrix of biopolymer chitosan on an Au-coated EQCM electrode [162]. This sensing platform was credited with a reliable method for estimating aspartame in real and commercial samples by hyphenating EQCM-MIP with DPV. This nanocomposite sensor endeavours a good immobilization of analyte along with imprinting process, providing a suitable way for detection. Such successful attempts bolster the hypothesis that molecular imprinting and electrochemistry will jointly pave a potential route to the generation of economic, miniaturized sensing tools.

MIP SENSOR BASED ON STARCH NANOPARTICLE/NANOCOMPOSITE

Starch nanoparticle (SNP) alone and/or nanocomposite with certain materials favourable for imprinting were reported. One example is a magnetic molecularly imprinted polymer (MMIPs) for protein amylase. These were synthesized utilizing poly(ethylene-co-vinyl liquor) (EVAL) with 27−44 mol % ethylene [163]. Routinely, conjugation and capture of amylase with biomaterials lessen the protein's activities. Attractive particles were additionally added into MIPs to empower the fast division of bound protein and reactants and products. MMIPs have the benefits of high surface territory, suspension, simple expulsion from the response, and quick reload of chemicals. Hexavalent chromium, Cr(VI), is a solid oxidizing specialist lethal to plants, creatures and people. Cr(VI) is found in effluents discharged from tanning, electroplating, paint and material ventures. Due to its lethal impacts, hexavalent chromium (Cr(VI)) must be removed from the modern effluents previously released into the earth [164]. ST with iron oxide NPs was used for the adsorption of Cr (VI) from contaminated water. ST functionalized iron oxide NPs (SIONPs) show critical enhancement in adsorption capacities for Cr(VI) as compared with IO. Adsorption limit is reliant on the pH; at a given Cr(VI) fixation, the adsorption limit is observed to be most at pH 2 for all SIO [165].

A method was developed for the extraction of three antimicrobial agents from bovine milk tests utilizing a GO–ST-based nanocomposite [166]. The prepared nanocomposites were utilized as an extractive stage for smaller-scale strong-stage extraction of antibiotics from dairy animals' milk tests. The extracted antimicrobials, for *e.g.,* amoxicillin, ampicillin and cloxacillin, were broken down by high-performance liquid chromatography– ultraviolet identification (HPLC– UV). Vital factors related to extraction and desorption productivity were streamlined. High extraction efficiencies for the chosen antimicrobials were accomplished utilizing the ST–based nanocomposite as the extractive stage [167].

An EQCM-MIP sensor for estimating transferrin in clinical samples was developed by grafting MIPs on gold-coated EQCM electrode surface using nanoparticles of biopolymer starch (Starch NP) - RGO nanocomposite as polymeric format (Fig. **12**) [168]. Electrodeposition of starch NP-RGO nano-composite on EQCM electrode produced the nano MIP sensor. As the imprinted sites are located on exposed surfaces and enormous specific surface areas, analyte molecules have better receptiveness with good binding affinity. The sensor was highly selective to analyte transferrin and was found to be very useful in terms of reusability and also a gain in adsorption capacity over other sensors was achieved. Another electrochemical sensor by utilizing starch nanoparticles as imprinting matrix for epinephrine (EP), a neurotransmitter, was prepared by grafting MIP on the surface of the Au-coated EQCM electrode (Fig. **13**) [169]. The starch NP-RGO composite was electrodeposited on the EQCM electrode in the presence of analyte EP. The location of imprinted sites on the exposed surface and high specific surface area induces the binding of analyte molecules to specific imprinted cavities. A biocompatible MIP-sensor was thus successfully prepared, which can easily be applied in 'real matrices' also.

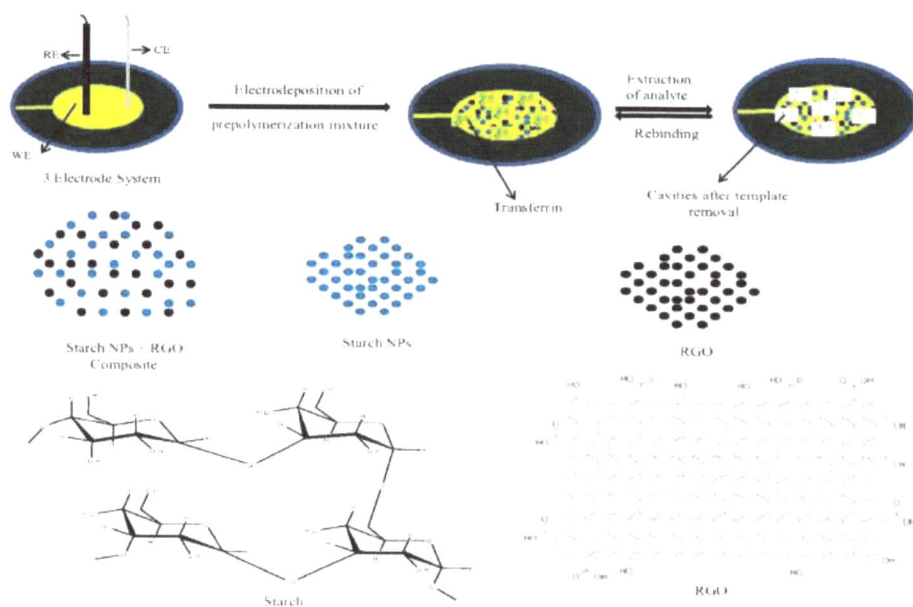

Fig. (12). Schematic representation for the fabrication of glycoprotein (Tf) imprinted sensor [168].

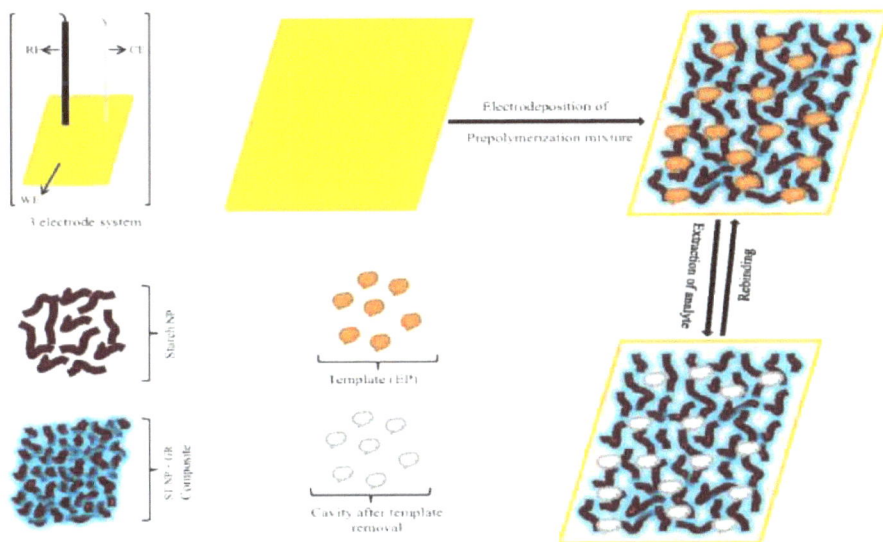

Fig. (13). Schematic diagram of the fabrication of SNP-RGO composite MIP sensor [169].

PERSPECTIVE

As per the literature survey cited above, the sensing industry needs to be transferred from bio-sensing (using biological components) to chemo-sensing and that biocompatible and eco-compatible variety is the need of the day, especially for 'real' samples involving (bio)molecules or samples from nature, both of which are aqueous-based media. In this course, nanocomposite matrices for imprinting are utilized as an imprinting matrix. Nanoparticles, which are particles smaller than a particular size, are used to meet a variety of societal needs, from tailoring unique substances/materials to reducing the time it takes for life-saving nanostructured drugs/pharmaceuticals to reach the body's biological fluids.MIPs have come a long way to serve science and society for various needs. Although the molecular imprinting technology for sensors is being developed, researchers are not content with the primary function of sensing. Many studies are trying to market prototypes into multi-functional kits that serve broad applications. For example, MIP sensors were linked with photothermal therapy to devise a three-i--one package for sensing, advancement, and wiping out bacteria. MIPs have outgrown from simple sensing, monitoring, tracking of (bio) molecules of interest to food assays and environmental monitoring. Innovations in all sectors are being incorporated for updating 2D MIP sensors to 3D bioMEMs. Newer devices based on MIP are appearing, such as MIP electronic tongues, handheld analyzers, *etc.* Electronic tongue on combining with imprinting technology permits sensing as well as separation of various analytes in solutions with better analytical signals.

Now, imprinting is not confined to research laboratories only; it is now being considered for commercial ventures also. Based on the scenario presented in this chapter, molecular imprinting, with its exemplary selectivity feature, appears to be waiting for more experimentation through the incorporation of newer technologies, hyphenations with sensitive detectors being developed, and trials for more and more applications. In the near future, this will pave the way for economic devices with efficacy parallel to biological counterparts.

CONSENT FOR PUBLICATION

Not applicable.

CONFLICT OF INTEREST

The author declares no conflict of interest, financial or otherwise.

ACKNOWLEDGEMENTS

The author acknowledges the Department of Chemistry, MMV and Institute of Science, Banaras Hindu University, for providing the facilities to carry out the research work. The author acknowledges the funding agencies (UGC, CSIR, SERB) for providing research grants.

REFERENCES

[1]   Turner, A.P.F. Biosensors: sense and sensibility. *Chem. Soc. Rev.,* **2013**, *42*(8), 3184-3196.
        [http://dx.doi.org/10.1039/c3cs35528d] [PMID: 23420144]

[2]   Alocilja, E.C.; Radke, S.M. Market analysis of biosensors for food safety. *Biosens. Bioelectron.,* **2003**, *18*(5-6), 841-846.
        [http://dx.doi.org/10.1016/S0956-5663(03)00009-5] [PMID: 12706600]

[3]   Van Dorst, B.; Mehta, J.; Bekaert, K.; Rouah-Martin, E.; De Coen, W.; Dubruel, P.; Blust, R.; Robbens, J. Recent advances in recognition elements of food and environmental biosensors: a review. *Biosens. Bioelectron.,* **2010**, *26*(4), 1178-1194.
        [http://dx.doi.org/10.1016/j.bios.2010.07.033] [PMID: 20729060]

[4]   Smith, R.G.; D'Souza, N.; Nicklin, S. A review of biosensors and biologically-inspired systems for explosives detection. *Analyst (Lond.),* **2008**, *133*(5), 571-584.
        [http://dx.doi.org/10.1039/b717933m] [PMID: 18427676]

[5]   Wanekaya, A.K.; Chen, W.; Mulchandani, A. Recent biosensing developments in environmental security. *J. Environ. Monit.,* **2008**, *10*(6), 703-712.
        [http://dx.doi.org/10.1039/b806830p] [PMID: 18528536]

[6]   Hulanicki, A.; Geab, S.; Ingman, F. Chemical sensors defnitions and classification. *Pure Appl. Chem.,* **1991**, *63*(9), 1247-1250.
        [http://dx.doi.org/10.1351/pac199163091247]

[7]   Thevenot, D.R.; Toth, K.; Durst, R.A.; Wilson, G.S. Electrochemical biosensors: recommended definitions and classification. *Pure Appl. Chem.,* **1999**, *7*(12), 2333-2348.
        [http://dx.doi.org/10.1351/pac199971122333]

[8]   Malitesta, C.; Mazzotta, E.; Picca, R.A.; Poma, A.; Chianella, I.; Piletsky, S.A. MIP sensors--the electrochemical approach. *Anal. Bioanal. Chem.,* **2012**, *402*(5), 1827-1846.
[http://dx.doi.org/10.1007/s00216-011-5405-5] [PMID: 21947439]

[9]   Sharma, P.S.; D'Souza, F.; Kutner, W. Molecular imprinting for selective chemical sensing of hazardous compounds and drugs of abuse. *Trends Analyt. Chem.,* **2012**, *34*, 59-77.
[http://dx.doi.org/10.1016/j.trac.2011.11.005]

[10]   Sharma, P.S.; Pietrzyk-Le, A.; D'Souza, F.; Kutner, W. Electrochemically synthesized polymers in molecular imprinting for chemical sensing. *Anal. Bioanal. Chem.,* **2012**, *402*(10), 3177-3204.
[http://dx.doi.org/10.1007/s00216-011-5696-6] [PMID: 22302165]

[11]   Wackerlig, J.; Lieberzeit, P.A. Molecularly imprinted polymer nanoparticles in chemical sensing — synthesis, characterisation and application. *Sens. Actuators B Chem.,* **2015**, *207*, 144-157.
[http://dx.doi.org/10.1016/j.snb.2014.09.094]

[12]   Whitcombe, M.J.; Kirsch, N.; Nicholls, I.A. Molecular imprinting science and technology: a survey of the literature for the years 2004-2011. *J. Mol. Recognit.,* **2014**, *27*(6), 297-401.
[http://dx.doi.org/10.1002/jmr.2347] [PMID: 24700625]

[13]   Suriyanarayanan, S.; Cywinski, P.J.; Moro, A.J.; Mohr, G.J.; Kutner, W. Chemosensors based on molecularly imprinted polymers. *Top. Curr. Chem.,* **2012**, *325*, 165-265.
[http://dx.doi.org/10.1007/128_2010_92] [PMID: 21128065]

[14]   Haupt, K.; Mosbach, K. Molecularly imprinted polymers and their use in biomimetic sensors. *Chem. Rev.,* **2000**, *100*(7), 2495-2504.
[http://dx.doi.org/10.1021/cr990099w] [PMID: 11749293]

[15]   Wulff, G. Molecular imprinting in cross-linked materials with the aid of molecular templates-a way towards artificial antibodies. *Angew. Chem. Int. Ed. Engl.,* **1995**, *34*(17), 1812-1832.
[http://dx.doi.org/10.1002/anie.199518121]

[16]   Xu, S.F.; Li, J.H.; Song, X.L.; Liu, J.S.; Lu, H.Z.; Chen, L.X. Photonic and magnetic dual responsive molecularly imprinted polymers: preparation, recognition characteristics and properties as novel sorbent for caffeine in complicated samples. *Anal. Methods,* **2013**, *5*(1), 124-133.
[http://dx.doi.org/10.1039/C2AY25922B]

[17]   Zhou, C.; Wang, T.; Liu, J.; Guo, C.; Peng, Y.; Bai, J.; Liu, M.; Dong, J.; Gao, N.; Ning, B.; Gao, Z. Molecularly imprinted photonic polymer as an optical sensor to detect chloramphenicol. *Analyst (Lond.),* **2012**, *137*(19), 4469-4474.
[http://dx.doi.org/10.1039/c2an35617a] [PMID: 22870501]

[18]   Wang, C.; Javadi, A.; Ghaffari, M.; Gong, S. A pH-sensitive molecularly imprinted nanospheres/hydrogel composite as a coating for implantable biosensors. *Biomaterials,* **2010**, *31*(18), 4944-4951.
[http://dx.doi.org/10.1016/j.biomaterials.2010.02.073] [PMID: 20346500]

[19]   Zhang, X.; Xu, S.; Lim, J.M.; Lee, Y.I. Molecularly imprinted solid phase microextraction fiber for trace analysis of catecholamines in urine and serum samples by capillary electrophoresis. *Talanta,* **2012**, *99*, 270-276.
[http://dx.doi.org/10.1016/j.talanta.2012.05.050] [PMID: 22967551]

[20]   Chen, L.; Xu, S.; Li, J. Recent advances in molecular imprinting technology: current status, challenges and highlighted applications. *Chem. Soc. Rev.,* **2011**, *40*(5), 2922-2942.
[http://dx.doi.org/10.1039/c0cs00084a] [PMID: 21359355]

[21]   Uzuriaga-Sánchez, R.J.; Khan, S.; Wong, A.; Picasso, G.; Pividori, M.I.; Sotomayor, M.D.P.T. Magnetically separable polymer (Mag-MIP) for selective analysis of biotin in food samples. *Food Chem.,* **2016**, *190*, 460-467.
[http://dx.doi.org/10.1016/j.foodchem.2015.05.129] [PMID: 26212997]

[22]   Ahmad, O.S.; Bedwell, T.S.; Esen, C.; Garcia-Cruz, A.; Piletsky, S.A. Molecularly imprinted

polymers in electrochemical and optical sensors. *Trends Biotechnol.,* **2019**, *37*(3), 294-309.
[http://dx.doi.org/10.1016/j.tibtech.2018.08.009] [PMID: 30241923]

[23]   BelBruno, J.J. Molecularly Imprinted Polymers. *Chem. Rev.,* **2019**, *119*(1), 94-119.
[http://dx.doi.org/10.1021/acs.chemrev.8b00171] [PMID: 30246529]

[24]   Zhang, H. Water-compatible molecularly imprinted polymers: Promising synthetic substitutes for biological receptors. *Polymer (Guildf.),* **2014**, *55*(3), 699-714.
[http://dx.doi.org/10.1016/j.polymer.2013.12.064]

[25]   Nicholls, I.A.; Ramstrom, O.; Mosbach, K. Insights into the role of the hydrogen bond and hydrophobic effect on recognition in molecularly imprinted polymer synthetic peptide receptor mimics. *J. Chromatogr. A,* **1995**, *691*(1-2), 349-353.
[http://dx.doi.org/10.1016/0021-9673(94)01067-O]

[26]   Nicholls, I.A. Towards the rational design of molecularly imprinted polymers. *J. Mol. Recognit.,* **1998**, *11*(1-6), 79-82.
[http://dx.doi.org/10.1002/(SICI)1099-1352(199812)11:1/6<79::AID-JMR394>3.0.CO;2-B] [PMID: 10076811]

[27]   Haupt, K.; Linares, A.V.; Bompart, M.; Bui, B.T. Molecularly imprinted polymers. *Top. Curr. Chem.,* **2012**, *325*, 1-28.
[http://dx.doi.org/10.1007/978-3-642-28421-2] [PMID: 22183146]

[28]   Bowen, J.L.; Manesiotis, P.; Allender, C.J. Twenty years since 'antibody mimics' by molecular imprinting were first proposed: a critical perspective. *Mol. Impr.,* **2013**, *1*, 35-40.
[http://dx.doi.org/10.2478/molim-2013-0001]

[29]   Shen, X.; Xu, C.; Ye, L. Molecularly imprinted polymers for clean water: analysis and purification. *Ind. Eng. Chem. Res.,* **2012**, *52*(39), 13890-13899.
[http://dx.doi.org/10.1021/ie302623s]

[30]   Lemieux, R.U. How Water provides the impetus for molecular eecognition in aqueous solution. *Acc. Chem. Res.,* **1996**, *29*(8), 373-380.
[http://dx.doi.org/10.1021/ar9600087]

[31]   Guan, W.; Pan, J.; Ou, H.; Wang, X.; Zou, X.; Hu, W.; Li, C.; Wu, X. Removal of strontium(II) ions by potassium tetratitanate whisker and sodium trititanate whisker from aqueous solution: equilibrium, kinetics and thermodynamics. *Chem. Eng. J.,* **2011**, *167*(1), 215-222.
[http://dx.doi.org/10.1016/j.cej.2010.12.025]

[32]   Wang, H.J.; Zhou, W.H.; Yin, X.F.; Zhuang, Z.X.; Yang, H.H.; Wang, X.R. Template synthesized molecularly imprinted polymer nanotube membranes for chemical separations. *J. Am. Chem. Soc.,* **2006**, *128*(50), 15954-15955.
[http://dx.doi.org/10.1021/ja065116v] [PMID: 17165706]

[33]   Novoselov, K.S.; Geim, A.K.; Morozov, S.V.; Jiang, D.; Katsnelson, M.I.; Grigorieva, I.V.; Dubonos, S.V.; Firsov, A.A. Two-dimensional gas of massless Dirac fermions in graphene. *Nature,* **2005**, *438*(7065), 197-200.
[http://dx.doi.org/10.1038/nature04233] [PMID: 16281030]

[34]   Zhu, G.; Fan, J.; Gao, Y.; Gao, X.; Wang, J. Synthesis of surface molecularly imprinted polymer and the selective solid phase extraction of imidazole from its structural analogs. *Talanta,* **2011**, *84*(4), 1124-1132.
[http://dx.doi.org/10.1016/j.talanta.2011.03.015] [PMID: 21530788]

[35]   Zhang, W.; Qin, L.; He, X.W.; Li, W.Y.; Zhang, Y.K. Novel surface modified molecularly imprinted polymer using acryloyl-β-cyclodextrin and acrylamide as monomers for selective recognition of lysozyme in aqueous solution. *J. Chromatogr. A,* **2009**, *1216*(21), 4560-4567.
[http://dx.doi.org/10.1016/j.chroma.2009.03.056] [PMID: 19361806]

[36]   Zhu, R.; Zhao, W.; Zhai, M.; Wei, F.; Cai, Z.; Sheng, N.; Hu, Q. Molecularly imprinted layer-coated

silica nanoparticles for selective solid-phase extraction of bisphenol A from chemical cleansing and cosmetics samples. *Anal. Chim. Acta,* **2010**, *658*(2), 209-216.
[http://dx.doi.org/10.1016/j.aca.2009.11.008] [PMID: 20103097]

[37]   Guo, X.; Liu, X.; Xu, B.; Dou, T. Synthesis and characterization of carbon sphere-silica core–shell structure and hollow silica spheres. *Colloids Surf. A Physicochem. Eng. Asp.,* **2009**, *345*(1-3), 141-146.
[http://dx.doi.org/10.1016/j.colsurfa.2009.04.048]

[38]   Ji, J.; Sun, X.; Tian, X.; Li, Z.; Zhang, Y. Synthesis of acrylamide molecularly imprinted polymers immobilized on graphite oxide through surface-initiated atom transfer radical polymerization. *Anal. Lett.,* **2012**, *46*(6), 969-981.
[http://dx.doi.org/10.1080/00032719.2012.745085]

[39]   Liu, Y.; Huang, Y.; Liu, J.; Wang, W.; Liu, G.; Zhao, R. Superparamagnetic surface molecularly imprinted nanoparticles for water-soluble pefloxacin mesylate prepared *via* surface initiated atom transfer radical polymerization and its application in egg sample analysis. *J. Chromatogr. A,* **2012**, *1246*, 15-21.
[http://dx.doi.org/10.1016/j.chroma.2012.01.045] [PMID: 22321951]

[40]   Schirhagl, R.; Latif, U.; Dickert, F.L. Atrazine detection based on antibody replicas. *J. Mater. Chem.,* **2011**, *21*(38), 14594-14598.
[http://dx.doi.org/10.1039/c1jm11576f]

[41]   Schirhagl, R.; Latif, U.; Podlipna, D.; Blumenstock, H.; Dickert, F.L. Natural and biomimetic materials for the detection of insulin. *Anal. Chem.,* **2012**, *84*(9), 3908-3913.
[http://dx.doi.org/10.1021/ac201687b] [PMID: 22468696]

[42]   Wang, J.; Dai, J.; Meng, M.; Song, Z.; Pan, J.; Yan, Y.; Li, C. Surface molecularly imprinted polymers based on yeast prepared by atom transfer radical emulsion polymerization for selective recognition of ciprofloxacin from aqueous medium. *J. Appl. Polym. Sci.,* **2014**, *131*(11), 40310-40319.
[http://dx.doi.org/10.1002/app.40310]

[43]   Niu, M.; Pham-Huy, C.; He, H. Core-shell nanoparticles coated with molecularly imprinted polymers: a review. *Mikrochim. Acta,* **2016**, *183*(10), 2677-2695.
[http://dx.doi.org/10.1007/s00604-016-1930-4]

[44]   Hart, B.R.; Shea, K.J. Synthetic peptide receptors: molecularly imprinted polymers for the recognition of peptides using peptide-metal interactions. *J. Am. Chem. Soc.,* **2001**, *123*(9), 2072-2073.
[http://dx.doi.org/10.1021/ja005661a] [PMID: 11456837]

[45]   Dirion, B.; Cobb, Z.; Schillinger, E.; Andersson, L.I.; Sellergren, B. Water-compatible molecularly imprinted polymers obtained *via* high-throughput synthesis and experimental design. *J. Am. Chem. Soc.,* **2003**, *125*(49), 15101-15109.
[http://dx.doi.org/10.1021/ja0355473] [PMID: 14653745]

[46]   Manesiotis, P.; Hall, A.J.; Courtois, J.; Irgum, K.; Sellergren, B. An artificial riboflavin receptor prepared by a template analogue imprinting strategy. *Angew. Chem. Int. Ed.,* **2005**, *44*(25), 3902-3906.
[http://dx.doi.org/10.1002/anie.200500342] [PMID: 15892138]

[47]   Urraca, J.L.; Moreno-Bondi, M.C.; Orellana, G.; Sellergren, B.; Hall, A.J. Molecularly imprinted polymers as antibody mimics in automated on-line fluorescent competitive assays. *Anal. Chem.,* **2007**, *79*(13), 4915-4923.
[http://dx.doi.org/10.1021/ac070277i] [PMID: 17550229]

[48]   Hoshino, Y.; Kodama, T.; Okahata, Y.; Shea, K.J. Peptide imprinted polymer nanoparticles: a plastic antibody. *J. Am. Chem. Soc.,* **2008**, *130*(46), 15242-15243.
[http://dx.doi.org/10.1021/ja8062875] [PMID: 18942788]

[49]   Helling, S.; Shinde, S.; Brosseron, F.; Schnabel, A.; Müller, T.; Meyer, H.E.; Marcus, K.; Sellergren, B. Ultratrace enrichment of tyrosine phosphorylated peptides on an imprinted polymer. *Anal. Chem.,* **2011**, *83*(5), 1862-1865.
[http://dx.doi.org/10.1021/ac103086v] [PMID: 21306124]

[50]   Manesiotis, P.; Borrelli, C.; Aureliano, C.S.A.; Svensson, C.; Sellergren, B. Water-compatible imprinted polymers for selective depletion of riboflavine from beverages. *J. Mater. Chem.*, **2009**, *19*(34), 6185-6193.
[http://dx.doi.org/10.1039/b906117g]

[51]   Ton, X.A.; Acha, V.; Haupt, K.; Tse Sum Bui, B. Direct fluorimetric sensing of UV-excited analytes in biological and environmental samples using molecularly imprinted polymer nanoparticles and fluorescence polarization. *Biosens. Bioelectron.*, **2012**, *36*(1), 22-28.
[http://dx.doi.org/10.1016/j.bios.2012.03.033] [PMID: 22541891]

[52]   Wagner, R.; Wan, W.; Biyikal, M.; Benito-Peña, E.; Moreno-Bondi, M.C.; Lazraq, I.; Rurack, K.; Sellergren, B. Synthesis, spectroscopic, and analyte-responsive behavior of a polymerizable naphthalimide-based carboxylate probe and molecularly imprinted polymers prepared thereof. *J. Org. Chem.*, **2013**, *78*(4), 1377-1389.
[http://dx.doi.org/10.1021/jo3019522] [PMID: 23356385]

[53]   Wang, Y.T.; Zhang, Z.Q.; Jain, V.; Yi, J.J.; Mueller, S.; Sokolov, J.; Liu, Z.X.; Levon, K.; Rigas, B.; Rafailovich, M.H. Potentiometric sensors based on surface molecular imprinting: detection of cancer biomarkers and viruses. *Sens. Actuators B Chem.*, **2010**, *146*(1), 381-387.
[http://dx.doi.org/10.1016/j.snb.2010.02.032]

[54]   Ma, L.; Tang, L.; Li, R-S.; Huang, Y-P.; Liu, Z-S. Water-compatible molecularly imprinted polymers prepared using metal–organic gel as porogen. *RSC Advances*, **2015**, *5*(103), 84601-84609.
[http://dx.doi.org/10.1039/C5RA16029D]

[55]   Duan, F.; Chen, C.; Zhao, X.; Yang, Y.; Liu, X.; Qin, Y. Water-compatible surface molecularly imprinted polymers with synergy of bi-functional monomers for enhanced selective adsorption of bisphenol A from aqueous solution. *Environ. Sci. Nano*, **2016**, *3*(1), 213-222.
[http://dx.doi.org/10.1039/C5EN00198F]

[56]   Singh, A.K.; Singh, M. Molecularly imprinted Au-nanoparticle composite-functionalized EQCM sensor for L-serine. *J. Electroanal. Chem. (Lausanne)*, **2016**, *780*, 169-175.
[http://dx.doi.org/10.1016/j.jelechem.2016.09.021]

[57]   Sanati, A; Moakhar, RS; Hosseini, I I; Raeissi, K; Karimzadeh, F; Jalali, M; Kharaziha, M; Sheibani, S; Shariati, L; Presley, JF; Vali, H; Mahshid, S Gold nano/micro-islands overcome the molecularly imprinted polymer limitations to achieve ultrasensitive protein detection
[http://dx.doi.org/10.1021/acssensors.0c01701]

[58]   Du, Q.; Wu, P.; Sun, Y.; Zhang, J.; He, H. Selective photodegradation of tetracycline by molecularly imprinted $ZnO@NH_2$-UiO-66 composites. *Chem. Eng. J.*, **2020**, *390*, 124614.
[http://dx.doi.org/10.1016/j.cej.2020.124614]

[59]   Guo, Q.; Hu, F.; Yang, X.; Yang, J.; Yang, S.; Chen, X.; Wu, F.; Minteer, S.D. HuF,Yang X, Yang J, Yang S, Chen X, Wu F, MinteerSD; *In-situ* and controllable synthesis of graphene-gold nanoparticles/molecularly imprinted polymers composite modified electrode for sensitive and selective rutin detection. *Microchem. J.*, **2020**, *158*, 105254.
[http://dx.doi.org/10.1016/j.microc.2020.105254]

[60]   Yuan, X.; Gao, X.; Yuan, Y.; Ji, Y.; Xiong, Z.; Zhao, L. Fe_3O_4/graphene molecularly imprinted composite for selective separation of catecholamine neurotransmitters and their analysis in rat brain tissues. *Talanta*, **2021**, *224*, 121843.
[http://dx.doi.org/10.1016/j.talanta.2020.121843] [PMID: 33379061]

[61]   Yang, Y.; Yan, W.; Wang, X.; Yu, L.; Zhang, J.; Bai, B.; Guo, C.; Fan, S. Development of a molecularly imprinted photoelectrochemical sensing platform based on NH_2-MIL-125(Ti)-TiO_2 composite for the sensitive and selective determination of oxtetracycline. *Biosens. Bioelectron.*, **2021**, *177*, 113000.
[http://dx.doi.org/10.1016/j.bios.2021.113000] [PMID: 33485152]

[62]   Özcan, N.; Medetalibeyoglu, H.; Akyıldırım, O.; Atar, N.; Yola, M.L. Electrochemical detection of

amyloid-β protein by delaminated titanium carbide MXene/multi-walled carbon nanotubes composite with molecularly imprinted polymer. *Mater. Today Commun.,* **2020**, *23*, 101097.
[http://dx.doi.org/10.1016/j.mtcomm.2020.101097]

[63]   Boke, C.P.; Karaman, O.; Medetalibeyoglu, H.; Karaman, C.; Atar, N.; Yola, M.L. A new approach for electrochemical detection of organochlorine compound lindane: Development of molecular imprinting polymer with polyoxometalate/carbon nitride nanotubes composite and validation. *Microchem. J.,* **2020**, *157*, 105012.
[http://dx.doi.org/10.1016/j.microc.2020.105012]

[64]   Yola, M.L.; Atar, N. Development of molecular imprinted sensor including graphitic carbon nitride/N-doped carbon dots composite for novel recognition of epinephrine;*Composites B. Engineering,* **2019**, *175*, 107113.

[65]   Zhou, T.; Ding, J.; He, Z.; Li, J.; Liang, Z.; Li, C.; Li, Y.; Chen, Y.; Ding, L. Preparation of magnetic superhydrophilic molecularly imprinted composite resin based on multi-walled carbon nanotubes to detect triazines in environmental water. *Chem. Eng. J.,* **2018**, *334*, 2293-2302.
[http://dx.doi.org/10.1016/j.cej.2017.11.185]

[66]   Gu, Y.; Wang, Y.; Wu, X.; Pan, M.; Hu, N.; Wang, J.; Wang, S. WangY,Wu X, Pan M, Hu N, Wang J,Wang S; Quartz crystal microbalance sensor based on covalent organic framework composite and molecularly imprinted polymer of poly(*o*-aminothiophenol) with gold nanoparticles for the determination of aflatoxin B1. *Sens. Actuators B Chem.,* **2019**, *291*, 293-297.
[http://dx.doi.org/10.1016/j.snb.2019.04.092]

[67]   Yang, L.; Xu, B.; Ye, H.; Zhao, F.; Zeng, B. A novel quercetin electrochemical sensor based on molecularly imprinted poly(para-aminobenzoic acid) on 3D Pd nanoparticles-porous graphene-carbon nanotubes composite. *Sens. Actuators B Chem.,* **2017**, *251*, 601-608.
[http://dx.doi.org/10.1016/j.snb.2017.04.006]

[68]   Huang, Q.; Lv, C.; Yuan, X.; He, M.; Lai, J.; Sun, H. A novel fluorescent optical fiber sensor for highly selective detection of antibiotic ciprofloxacin based on replaceable molecularly imprinted nanoparticles composite hydrogel detector. *Sens. Actuators B Chem.,* **2021**, *328*, 129000.
[http://dx.doi.org/10.1016/j.snb.2020.129000]

[69]   Wang, H.; Qian, D.; Xiao, X.; Deng, C.; Liao, L.; Deng, J.; Lin, Y-W. Preparation and application of a carbon paste electrode modified with multi-walled carbon nanotubes and boron-embedded molecularly imprinted composite membranes. *Bioelectrochemistry,* **2018**, *121*, 115-124.
[http://dx.doi.org/10.1016/j.bioelechem.2018.01.006] [PMID: 29413861]

[70]   Wang, S.; Sun, G.; Chen, Z.; Liang, Y.; Zhou, Q.; Pan, Y.; Zhai, H. Constructing a novel composite of molecularly imprinted polymer-coated AuNPs electrochemical sensor for the determination of 3-nitrotyrosine. *Electrochim. Acta,* **2018**, *259*, 893-902.
[http://dx.doi.org/10.1016/j.electacta.2017.11.033]

[71]   Yang, L.; Zhang, B.; Xu, B.; Zhao, F.; Zeng, B. Ionic liquid functionalized 3D graphene-carbon nanotubes–AuPd nanoparticles–molecularly imprinted copolymer based paracetamol electrochemical sensor: Preparation, characterization and application. *Talanta,* **2021**, *224*, 121845.
[http://dx.doi.org/10.1016/j.talanta.2020.121845] [PMID: 33379063]

[72]   Wang, X.; Huang, K.; Cui, H.; Zhang, H.; Zeng, L.; Zhou, Y.; Jing, T. Label-free determination of thyroglobulin using template-probe double imprinted composites. *Sens. Actuators B Chem.,* **2020**, *313*, 128028.
[http://dx.doi.org/10.1016/j.snb.2020.128028]

[73]   Duan, D.; Yang, H.; Ding, Y.; Li, L.; Ma, G. A three-dimensional conductive molecularly imprinted electrochemical sensor based on MOF derived porous carbon/carbon nanotubes composites and prussian blue nanocubes mediated amplification for chiral analysis of cysteine enantiomers. *Electrochim. Acta,* **2019**, *302*, 137-144.
[http://dx.doi.org/10.1016/j.electacta.2019.02.028]

[74] Jalili, R.; Amjadi, M. Bio-inspired molecularly imprinted polymer–green emitting carbon dot composite for selective and sensitive detection of 3-nitrotyrosine as a biomarker. *Sens. Actuators B Chem.,* **2018**, *255*, 1072-1078.
[http://dx.doi.org/10.1016/j.snb.2017.08.145]

[75] Si, Y.; Samulski, E.T. Synthesis of water soluble graphene. *Nano Lett.,* **2008**, *8*(6), 1679-1682.
[http://dx.doi.org/10.1021/nl080604h] [PMID: 18498200]

[76] Li, B.; Wang, L.; Li, D.; Chiu, Y.L.; Zhang, Z.; Shi, J.; Chen, X.D.; Mao, Z. Physical properties and loading capacity of starch-based microparticles crosslinked with trisodium trimetaphosphate. *J. Food Eng.,* **2009**, *92*(3), 255-260.
[http://dx.doi.org/10.1016/j.jfoodeng.2008.10.008]

[77] Bai, H.; Li, C.; Shi, G. Functional composite materials based on chemically converted graphene. *Adv. Mater.,* **2011**, *23*(9), 1089-1115.
[http://dx.doi.org/10.1002/adma.201003753] [PMID: 21360763]

[78] Low, C.T.J.; Walsh, F.C.; Chakrabarti, M.H.; Hashim, M.A.; Hussain, M.A. Electrochemical approaches to the production of graphene flakes and their potential applications. *Carbon,* **2013**, *54*, 1-21.
[http://dx.doi.org/10.1016/j.carbon.2012.11.030]

[79] Compton, O.C.; Nguyen, S.T. Graphene oxide, highly reduced graphene oxide, and graphene: versatile building blocks for carbon-based materials. *Small,* **2010**, *6*(6), 711-723.
[http://dx.doi.org/10.1002/smll.200901934] [PMID: 20225186]

[80] Sahoo, P.K.; Panigrahy, B.; Sahoo, S.; Satpati, A.K.; Li, D.; Bahadur, D. *In situ* synthesis and properties of reduced graphene oxide/Bi nanocomposites: as an electroactive material for analysis of heavy metals. *Biosens. Bioelectron.,* **2013**, *43*, 293-296.
[http://dx.doi.org/10.1016/j.bios.2012.12.031] [PMID: 23334218]

[81] Kuila, T.; Bose, S.; Khanra, P.; Mishra, A.K.; Kim, N.H.; Lee, J.H. Recent advances in graphene-based biosensors. *Biosens. Bioelectron.,* **2011**, *26*(12), 4637-4648.
[http://dx.doi.org/10.1016/j.bios.2011.05.039] [PMID: 21683572]

[82] Tran, P.D.; Batabyal, S.K.; Pramana, S.S.; Barber, J.; Wong, L.H.; Loo, S.C.J. A cuprous oxide-reduced graphene oxide (Cu2O-rGO) composite photocatalyst for hydrogen generation: employing rGO as an electron acceptor to enhance the photocatalytic activity and stability of Cu$_2$O. *Nanoscale,* **2012**, *4*(13), 3875-3878.
[http://dx.doi.org/10.1039/c2nr30881a] [PMID: 22653156]

[83] Shin, M.K.; Lee, B.; Kim, S.H.; Lee, J.A.; Spinks, G.M.; Gambhir, S.; Wallace, G.G.; Kozlov, M.E.; Baughman, R.H.; Kim, S.J. Synergistic toughening of composite fibres by self-alignment of reduced graphene oxide and carbon nanotubes. *Nat. Commun.,* **2012**, *3*(1), 650-657.
[http://dx.doi.org/10.1038/ncomms1661] [PMID: 22337128]

[84] Deng, Y.; Li, Y.J.; Dai, J.; Lang, M.D.; Huang, X.Y. An efficient way to functionalize graphene sheets with presynthesized polymer *via* ATNRC chemistry. *J. Polym. Sci. A Polym. Chem.,* **2011**, *49*(7), ·1582-1590.
[http://dx.doi.org/10.1002/pola.24579]

[85] Park, O.K.; Hahm, M.G.; Lee, S.; Joh, H.I.; Na, S.I.; Vajtai, R.; Lee, J.H.; Ku, B.C.; Ajayan, P.M. *In situ* synthesis of thermochemically reduced graphene oxide conducting nanocomposites. *Nano Lett.,* **2012**, *12*(4), 1789-1793.
[http://dx.doi.org/10.1021/nl203803d] [PMID: 22260510]

[86] Wu, N.; She, X.L.; Yang, D.J.; Wu, X.F.; Su, F.B.; Chen, Y.F. Synthesis of network reduced graphene oxide in polystyrene matrix by a two-step reduction method for superior conductivity of the composite. *J. Mater. Chem.,* **2012**, *22*(33), 17254-17261.
[http://dx.doi.org/10.1039/c2jm33114d]

[87]   Schedin, F.; Geim, A.K.; Morozov, S.V.; Hill, E.W.; Blake, P.; Katsnelson, M.I.; Novoselov, K.S. Detection of individual gas molecules adsorbed on graphene. *Nat. Mater.,* **2007**, *6*(9), 652-655.
[http://dx.doi.org/10.1038/nmat1967] [PMID: 17660825]

[88]   Robinson, J.T.; Perkins, F.K.; Snow, E.S.; Wei, Z.; Sheehan, P.E. Reduced graphene oxide molecular sensors. *Nano Lett.,* **2008**, *8*(10), 3137-3140.
[http://dx.doi.org/10.1021/nl8013007] [PMID: 18763832]

[89]   Gautam, M.; Jayatissa, A.H. Gas sensing properties of graphene synthesized by chemical vapor deposition. *Mater. Sci. Eng. C,* **2011**, *31*(7), 1405-1411.
[http://dx.doi.org/10.1016/j.msec.2011.05.008]

[90]   Ko, G.; Kim, H.Y.; Ahn, J.; Park, Y.M.; Lee, K.Y.; Kim, J. Graphene-based nitrogen dioxide gas sensors. *Curr. Appl. Phys.,* **2010**, *10*(4), 1002-1004.
[http://dx.doi.org/10.1016/j.cap.2009.12.024]

[91]   Yoon, H.J.; Jun, D.H.; Yang, J.H.; Zhou, Z.; Yang, S.S.; Cheng, M.M.C. Carbon dioxide gas sensor using a graphene sheet. *Sens. Actuators B Chem.,* **2011**, *157*(1), 310-313.
[http://dx.doi.org/10.1016/j.snb.2011.03.035]

[92]   Dua, V.; Surwade, S.P.; Ammu, S.; Agnihotra, S.R.; Jain, S.; Roberts, K.E.; Park, S.; Ruoff, R.S.; Manohar, S.K. All-organic vapor sensor using inkjet-printed reduced graphene oxide. *Angew. Chem. Int. Ed. Engl.,* **2010**, *49*(12), 2154-2157.
[http://dx.doi.org/10.1002/anie.200905089] [PMID: 20187049]

[93]   Zhou, X.; Yuan, C.; Qin, D.; Xue, Z.; Wang, Y.; Du, J.; Ma, L.; Ma, L.; Lu, X. Pd Nanoparticles on functionalized graphene for excellent detection of nitro aromatic compounds. *Electrochim. Acta,* **2014**, *119*, 243-250.
[http://dx.doi.org/10.1016/j.electacta.2013.10.197]

[94]   Alizadeh, T.; Akhoundian, M. Graphene/nano-sized imprinted polymer/poly(methyl methacrylate) nanocomposite as a new gas sensor for the determination of nitrobenzene vapors. *Anal Bioanal Electrochem,* **2017**, *9*, 1070-1079.

[95]   Broznić, D.; Milin, C. Effects of temperature on sorption-desorption processes of imidacloprid in soils of Croatian coastal regions. *J. Environ. Sci. Health B,* **2012**, *47*(8), 779-794.
[http://dx.doi.org/10.1080/03601234.2012.676413] [PMID: 22575005]

[96]   Guzsvany, V.J.; Gaal, F.F.; Bjelica, L.J.; Okresz, S.N. Voltammetric determination of imidacloprid and thiamethoxam. *J. Serb. Chem. Soc.,* **1995**, *70*(5), 735-743.
[http://dx.doi.org/10.2298/JSC0505735G]

[97]   Kong, L.; Jiang, X.; Zeng, Y.; Zhou, T.; Shi, G. Molecularly imprinted sensor based on electropolmerized poly(o-phenylenediamine) membranes at reduced graphene oxide modified electrode for imidacloprid determination. *Sens. Actuators B Chem.,* **2013**, *185*, 424-431.
[http://dx.doi.org/10.1016/j.snb.2013.05.033]

[98]   Zeng, Y.; Zhou, Y.; Zhou, T.; Shi, G. A novel composite of reduced graphene oxide and molecularly imprinted polymer for electrochemical sensing of 4-nitrophenol. *Electrochim. Acta,* **2014**, *130*, 504-511.
[http://dx.doi.org/10.1016/j.electacta.2014.02.130]

[99]   Tan, X.; Hu, Q.; Wu, J.; Li, X.; Li, P.; Yu, H.; Li, X.; Lei, F. Electrochemical sensor based on molecularly imprinted polymer reduced graphene oxide and gold nanoparticles modified electrode for detection of carbofuran. *Sens. Actuators B Chem.,* **2015**, *220*, 216-221.
[http://dx.doi.org/10.1016/j.snb.2015.05.048]

[100]   Zhang, Y.; Zheng, J.; Guo, M. Preparation of molecularly imprinted electrochemical sensor for detection of vincristine based on reduced graphene oxide/gold nanoparticle composite film. *Chin. J. Chem.,* **2016**, *34*(12), 1268-1276.
[http://dx.doi.org/10.1002/cjoc.201600582]

[101]   Latifoglu, A.; Gurol, M.D. The effect of humic acids on nitrobenzene oxidation by ozonation and O$_3$/UV processes. *Water Res.,* **2003**, *37*(8), 1879-1889.
[http://dx.doi.org/10.1016/S0043-1354(02)00583-3] [PMID: 12697231]

[102]   Yang, S.; Yang, J.; Cao, Q.; Zheng, Y.; Bai, C.; Teng, Y.; Xu, W. A molecularly imprinted polythionine-modified electrode based on a graphene-gold nanoparticle composite (MIP/AuNPs/RGO/GCE) for the determination of thrombin. *Int. J. Electrochem. Sci.,* **2018**, *13*, 9333-9345.
[http://dx.doi.org/10.20964/2018.10.53]

[103]   Kadivar, M.; Aliakbar, A. A molecularly imprinted poly 2-aminophenol-gold nanoparticle-reduced graphene oxide composite for electrochemical determination of flutamide in environmental and biological samples. *Anal. Methods,* **2021**, *13*(4), 536-551.
[http://dx.doi.org/10.1039/D0AY01812K] [PMID: 33449062]

[104]   Zeng, G.; Liu, Y.; Ma, X.; Fan, Y. Fabrication of magnetic multi-template molecularly imprinted polymer composite for the selective and efficient removal of tetracyclines from water. *Front. Environ. Sci. Eng.,* **2021**, *15*(5), 107.
[http://dx.doi.org/10.1007/s11783-021-1395-5]

[105]   Shen, M.; Kan, X. Aptamer and molecularly imprinted polymer: Synergistic recognition and sensing of dopamine. *Electrochim. Acta,* **2021**, *367*, 137433.
[http://dx.doi.org/10.1016/j.electacta.2020.137433]

[106]   Kumar, D.R.; Dhakal, G.; Nguyen, V.Q.; Shim, J.J. Molecularly imprinted hornlike polymer@electrochemically reduced graphene oxide electrode for the highly selective determination of an antiemetic drug. *Anal. Chim. Acta,* **2021**, *1141*, 71-82.
[http://dx.doi.org/10.1016/j.aca.2020.10.014] [PMID: 33248664]

[107]   Zhou, X.; Gao, X.; Song, F.; Wang, C.; Chu, F.; Wu, S. A sensing approach for dopamine determination by boronic acid-functionalized molecularly imprinted graphene quantum dots composite. *Appl. Surf. Sci.,* **2017**, *423*, 810-816.
[http://dx.doi.org/10.1016/j.apsusc.2017.06.199]

[108]   Mostafavi, M.; Yaftian, M.R.; Piri, F.; Shayani-Jam, H. A new diclofenac molecularly imprinted electrochemical sensor based upon a polyaniline/reduced graphene oxide nano-composite. *Biosens. Bioelectron.,* **2018**, *122*, 160-167.
[http://dx.doi.org/10.1016/j.bios.2018.09.047] [PMID: 30265965]

[109]   Zhu, X.; Zeng, Y.; Zhang, Z.; Yang, Y.; Zhai, Y.; Wang, H.; Liu, L.; Hu, J.; Li, L. A new composite of graphene and molecularly imprinted polymer based on ionic liquids as functional monomer and cross-linker for electrochemical sensing 6-benzylaminopurine. *Biosens. Bioelectron.,* **2018**, *108*, 38-45.
[http://dx.doi.org/10.1016/j.bios.2018.02.032] [PMID: 29499557]

[110]   Zhao, X.; Chen, L.; Li, B. Magnetic molecular imprinting polymers based on three-dimensional (3D) graphene-carbon nanotube hybrid composites for analysis of melamine in milk powder. *Food Chem.,* **2018**, *255*, 226-234.
[http://dx.doi.org/10.1016/j.foodchem.2018.02.078] [PMID: 29571470]

[111]   Chen, X.; Yuan, Y.; Yan, H.; Shen, S. Selective, sensitive, and miniaturized analytical method based on molecularly imprinted graphene oxide composites for the determination of naphthalene-derived plant growth regulators in apples. *Food Chem.,* **2021**, *349*, 128982.
[http://dx.doi.org/10.1016/j.foodchem.2020.128982] [PMID: 33561797]

[112]   Zheng, W.; Zhao, M.; Liu, W.; Yu, S.; Niu, L.; Li, G.; Li, H.; Liu, W. Electrochemical sensor based on molecularly imprinted polymer/reduced graphene oxide composite for simultaneous determination of uric acid and tyrosine. *J. Electroanal. Chem. (Lausanne),* **2018**, *813*, 75-82.
[http://dx.doi.org/10.1016/j.jelechem.2018.02.022]

[113]   Kuceki, M.; Oliveira, F.M.; Segatelli, M.G.; Coelho, M.K.L.; Pereira, A.C.; Rocha, L.R.; Mendonça, J.C.; Tarley, C.R.T. Selective and sensitive voltammetric determination of folic acid using

graphite/restricted access molecularly imprinted poly(methacrylic acid)/ SiO_2 composite. *J. Electroanal. Chem. (Lausanne),* **2018**, *818*, 223-230.
[http://dx.doi.org/10.1016/j.jelechem.2018.04.043]

[114] Liu, Y; Zhu, L; Zhang, Y; Tang, H Electrochemical sensoring of 2,4-dinitrophenol by using composites of graphene oxide with surface molecular imprinted polymer; *Sens Actuat B Chem,* **2012**, *171– 172*(1151– 1158)

[115] Luo, J.; Jiang, S.; Liu, X. Electrochemical sensor for bovine hemoglobin based on a novel graphene-molecular imprinted polymers composite as recognition element. *Sens. Actuators B Chem.,* **2014**, *203*, 782-789.
[http://dx.doi.org/10.1016/j.snb.2014.07.061]

[116] Wang, H.; He, Y.; He, X.; Li, W.; Chen, L.; Zhang, Y. BSA-imprinted synthetic receptor for reversible template recognition. *J. Sep. Sci.,* **2009**, *32*(11), 1981-1986.
[http://dx.doi.org/10.1002/jssc.200800562] [PMID: 19479780]

[117] Chang, P.R.; Jian, R.; Zheng, P.; Yu, J.; Ma, X. Preparation and proper-ties of glycerol plasticized-starch (GPS)/cellulose nanoparticle (CN) composites. *Carbohydr. Polym.,* **2010**, *79*(2), 301-305.
[http://dx.doi.org/10.1016/j.carbpol.2009.08.007]

[118] Aronson, D.; Mittleman, M.A.; Burger, A.J. Elevated blood urea nitrogen level as a predictor of mortality in patients admitted for decompensated heart failure. *Am. J. Med.,* **2004**, *116*(7), 466-473.
[http://dx.doi.org/10.1016/j.amjmed.2003.11.014] [PMID: 15047036]

[119] Chen, J.; Lian, H.; Sun, X.; Liu, B. Development of a chitosan molecularly imprinted electrochemical sensor for trichlorphon determination. *Int. J. Environ. Anal. Chem.,* **2012**, *92*(9), 1046-1058.
[http://dx.doi.org/10.1080/03067319.2010.496054]

[120] Tokonami, S.; Shiigi, H.; Nagaoka, T. Review: micro- and nanosized molecularly imprinted polymers for high-throughput analytical applications. *Anal. Chim. Acta,* **2009**, *641*(1-2), 7-13.
[http://dx.doi.org/10.1016/j.aca.2009.03.035] [PMID: 19393361]

[121] Hutchinson, M.J.; Young, P.B.; Kennedy, D.G. Confirmation of carbadox and olaquindox metabolites in porcine liver using liquid chromatography-electrospray, tandem mass spectrometry. *J. Chromatogr. B Analyt. Technol. Biomed. Life Sci.,* **2005**, *816*(1-2), 15-20.
[http://dx.doi.org/10.1016/j.jchromb.2004.09.024] [PMID: 15664328]

[122] Yang, Y.; Fang, G.; Liu, G.; Pan, M.; Wang, X.; Kong, L.; He, X.; Wang, S. Electrochemical sensor based on molecularly imprinted polymer film *via* sol-gel technology and multi-walled carbon nanotubes-chitosan functional layer for sensitive determination of quinoxaline-2-carboxylic acid. *Biosens. Bioelectron.,* **2013**, *47*, 475-481.
[http://dx.doi.org/10.1016/j.bios.2013.03.054] [PMID: 23624016]

[123] Li, H.X.; Yao, W.; Wu, Q.; Xia, W.S. Glucose molecularly imprinted electrochemical sensor based on chitosan and nickel oxide electrode. *Adv. Mat. Res.,* **2014**, *1052*, 215-219.
[http://dx.doi.org/10.4028/www.scientific.net/AMR.1052.215]

[124] Zhang, L.; Zhong, L.; Yang, S.; Liu, D.; Wang, Y.; Wang, S.; Han, X.; Zhang, X. Adsorption of Ni(II) ion on Ni(II) ion-imprinted magnetic chitosan/poly(vinyl alcohol) composite. *Colloid Polym. Sci.,* **2015**, *293*(9), 2497-2506.
[http://dx.doi.org/10.1007/s00396-015-3626-4]

[125] Lu, B.J.; Liu, M.C.; Shi, H.J.; Huang, X.F.; Zhao, G.H. A Novel Photoelectrochemical sensor for bisphenol A with high sensitivity and selectivity based on surface molecularly imprinted polypyrrole modified TiO_2 nanotubes. *Electroanalysis,* **2013**, *25*(3), 771-779.
[http://dx.doi.org/10.1002/elan.201200585]

[126] Keri, R.A.; Ho, S.M.; Hunt, P.A.; Knudsen, K.E.; Soto, A.M.; Prins, G.S. An evaluation of evidence for the carcinogenic activity of bisphenol A. *Reprod. Toxicol.,* **2007**, *24*(2), 240-252.
[http://dx.doi.org/10.1016/j.reprotox.2007.06.008] [PMID: 17706921]

[127]   Tan, Y.; Jin, J.; Zhang, S.; Shi, Z.; Wang, J.; Zhang, J.; Pu, W.; Yang, C. Electrochemical determination of bisphenol A using a molecularly imprinted chitosan-acetylene black composite film modified glassy carbon electrode. *Electroanalysis,* **2016**, *28*(1), 189-196.
[http://dx.doi.org/10.1002/elan.201500533]

[128]   Hain, R.D.; Hardcastle, A.; Pinkerton, C.R.; Aherne, G.W. Morphine and morphine-6-glucuronide in the plasma and cerebrospinal fluid of children. *Br. J. Clin. Pharmacol.,* **1999**, *48*(1), 37-42.
[http://dx.doi.org/10.1046/j.1365-2125.1999.00948.x] [PMID: 10383558]

[129]   Meineke, I.; Freudenthaler, S.; Hofmann, U.; Schaeffeler, E.; Mikus, G.; Schwab, M.; Prange, H.W.; Gleiter, C.H.; Brockmöller, J. Pharmacokinetic modelling of morphine, morphine-3-glucuronide and morphine-6-glucuronide in plasma and cerebrospinal fluid of neurosurgical patients after short-term infusion of morphine. *Br. J. Clin. Pharmacol.,* **2002**, *54*(6), 592-603.
[PMID: 12492606]

[130]   Sverrisdóttir, E.; Lund, T.M.; Olesen, A.E.; Drewes, A.M.; Christrup, L.L.; Kreilgaard, M. A review of morphine and morphine-6-glucuronide's pharmacokinetic-pharmacodynamic relationships in experimental and clinical pain. *Eur. J. Pharm. Sci.,* **2015**, *74*, 45-62.
[http://dx.doi.org/10.1016/j.ejps.2015.03.020] [PMID: 25861720]

[131]   Klimas, R.; Mikus, G. Morphine-6-glucuronide is responsible for the analgesic effect after morphine administration: a quantitative review of morphine, morphine-6-glucuronide, and morphine---glucuronide. *Br. J. Anaesth.,* **2014**, *113*(6), 935-944.
[http://dx.doi.org/10.1093/bja/aeu186] [PMID: 24985077]

[132]   Keil, G.J., II; Delander, G.E. Time-dependent antinociceptive interactions between opioids and nucleoside transport inhibitors. *J. Pharmacol. Exp. Ther.,* **1995**, *274*(3), 1387-1392.
[PMID: 7562512]

[133]   Hassanzadeh, M.; Ghaemy, M.; Ahmadi, S. Extending time profile of morphine-induced analgesia using a chitosan-based molecular imprinted polymer nanogel. *Macromol. Biosci.,* **2016**, *16*(10), 1515-1523.
[http://dx.doi.org/10.1002/mabi.201600177] [PMID: 27411404]

[134]   Tabery, H.M. Toxic effect of rose bengal dye on the living human corneal epithelium. *Acta Ophthalmol. Scand.,* **1998**, *76*(2), 142-145.
[http://dx.doi.org/10.1034/j.1600-0420.1998.760203.x] [PMID: 9591941]

[135]   Ahmed, M.A.; Abdelbar, N.M.; Mohamed, A.A. Molecular imprinted chitosan-TiO$_2$ nanocomposite for the selective removal of Rose Bengal from wastewater. *Int. J. Biol. Macromol.,* **2018**, *107*(Pt A), 1046-1053.
[http://dx.doi.org/10.1016/j.ijbiomac.2017.09.082] [PMID: 28943440]

[136]   Liu, Y.; Song, Q.; Wang, L. Development and characterization of an amperometric sensor for triclosan detection based on electropolymerized molecularly imprinted polymer. *Microchem. J.,* **2009**, *91*(2), 222-226.
[http://dx.doi.org/10.1016/j.microc.2008.11.007]

[137]   von der Ohe, P.C.; Schmitt-Jansen, M.; Slobodnik, J.; Brack, W. Triclosan--the forgotten priority substance? *Environ. Sci. Pollut. Res. Int.,* **2012**, *19*(2), 585-591.
[http://dx.doi.org/10.1007/s11356-011-0580-7] [PMID: 21833630]

[138]   Chen, Y.; Lei, X.; Dou, R.; Chen, Y.; Hu, Y.; Zhang, Z. Selective removal and preconcentration of triclosan using a water-compatible imprinted nano-magnetic chitosan particles. *Environ. Sci. Pollut. Res. Int.,* **2017**, *24*(22), 18640-18650.
[http://dx.doi.org/10.1007/s11356-017-9467-6] [PMID: 28647880]

[139]   Shahdost-fard, F.; Salimi, A.; Sharifi, E.; Korani, A. Fabrication of a highly sensitive adenosine aptasensor based on covalent attachment of aptamer onto chitosan-carbon nanotubes-ionic liquid nanocomposite. *Biosens. Bioelectron.,* **2013**, *48*, 100-107.
[http://dx.doi.org/10.1016/j.bios.2013.03.060] [PMID: 23660341]

[140] Ghanbari, K.; Roushani, M. A nanohybrid probe based on double recognition of an aptamer MIP grafted onto a MWCNTs-Chit nanocomposite for sensing hepatitis C virus core antigen. *Sens. Actuators B Chem.,* **2018**, *258*, 1066-1071.
[http://dx.doi.org/10.1016/j.snb.2017.11.145]

[141] Houde, M.; Czub, G.; Small, J.M.; Backus, S.; Wang, X.; Alaee, M.; Muir, D.C. Fractionation and bioaccumulation of perfluorooctane sulfonate (PFOS) isomers in a Lake Ontario food web. *Environ. Sci. Technol.,* **2008**, *42*(24), 9397-9403.
[http://dx.doi.org/10.1021/es800906r] [PMID: 19174922]

[142] Loos, R.; Wollgast, J.; Huber, T.; Hanke, G. Polar herbicides, pharmaceutical products, perfluorooctanesulfonate (PFOS), perfluorooctanoate (PFOA), and nonylphenol and its carboxylates and ethoxylates in surface and tap waters around Lake Maggiore in Northern Italy. *Anal. Bioanal. Chem.,* **2007**, *387*(4), 1469-1478.
[http://dx.doi.org/10.1007/s00216-006-1036-7] [PMID: 17200857]

[143] Tittlemier, S.A.; Pepper, K.; Seymour, C.; Moisey, J.; Bronson, R.; Cao, X.L.; Dabeka, R.W. Dietary exposure of Canadians to perfluorinated carboxylates and perfluorooctane sulfonate *via* consumption of meat, fish, fast foods, and food items prepared in their packaging. *J. Agric. Food Chem.,* **2007**, *55*(8), 3203-3210.
[http://dx.doi.org/10.1021/jf0634045] [PMID: 17381114]

[144] Olsen, G.W.; Mair, D.C.; Reagen, W.K.; Ellefson, M.E.; Ehresman, D.J.; Butenhoff, J.L.; Zobel, L.R. Preliminary evidence of a decline in perfluorooctanesulfonate (PFOS) and perfluorooctanoate (PFOA) concentrations in American Red Cross blood donors. *Chemosphere,* **2007**, *68*(1), 105-111.
[http://dx.doi.org/10.1016/j.chemosphere.2006.12.031] [PMID: 17267015]

[145] Jiao, Z.; Li, J.; Mo, L.; Liang, J.; Fan, H. A molecularly imprinted chitosan doped with carbon quantum dots for fluorometric determination of perfluorooctane sulfonate. *Mikrochim. Acta,* **2018**, *185*(10), 473-481.
[http://dx.doi.org/10.1007/s00604-018-2996-y] [PMID: 30242509]

[146] Dranove, J.; Horn, D.; Reddy, S.N.; Croffie, J. Effect of intravenous erythromycin on the colonic motility of children and young adults during colonic manometry. *J. Pediatr. Surg.,* **2010**, *45*(4), 777-783.
[http://dx.doi.org/10.1016/j.jpedsurg.2009.07.039] [PMID: 20385286]

[147] Wu, L.; Lin, J.H.; Bao, K.; Li, P.F.; Zhang, W.G. *In vitro* effects of erythromycin on RANKL and nuclear factor-kappa B by human TNF-α stimulated Jurkat cells. *Int. Immunopharmacol.,* **2009**, *9*(9), 1105-1109.
[http://dx.doi.org/10.1016/j.intimp.2009.05.008] [PMID: 19500694]

[148] Pendela, M.; Van den Bossche, L.; Hoogmartens, J.; Van Schepdael, A.; Adams, E. Combination of a liquid chromatography-ultraviolet method with a non-volatile eluent, peak trapping and a liquid chromatography-mass spectrometry method with a volatile eluent to characterise erythromycin related substances. *J. Chromatogr. A,* **2008**, *1180*(1-2), 108-121.
[http://dx.doi.org/10.1016/j.chroma.2007.11.079] [PMID: 18177878]

[149] Hu, X.; Wang, P.; Yang, J.; Zhang, B.; Li, J.; Luo, J.; Wu, K. Enhanced electrochemical detection of erythromycin based on acetylene black nanoparticles. *Colloids Surf. B Biointerfaces,* **2010**, *81*(1), 27-31.
[http://dx.doi.org/10.1016/j.colsurfb.2010.06.018] [PMID: 20643533]

[150] Lian, W.; Liu, S.; Yu, J.; Xing, X.; Li, J.; Cui, M.; Huang, J. Electrochemical sensor based on gold nanoparticles fabricated molecularly imprinted polymer film at chitosan-platinum nanoparticles/graphene-gold nanoparticles double nanocomposites modified electrode for detection of erythromycin. *Biosens. Bioelectron.,* **2012**, *38*(1), 163-169.
[http://dx.doi.org/10.1016/j.bios.2012.05.017] [PMID: 22683249]

[151] Zhu, Y.; Son, J.I.K.; Shim, Y.B. Amplification strategy based on gold nanoparticle-decorated carbon

nanotubes for neomycin immunosensors. *Biosens. Bioelectron.,* **2010**, *26*(3), 1002-1008.
[http://dx.doi.org/10.1016/j.bios.2010.08.023] [PMID: 20869230]

[152]   Zawilla, N.H.; Diana, J.; Hoogmartens, J.; Adams, E. Analysis of neomycin using an improved liquid chromatographic method combined with pulsed electrochemical detection. *J. Chromatogr. B Analyt. Technol. Biomed. Life Sci.,* **2006**, *833*(2), 191-198.
[http://dx.doi.org/10.1016/j.jchromb.2006.01.034] [PMID: 16504608]

[153]   Gaudin, V.; Cadieu, N.; Sanders, P. Results of a european proficiency test for the detection of streptomycin/dihydrostreptomycin, gentamicin and neomycin in milk by ELISA and biosensor methods. *Anal. Chim. Acta,* **2005**, *529*(1-2), 273-283.
[http://dx.doi.org/10.1016/j.aca.2004.06.058]

[154]   Wang, S.; Xu, B.; Zhang, Y.; He, J.X. Development of enzyme-linked immunosorbent assay (ELISA) for the detection of neomycin residues in pig muscle, chicken muscle, egg, fish, milk and kidney. *Meat Sci.,* **2009**, *82*(1), 53-58.
[http://dx.doi.org/10.1016/j.meatsci.2008.12.003] [PMID: 20416595]

[155]   Lian, W.; Liu, S.; Yu, J.; Li, J.; Cui, M.; Xu, W.; Huang, J. Electrochemical sensor using neomycin-imprinted film as recognition element based on chitosan-silver nanoparticles/graphene-multiwalled carbon nanotubes composites modified electrode. *Biosens. Bioelectron.,* **2013**, *44*, 70-76.
[http://dx.doi.org/10.1016/j.bios.2013.01.002] [PMID: 23395725]

[156]   Hu, J.Y.; Aizawa, T.; Ookubo, S. Products of aqueous chlorination of bisphenol A and their estrogenic activity. *Environ. Sci. Technol.,* **2002**, *36*(9), 1980-1987.
[http://dx.doi.org/10.1021/es011177b] [PMID: 12026981]

[157]   Muñoz-de-Toro, M.; Markey, C.M.; Wadia, P.R.; Luque, E.H.; Rubin, B.S.; Sonnenschein, C.; Soto, A.M. Perinatal exposure to bisphenol-A alters peripubertal mammary gland development in mice. *Endocrinology,* **2005**, *146*(9), 4138-4147.
[http://dx.doi.org/10.1210/en.2005-0340] [PMID: 15919749]

[158]   Deng, P.; Xu, Z.; Kuang, Y. Electrochemical determination of bisphenol A in plastic bottled drinking water and canned beverages using a molecularly imprinted chitosan-graphene composite film modified electrode. *Food Chem.,* **2014**, *157*, 490-497.
[http://dx.doi.org/10.1016/j.foodchem.2014.02.074] [PMID: 24679809]

[159]   Xia, J.; Cao, X.; Wang, Z.; Yang, M.; Zhang, F.; Bing, L.; Li, F.; Xia, L.; Li, Y.; Xia, Y. Molecularly imprinted electrochemical biosensor based on chitosan/ionic liquid–graphene composites modified electrode for determination of bovine serum albumin. *Sens. Actuators B Chem.,* **2016**, *225*, 305-311.
[http://dx.doi.org/10.1016/j.snb.2015.11.060]

[160]   Kazemi, E.; Dadfarnia, S.; Haji Shabani, A.M.; Ranjbar, M. Synthesis, characterization, and application of a Zn (II)-imprinted polymer grafted on graphene oxide/magnetic chitosan nanocomposite for selective extraction of zinc ions from different food samples. *Food Chem.,* **2017**, *237*, 921-928.
[http://dx.doi.org/10.1016/j.foodchem.2017.06.053] [PMID: 28764087]

[161]   Wu, S.; Dai, X.; Cheng, T.; Li, S. Highly sensitive and selective ion-imprinted polymers based on one-step electrodeposition of chitosan-graphene nanocomposites for the determination of Cr(VI). *Carbohydr. Polym.,* **2018**, *195*, 199-206.
[http://dx.doi.org/10.1016/j.carbpol.2018.04.077] [PMID: 29804969]

[162]   Srivastava, J.; Gupta, N.; Kushwaha, A.; Umrao, S.; Srivastava, A.; Singh, M. Highly sensitive and selective estimation of aspartame by chitosan nanoparticle – graphene nanocomposite tailored EQCM-MIP sensor. *Polym. Bull.,* **2018**, *76*(9), 4431-4449.
[http://dx.doi.org/10.1007/s00289-018-2597-2]

[163]   Lee, M.H.; Thomas, J.L.; Chen, Y.C.; Wang, H.Y.; Lin, H.Y. Hydrolysis of magnetic amylase-imprinted poly(ethylene-co-vinyl alcohol) composite nanoparticles. *ACS Appl. Mater. Interfaces,* **2012**, *4*(2), 916-921.

[http://dx.doi.org/10.1021/am201576y] [PMID: 22276908]

[164] Wang, P.; Lo, I.M.C. Synthesis of mesoporous magnetic γ-Fe$_2$O$_3$ and its application to Cr(VI) removal from contaminated water. *Water Res.,* **2009**, *43*(15), 3727-3734.
[http://dx.doi.org/10.1016/j.watres.2009.05.041] [PMID: 19559458]

[165] Singh, P.N.; Tiwary, D.; Sinha, I. Improved removal of Cr (VI) by starch functionalized iron oxide nanoparticles. *J. Environ. Chem. Eng.,* **2014**, *2*(4), 2252-2258.
[http://dx.doi.org/10.1016/j.jece.2014.10.003]

[166] Rinken, T.; Riik, H. Determination of antibiotic residues and their interaction in milk with lactate biosensor. *J. Biochem. Biophys. Methods,* **2006**, *66*(1-3), 13-21.
[http://dx.doi.org/10.1016/j.jbbm.2005.04.009] [PMID: 16423405]

[167] Aqda, T.G.; Raoofi, M.; Behkami, S.; Bagheri, H. Graphene oxide-starch-based micro-solid phase extraction of antibiotic residues from milk samples. *J. Chromatogr. A,* **2018**.
[http://dx.doi.org/10.1016/j.chroma.2018.11.069] [PMID: 30503697]

[168] Srivastava, J.; Kushwaha, A.; Srivastava, M.; Srivastava, A.; Singh, M. Glycoprotein imprinted RGO-starch nanocomposite modified EQCM sensor for sensitive and specific detection of transferrin. *J. Electroanal. Chem. (Lausanne),* **2019**, *835*, 169-177.
[http://dx.doi.org/10.1016/j.jelechem.2019.01.033]

[169] Srivastava, J.; Kushwaha, A.; Singh, M. Imprinted graphene-starch nanocomposite matrix-anchored eqcm platform for highly selective sensing of epinephrine. *Nano,* **2018**, *13*(11), 1850131.
[http://dx.doi.org/10.1142/S179329201850131X]

CHAPTER 9

Advancement in Nanocomposites for Explosive Sensing

V Dhinakaran[1,*], **M Varsha Shree**[1] and **M Swapna Sai**[1]

[1] *Centre for Applied Research, Chennai Institute of Technology, Chennai-600069, India*

Abstract: In the research of nanocomposite, its selective and sensitive explosive detection is very critical. Due to a series of causes, including the vast collection of materials that can be used as explosives, lack of easy to detect signatures and broad ranges of means to deploy such weapons, and lack of affordable sensors of great sensibility and selectivity, explosive trace detection has been exceedingly difficult and costly. The fight against explosives needs a high resilience and selectivity coupled with the potential to lower deployment costs of sensors using mass processing. Nanosensors should fulfill the criteria of an efficient framework for explosive trace detection. In this research work, we confer about the ability of nanosensors to detect trace explosions, based upon high sensitivity and selectivity on nano mechanical sensors for both receiver and receptor-free sensing, which can be used because of their versatility and are incorporated into a multimodal sensor system.

Keywords: Explosives, Nanocomposites, Nanosensors, Polymeric Nanocomposites, Nitramines.

INTRODUCTION

Explosives usually need to be tracked through the processing or examination of vapours or particulate samples using a responsive sensor device. There are actually many approaches used for trace explosives identification [1]. Most generally, the spectrometry of ion mobility (IMS), mass spectrometry (MS) and gas chromatography (GC) are accompanied by responsive sensors identification. However, most of these instruments are very heavy, pricey and take time [2]. Owing to these drawbacks, systems like airports and government buildings are being sparsely built in sensitive areas. Complications not only occur at airports (where sensing and detecting environments are relatively controllable) but at practically unregulated points of entry to public areas, transit networks, infrastructures, or road networks with erratic cars and pedestrian traffic [3]. Ther-

* **Corresponding author V. Dhinakaran**, Centre for Applied Research, Chennai, Institute of Technology, Chennai-600069, India; Email: dhinakaranv@citchennai.net

Manorama Singh, Vijai K Rai and Ankita Rai (Eds.)

efore, protection against explosives can only be accomplished by the use of miniaturized sensors in mass deployment that are adaptive, selective, and economical enough to manufacture in mass. For detection, it is important to detect chemicals using a selective agent and signal transducer. The signal is then sent to an electronic computer for reporting. High sensitivity, selectivity, reversibility and function in real time are the main features of sensors for trace level explosive detection [4]. Excellent sensitivity and low detection limits are needed for trace explosive detection due to the comparatively small number of molecules and vapour pressures. If false-positive rates are to be adequate, high selection is essential. To allow for continuous operation, the sensor should be reversible at room temperature. This sensor must also be easily detected and regenerated to function efficiently [5]. Finally, due to the vast number of terrorism attacks, including bombs, adequate sensors should be deployed. These rigid specifications cannot be met by currently available sensor platforms. Nevertheless, nanoscience-based sensors have a straightforward way to build trace explosive sensors that follow those requirements [6].

EXPLOSIVES

Explosives are chemical compounds that can immediately disperse heat and pressure. Explosives are known to be medium or high explosives because of their burn rates. The propellant, black powder, *etc.,* are low explosive burning (centimetres per second) [7]. In addition, high explosives, which rise at kilometres per second, are divided according to stability into main and secondary explosives. Core explosives such as plumb azides are very vulnerable to environmental stimuli, such as explosives and vibrations, thermal or electrical. Secondary explosives, such as 2,4,6 trinitrotoluene (TNT), hexogen (RDX), and other large explosives, are extremely stable [8].

Various typical explosives are organic compounds and can be classified according to their chemistry in six major classes:

- Acid salts (*e.g.,* Ammonium nitrate)
- Aliphatic nitro compound (*e.g.,* Nitromethane, Hydrazine nitrate)
- Nitramines or nitrosamines (*e.g.,* Octogen (HMX) or RDX)
- Nitrate esters (*e.g.,* Pentrite (PETN), Ethylene glycol dinitrate (EDGN), Nitroguanidine (NQ) and Nitroglycerine)
- Nitroaromatic compounds (*e.g.,* TNT, Dinitrobenzene (DNB), Hexanitrostil-bene, Picric acid)
- Organic peroxides (*e.g.,* Triacetone triperoxide (TATP) and Hexamethylene triperoxide diamine (HMTD).

Last-group explosives called household-produced explosive materials (HME), owing to the presence of volatile organic compounds (VOC), such as acetone, are at very high vapour pressure. HME-based terrorism has recently been quickly extended due to the ease with which it can be made [9]. However, HMEs are highly unstable and require careful treatment in order to prevent detonation. Most typical explosives have a very low ambient temperature vapour pressure. These molecules are very rigid and adsorbed very quickly by surfaces due to low vapour pressure. It is worth noting that explosive vapour pressure rises with temperature rapidly [10]. During sample heating, explosive heating steams quickly condense in colder locations. The adsorption will be stronger on elevated surfaces, such as metal oxides, *etc.*

Materials like polymers, fibres, *etc.*, are compared to low-power surfaces. Sometimes, since explosive molecules are sticky on surfaces at room temperature, they can be condensed into sensor structures by supply chains. Trace sampling of these families of very low vapour explosives is often difficult since the sampled volume includes such a minimal number of molecules [11].

Π-ELECTRON RICH LUMINESCENT POLYMERIC NANOCOMPOSITES

In the modern scientific world, sensitive identification of trace-level explosive quantities is considered imperative because of safety issues and ecological contamination problems. By a single, phase-free radical polymerisation reaction, impregnated polymer vinyl alcohol and graft polymers for the detection of fluorescence compounds with carbon and silver nanoparticles are synthesised. The key function of the leading nanoparticles is to increase the luminous polymer's β-electron density [12]. The highly selective, ultrasensitive identification of nitro-aromatic explosives, based on a fluorescence scan process, by these electron-rich nanocomposites is successfully performed, as shown in Fig. (**1**) [13]. The quenching ability of carbon and silver nano-composites for strongly electron-deficient picric acid (PA) and 2, 4, 6-trinitrotoluene (TNT) was found to be 99 and 95%, respectively. Mechanical research has been studied in-depth and the combination of Forster Resonant Energy Transfer (FRET) and Phenogenic Electron Transfer (PET) in most analytics, like picric acid, results in fluorescence quenching. The analysis has been carried out in detail. On the other hand, PET is the only explanation behind fluorescence quenching in analytes such as 2, 4, 6-trinitrotoluen, 1,4-dinitrobenzene and nitrobenzene. Carbon and silver nanocomposite detection limits are located at the nM level [14]. Fast synthesis and scalable development in bulk volumes ensure their wide future application for real-time experience in defence and the industrial sectors for real-time explosive detection. Fluorescence Quenching Phenomena decides the excellent sensing properties of electron-rich nanocomposites for NACs. The nonlinear SV plots demonstrate both static and dynamic processes of quenching coexistence [15]. The carbon and silver nanocomposite detection limit is observed to be much smaller for both TNT and PA and is considerably lower than other works reported earlier. The limit for nanocomposites identification of TNT falls below the permitted levels for TNT exposure in drinking water. The fast development of clear synthetic paths, flexible high quantity production, good thermal properties, and high sensitivity ensures explosive detection in defence sectors and industries in real time [16]. For the protection of human health and the environment, the identification of these harmful explosives and contaminants is of paramount importance. The selective detection of nitroaromatic explosive sensors from this new series of electro-rich, fluorescent polymer nanocomposites could allow the future production of improved electron-rich sensors [17].

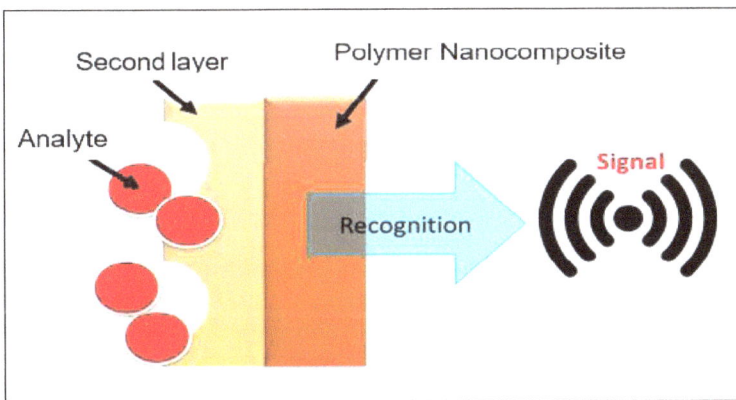

Fig.(1). Luminescent Polymeric Nanocomposites [18].

CDSE QUANTUM DOTS CAPPED PAMAM DENDRIMER NANOCOMPOSITES

In certain areas, ecological research, public protection and forensics include the discovery of nitro-aromatic compounds best used as raw material in explosive preparations. CdSe-quants are used to synthesize the PAMAM-G4 dendrimer in water as traces of three nitro-aromatic compounds: methoxyl-2-nitrofenolic acid (MNP), 2-amino-5 chloro-1,3-dinitrobenzoic acid (ACNB), and 3-methoxy-4-nitrobenzoic acid (CdSe) (MNB [19]. For the intention of encouraging the development of inclusion complexes, the apparent water solubility of these compounds was improved through cyclodextrin (-CD). With the linear Stern/Volmert components, the fluorescence of the nanocomposite was greatly quenched by the analysed nitroaromatic compounds (plus-CD). The study of nitrated aromatics is an important topic in environmental forensics [20]. This thesis has identified a new sensor that demonstrates increased sensitivity to the identification of some of these substances. In this article, the sensor mentioned is mainly comprised of very fluorescent nanomaterials (QP – CdSe). These nanomaterials provide the sensor with a relatively large analytical signal. In addition, a CD is used by the inclusion complex formation to improve the apparent solubility of nitroaromatic compounds [21]. The existence of such inclusion complexes with nitroaromatic compounds causes an interaction with the sensor, which causes fluorescence to quench.

SAMPLE COLLECTION AND PRE-CONCENTRATION

The processing of samples is the front end of any interconnected sensor device and is the most difficult job for trace explosion detection. Because most explosive vapour pressures are incredibly low, it is important to sample significant volumes of air to get adequate particulate molecules for the sensor element detection [22]. The more sensitive the sensor is, the fewer molecules are required for measurement. Sensor sensitivity is always inadequate in order to detect in real-time trace explosive vapours, hence pre-concentration instruments are critical for the identification of explosives. Usually, a large amount of air is gathered using a pump, and explosives and contaminants are trapped using special chemicals and are a combination of several molecules with very low explosive amounts [23]. The substance for the trapping of trapped molecules is then heated. The trapping process is based on the phenomenon of adsorption on large fields. The traditional bottlenecks are voluminous, require high power levels, and have long receptive periods [24]. Recent microfabrication advancements have allowed the development of thermally low thermal pre-concentrators, which can be heated quickly. These microfabricated pre concentrators are coated with high-affinity materials, which resist the cycling of temperature and can theoretically have higher efficiency in the collection of explosive molecules than their bulky predecessors [25]. In recent years, researchers of the U.S. Naval Research Laboratory (NRL) have produced a sorbent-capped micro fabricated pre-concentrator system, where the temperature of the device will rise to 180°C in 40 ms. Since the sorbent material has a high partition coefficient for explosives, a sensing device can detect all blast molecules obtained over a time span of 40 ms. The NRL pre-centred metallic oxide pre-concentrator consists of a hanging series of micro heating plates constructed using complementary metal-oxid--semiconductor (CMOS) technology [26].

SIGNAL TRANSDUCTION USING NANOSENSORS

Nanoscale effects provide outstanding prospects for the development of ideal explosive trace sensors that meet all of the above criteria. In this case, we define the general term "nano," where a size decrease leads to results that cannot be detected by macroscale materials or which can disproportionately enhance susceptibility [27]. Nanoscale measurements can increase physical and chemical characteristics in certain situations, and the change can come from multiple sources, such as expanded surface area, containment impact, *etc*. Micro- and nanotechnologies have been spurred on by recent developments in top-down production techniques. Related top-down methods are used to create micro and nanoelectromechanical structures (MEMS and NEMS) [28]. For certain types of signal transduction, the improved physical output can be predicted as manufacturing leads to smaller and smaller proportions. The sensitivity can be increased, for example, by reducing the sensor axis. Increased sensitivity cannot, however, fulfil all perfect sensor demands alone. The sensor must also be extremely chemically recognized [29]. Many nanosensor platforms suffer from low selectivity considering their high sensitivity. However, nanoscale-effect materials are still in the early phase of production and could lead to chemical speciation. An example is the use of polymers and nanoparticles, which, when combined with explosive molecules, modify one of their detectable features such that their specificity is clearly detected (see the section on receptor-free detection below [30]. The nanoscale effects are, however, used to enhance physical properties, including electrical, magnetic, mechanical and optical properties. By combining with the chemical selectivity of particular receptors, nano-sensors are highly selective, reactional and reversible miniatures. The nanoparticles, and functional groups use a range of methods that are used for molecular identification or assembly to build realistic structures using single atoms and molecules. These functional groups are then mixed with micro and nanodevices to maximize the performance of sensors and transduce the real-world signal [31]. By combining micro/nanoscale top-down and nano-level bottom-up processes, nanosensors can be generated easily by integrating a comprehensive variety of old and new concepts. There are several sensors/platforms known as nanosensors, such as micro- and nano-made machines, quantum points, nanowires, nano-bells, and so on. These sensors are based on chemically modified surfaces by molecular adsorption. For example, after the adsorption of molecules micro- and nano pharmaceutical resonant structures are subjected to a shift in their resonance frequency [32]. However, nanosensors are severely disadvantaged. Initially, sensors focused on recipients are also poorly selective. Wide surface sensors are safer than miniature molecules for the collection of explosive molecules. There is a lack of appropriate molecules for quick identification of increased sensitivity. Furthermore, because of the small adsorption sites in the sensor, nanosensors usually do not have a very wide dynamic range [33].

RECEPTOR-BASED TRACE EXPLOSIVE DETECTION

Nano-sensors may be labelled as non-receptor or receptor-based. Receptor-free nanosensors assess the physical features of explosive compounds, such as thermodynamic, biological, or optical properties (RNs). The exact coordination between the receptor molecule and the electrode sensor of the explosive analyte increases selectivity. A measurable change in weight, conductivity and absorption is then employed to assess the relationship. However, as a certain number of reversible receptors do not exist, they are used for speciation in electronic nose [34]. Based on the name of the receptor blast, the chemical reaction between explosive molecules and receptor molecules leads to chemical selectivity. Explosive molecules must bundle with receptors with weak chemical relations, including van der Waals' interactions, hydrogen bonding, *etc.* Chemical selectivity based on weak interactions is extremely selective. It can also be used in array formats such as an optical nose [35]. Any coatings may be used for the transmission of a partial selectivity, such as SAMs, polymers, metal oxides, single beach DNA, *etc.* The option of the layer must take the reaction time and the recovery time into consideration. A SAM of 4-mercaptobenzoic acid is one example of a partially selective explosives coating (4-MBA; also known as thio salicylic acid). 4-MBA SAMs include carboxylic end groups with explosive molecules for acid-base reactions [36]. An Au-coat SAM with 4 MBA SAM is 5–7 pKa and firmly attaches to essential groups, such as nitro substituted explosive vapour molecules [37]. 6-mercaptonicotinic acid (6-MNA) monolayers are used to improve TNT susceptibility. In addition, there are several polymers with high explosive partition coefficients. Another means of achieving chemical speciation is to acquire molecularly imprinted polymers [38].

ELECTRONIC NOSE APPROACH

Although the most reversible receptors are not selective, a reply pattern similar to artificial olfaction systems may be used in an array format. For designing such an electronic nose concept, a wide range of highly sensible physical sensor platforms is available. Pattern recognition algorithms evaluate the responses in the series [39], although the pattern recognition must be transmitted by each receiver group in the array to a single signal. Currently, the array format will classify a single chemical using a pattern detection algorithm. The confidence level of the pattern recognition decreases in binary mixtures, however, for mixtures made up of three or more components. By actually increasing the size of the set, selectivity does not increase. This possibly is due to the limited number of interactions on which one particular interaction can be based [40]. Therefore, it is uncertain that all techniques based on sensor arrays and pattern detection achieve the selectiveness of small molecules, particularly explosives while relying on simple interactions for speciation. Construction of highly selective receptors is also being undertaken. As mentioned above, there are currently no coatings capable of detecting volatile molecules without intervention. This will include the use of partially selective coatings in tablets that may have orthogonal or orthogonal responses (see the section on nanomechanical sensor platforms) [41]. An optimal detector array will generate a special chemical reaction signature for each vapour analysis (explosive and interfering molecules). For a significant number of analysts of interest, a database of such signatures may be generated and then used to classify unknown vapours. This job is relatively easy for single component analysts, but a complex computational algorithm is essential for mixtures of many different analysts [42].

NANOWIRE PLATFORM

A highly sensitive surface detection method is provided for molecular adsorption by nanostructures like nanowire and nanocrystals. Nanotubes, such as carbon nanotubes, have sizes in the nm range. Nanowires and nanocrystals are distinguished by distinct electrical and optical properties for detecting highly sensitive molecular adsorption. Molecular adsorption can modulate the optical characteristics of the nanocrystals [43]. Due to molecular adsorption, the electrical conduction in the nanowire is considerably different by field effects. An updated, single-walled carbon nanotubes (SWNTs) electrical condenser was demonstrated by Ski *et al*. to detect the sensitivity of groups of chemical vapours. The large electrical field that radiates from the SWNTs affects the polarization of the absorbed molecules while the electrical potential is used, impacting the capacitance measured. The improvement in ability can be partly selective and depend on the molecule type. However, it is possible to obtain much greater selectivity by shielding the SWNTs by chemically selective layers [44]. Nanowires can be constructed from many diverse components, such as Si, ZnO and many others, and thus have numerous surface characteristics for chemical sensing and functional groups alteration [45].

NANOMECHANICAL SENSOR PLATFORM

The nanomechanical effects caused by molecular adsorption give unmatched trace detection opportunities. There are several common operating modes for nano mechanic sensors, such as cantilever beams. For starters, a cantilever's resonance frequency varies with mass adsorption. The detector sensitivity depends on the cantilever's resonance frequency [46]. Antibodies with an incredibly high frequency of resonance nano cantilevers are very susceptible to mass adsorption. When molecular adsorption is confined to one side of the lever, the beam curves thanks to its adsorption powers. Bending sensitivity is high in cantilevers with very low spring constants for bending mode (low resonance frequency). The lifting force relies solely on the adsorbed weight and only on the adsorbed surface's binding energy. Often cantilevers beams are made by conventional picture and etching techniques. Depending on operating mode, standard measurements of the lifting system can range from 200 μm to a few microns in duration. Detecting cantilever reactions with variance in piezoresistance is very desirable for miniature sensor systems. Some changes in the deflection of the cantilever contribute to a difference in the lifting resistance in piezoresistive cantilevers [47]. Boisen's Si piezoresistive engineering equipment manufactures high sensitivity, low electronic noise and very minimal drift cantilevers. The poor bonding between analytes and flow coats allows the analytes to be desorbed by a flow of nitrogen and the detector is then regenerated. Pattern recognition algorithms can then be used to evaluate array response patterns to speciate the analytes.

RECEPTOR-FREE TRACE EXPLOSIVE DETECTION

Due to its low thermal weight, a cantilever platform can be used for detection procedures based on thermal processes. If the adsorbed explosive micro beam heats up easily at 105°C/s, the blast is subjected to thermal deflation and decomposition. An existing pulse means that a cantilever has a high degree of heating (10 V, 10 ms). The temperature of the cantilever is raised during heating from room temperature to around 500°C. A TNT adsorbed deflagration reaction has an uneven bowing with and without adsorbed explosives. A sensitivity of 40 pg is shown for deflagration-based cantilevers. At the end of each heat cycle, the cantilever purifies itself with this method. The technology has the capacity to miniaturize. Another technology that uses the very high thermal sensitivity of a cantilever laser is photothermal deflection spectroscopy. Temperature variations were highly prone to bi-material cantilevers. Because of changes in temperature, such as the adsorption by cantilever or adsorbed molecules of infrasound (IR) power, a bi-material cantilever is twisted. Resonant absorption by an explosive molecule monolayer of IR radiation results in a non-radiative de-excitation of molecules resulting in a bi-material amount bending. The series of exposure of the cantilever to IR from a monochromator will generate a mechanical IR of adsorbed explosives. In order to bend the beam, the IR-absorbed molecules adsorbed on the cantilever surface are seen by the light wavelength bending [48]. This photothermal deflection signal is connected directly to the vibratory mode of adsorbate and the heat capacity and thermal conductivity of the cantilever beam, based on variations in temperature resulting from the absorption of IR radiation. The photothermal spectrum observed leads to the arousal of multiple explosive vibration modes. Then, the thermal energy is transmitted to the substrate that induces a lift bending. The bending degree is commensurate with the thermal energy released. Nearly all the picture-spectral peaks follow the conventional IR absorption spectrum very well.

With clear optical materials, fluorescence can be observed. For the identification of TNT vapour, the AFP signal transduction is extremely selective and sensitive. Nomadic, which uses thin AFP films, offers a consumer version of this sensor. There are also improvements to the platform for fluorescence. In the case of explosive molecular adsorption, nanocomposite films such as silicon doped with macrocyclic nitrogen molecules (porphyrin), for example, demonstrate fluorescence quenching [49].

COMBINED PLATFORMS

The combination of different sensor systems and different operating modes from the same sensor architecture allows true chemical selectivity to obtain orthogonal signals for explosive molecular detection. For *e.g.*, the mass, stress and thermal signals of explosives determined by means of cantilevers are truly autonomous. Enthalpies (H) are also available for coatings that bind explosive, high-performance molecules such as the Lewis Donor-Acceptor, Bronsted acid-base, and 100-400 kJ/mole charge transfer interactions. The sensor regeneration process can be conducted *via* the electrical current heating of the lift while the connecting structures are kept at room temperature. Selectivity data are also possible while such observers are orbiting at temperature. Furthermore, multiple platforms can also be used in one sensor system to test various chemical and physical properties by combining selective capacity and sensibilities [50]. Adsorbed explosives, for instance, can calculate the effects of electric polarization by means of weight and stress measurements as improvements to an SWNT sensor power. It may also be paired with methods of separation, such as gas chromatography, to achieve chemical selectivity on many of these ultrasensitive platforms. Separation methods can, therefore, increase the time for identification. Quick micro-GCs are being developed, which can detect explosive vapours in combination with ultrasensitive sensors.

CHALLENGES IN MOVING TOWARDS PRACTICAL APPLICATION

The first part of the challenge is to create a sensor that is adaptive and selective. The second portion involves the movement of explosives to the sensor element from the atmosphere. Nanosensor modules can be combined into a package that provides vapour accumulation and pre-concentration for the practical application of trace explosives detect sensors. For integration, certain issues have to be overcome. In an incredibly low vapour pressure, the quantity of explosive molecules which can be achieved in a reasonable time is decreased. Furthermore, the concentration of the explosive reduces exponentially based on the distance from the source. The false-positive rate can increase very low concentrations of explosive molecules. Therefore, the production of practical sensor systems is focused on effective techniques of samples acquisition and pre-concentration [51]. The pre-concentration stage increases the time of detection. Since surfaces stick to explosive molecules, supply lines and other materials must be held at a higher temperature. Susceptibility decreases with higher temperatures. However, it is desirable to compromise selectivity to minimize sensitivity, particularly when using receiver-based signal transduction, which depends on low-energy molecular interactions. For *e.g.*, water molecules with concentrations higher than explosives at a magnitude order may significantly limit the selectivity of certain receptors. Array-based techniques can provide partial selectivity. But a multi-modal array approach, which uses a mixture of receptor-based and receptive-free detection strategies, will lead to high explosive detection selective and sensitive [52]. Nanosensors give multimodal detection independence, making them the perfect medium for the detection of trace explosives.

CONCLUDING REMARKS

We analyse the ability of nanosensors to detect trace explosions based on high sensitivity and selectivity on nanomechanical sensors for both receiver and receptor-free sensing, which can be used because of their versatility and are incorporated into a multimodal sensor system in this research.

- Due to a multitude of structural issues, such as the extreme scarcity of explosion molecules, the lack of selectivity due to interference with other molecules, and the breadth of explosive chemicals and compounds, high-sensitivity and blast trace detection is a huge challenge.
- However, nanoscale results can be used to create sensors that meet all trace explosive detection requirements.
- Nanosensor with enhanced flexibility and selectivity, as well as the ability to function on a multimodal network, gives a viable model for a large number of sensors.
- Nanosensor can be used as a highly sensitive and selective signal transduction platform for an implanted explosive sensor device.

CONSENT FOR PUBLICATION

Not Applicable.

CONFLICTS OF INTEREST

The author declares no conflict of interest, financial or otherwise.

ACKNOWLEDGEMENTS

This work is partially funded by the Centre for Applied Research, Chennai Institute of Technology, with funding number CIT/CAR/2021/003.

REFERENCES

[1]   Ates, B.; Koytepe, S.; Ulu, A.; Gurses, C.; Thakur, V.K. Chemistry, structures, and advanced applications of nanocomposites from biorenewable resources. *Chem. Rev.,* **2020**, *120*(17), 9304-9362.
        [http://dx.doi.org/10.1021/acs.chemrev.9b00553] [PMID: 32786427]

[2]   Fu, S.; Sun, Z.; Huang, P.; Li, Y.; Hu, N. Some basic aspects of polymer nanocomposites. *Mater. Sci.,* **2019**, *1*, 2-30.

[3]   Liao, G.; Fang, J.; Li, Q.; Li, S.; Xu, Z.; Fang, B. Ag-Based nanocomposites: synthesis and applications in catalysis. *Nanoscale,* **2019**, *11*(15), 7062-7096.
        [http://dx.doi.org/10.1039/C9NR01408J] [PMID: 30931457]

[4]   Rao, C. N. R.; Pramoda, K. Borocarbonitrides, BxCyNz, 2D nanocomposites with novel properties Japan. **2019**, *92*, 441-468.

[5]   Xu, X.; Chen, J.; Zhou, J.; Li, B. Thermal conductivity of polymers and their nanocomposites. *Adv. Mater.,* **2018**, *30*(17), e1705544.
        [http://dx.doi.org/10.1002/adma.201705544] [PMID: 29573283]

[6]   Papageorgiou, D.G.; Kinloch, I.A.; Young, R.J. Mechanical properties of graphene and graphene-based nanocomposites. *Prog. Mater. Sci.,* **2017**, *90*, 75-127.
        [http://dx.doi.org/10.1016/j.pmatsci.2017.07.004]

[7]   Forbes, T.P.; Sisco, E. Recent advances in ambient mass spectrometry of trace explosives. *Analyst (Lond.),* **2018**, *143*(9), 1948-1969.
        [http://dx.doi.org/10.1039/C7AN02066J] [PMID: 29664070]

[8]   Sun, R.; Huo, X.; Lu, H.; Feng, S.; Wang, D.; Liu, H. Recyclable fluorescent paper sensor for visual detection of nitroaromatic explosives. *Sens. Actuators B Chem.,* **2018**, *265*, 476-487.
        [http://dx.doi.org/10.1016/j.snb.2018.03.072]

[9]   Wang, S.; Wang, Q.; Feng, X.; Wang, B.; Yang, L. Explosives in the cage: metal–organic frameworks for high-energy materials sensing and desensitization. *Adv. Mater.,* **2017**, *29*(36), 1701898.
        [http://dx.doi.org/10.1002/adma.201701898] [PMID: 28731218]

[10]   Jurcic, M.; Peveler, W.J.; Savory, C.N.; Bučar, D-K.; Kenyon, A.J.; Scanlon, D.O.; Parkin, I.P. Sensing and discrimination of explosives at variable concentrations with a large-pore MOF as part of a luminescent array. *ACS Appl. Mater. Interfaces,* **2019**, *11*(12), 11618-11626.
        [http://dx.doi.org/10.1021/acsami.8b22385] [PMID: 30830741]

[11]   Huang, W-H.; Ren, J.; Yang, Y-H.; Li, X-M.; Wang, Q.; Jiang, N.; Yu, J-Q.; Wang, F.; Zhang, J. Water-stable metal–organic frameworks with selective sensing on Fe3+ and nitroaromatic explosives, and stimuli-responsive luminescence on lanthanide encapsulation. *Inorg. Chem.,* **2019**, *58*(2), 1481-1491.
        [http://dx.doi.org/10.1021/acs.inorgchem.8b02994] [PMID: 30601003]

[12]   Wong, Y-C.; De Andrew Ng, J.; Tan, Z-K. Perovskite-initiated photopolymerization for singly dispersed luminescent nanocomposites. *Adv. Mater.,* **2018**, *30*(21), e1800774.
        [http://dx.doi.org/10.1002/adma.201800774] [PMID: 29638013]

[13]   Lyu, L.; Han, M.; Cao, W.; Gao, Y.; Zeng, Q.; Yu, G.; Huang, X.; Hu, C. Efficient Fenton-like process for organic pollutant degradation on Cu-doped mesoporous polyimide nanocomposites. *Nano,* **2019**, *6*, 798-808.

[14]   Adegoke, O.; Montaseri, H.; Nsibande, S.A.; Patricia, B.C. Forbes. Alloyed quaternary/binary

core/shell quantum dot-graphene oxide nanocomposite: Preparation, characterization and application as a fluorescence switch ON probe for environmental pollutants. *J. Alloys Compd.,* **2017**, *720*, 70-78.
[http://dx.doi.org/10.1016/j.jallcom.2017.05.242]

[15] Zabihi, O.; Ahmadi, M.; Abdollahi, T.; Nikafshar, S.; Naebe, M. Collision-induced activation: Towards industrially scalable approach to graphite nanoplatelets functionalization for superior polymer nanocomposites. *Sci. Rep.,* **2017**, *7*(1), 3560.
[http://dx.doi.org/10.1038/s41598-017-03890-8] [PMID: 28620178]

[16] Amini, A.; Kazemi, S.; Safarifard, V. Metal-organic framework-based nanocomposites for sensing applications–A review. *Polyhedron,* **2020**, *177*, 114260.
[http://dx.doi.org/10.1016/j.poly.2019.114260]

[17] Huang, D.; Wang, X.; Zhang, C.; Zeng, G.; Peng, Z.; Zhou, J.; Cheng, M.; Wang, R.; Hu, Z.; Qin, X. Sorptive removal of ionizable antibiotic sulfamethazine from aqueous solution by graphene oxide-coated biochar nanocomposites: Influencing factors and mechanism. *Chemosphere,* **2017**, *186*, 414-421.
[http://dx.doi.org/10.1016/j.chemosphere.2017.07.154] [PMID: 28802133]

[18] Dutta, P.; Chakravarty, S.; Sen Sarma, N. Detection of nitroaromatic explosives using π-electron rich luminescent polymeric nanocomposites. *RSC Advances,* **2016**, *6*(5), 3680-3689.
[http://dx.doi.org/10.1039/C5RA20347C]

[19] Mirzaei, J.; Urbanski, M.; Yu, K.; Heinz-S., K.; Hegmann, T. Nanocomposites of a nematic liquid crystal doped with magic-sized CdSe quantum dots. *J. Mater. Chem.,* **2011**, *21*(34), 12710-12716.
[http://dx.doi.org/10.1039/c1jm11832c]

[20] Liu, P.-Z.; Hu, X-W.; Mao, C-J.; Niu, H-L.; Song, J-M.; Jin, B-K.; Zhang, S-Y. Electrochemiluminescence immunosensor based on graphene oxide nanosheets/polyaniline nanowires/CdSe quantum dots nanocomposites for ultrasensitive determination of human interleukin-6. *Electrochim. Acta,* **2013**, *113*, 176-180.
[http://dx.doi.org/10.1016/j.electacta.2013.09.074]

[21] Shandryuk, G.A.; Matukhina, E.V.; Roman, B. Vasil'Ev, Alexander Rebrov, Galina N. Bondarenko, Alexey S. Merekalov, Alexander M. Gas' kov, and Raisa V. Talroze. Effect of H-bonded liquid crystal polymers on CdSe quantum dot alignment within nanocomposite. *Macromolecules,* **2008**, *41*(6), 2178-2185.
[http://dx.doi.org/10.1021/ma701983y]

[22] Munonde, T.S.; Maxakato, N.W.; Nomngongo, P.N. Preparation of magnetic Fe3O4 nanocomposites modified with MnO2, Al2O3, Au and their application for preconcentration of arsenic in river water samples. *J. Environ. Chem. Eng.,* **2018**, *6*(2), 1673-1681.
[http://dx.doi.org/10.1016/j.jece.2018.02.017]

[23] Taghizadeh, M.; Asgharinezhad, A.A.; Pooladi, M.; Barzin, M.; Abbaszadeh, A.; Tadjarodi, A. A novel magnetic metal organic framework nanocomposite for extraction and preconcentration of heavy metal ions, and its optimization *via* experimental design methodology. *Mikrochim. Acta,* **2013**, *180*(11-12), 1073-1084.
[http://dx.doi.org/10.1007/s00604-013-1010-y]

[24] Munonde, T.S.; Maxakato, N.W.; Nomngongo, P.N. Preconcentration and speciation of chromium species using ICP-OES after ultrasound-assisted magnetic solid phase extraction with an amino-modified magnetic nanocomposite prepared from Fe_3O_4, MnO_2 and Al_2O_3. *Mikrochim. Acta,* **2017**, *184*(4), 1223-1232.
[http://dx.doi.org/10.1007/s00604-017-2126-2]

[25] Bagheri, H.; Afkhami, A.; Saber-Tehrani, M.; Khoshsafar, H. Preparation and characterization of magnetic nanocomposite of Schiff base/silica/magnetite as a preconcentration phase for the trace determination of heavy metal ions in water, food and biological samples using atomic absorption spectrometry. *Talanta,* **2012**, *97*, 87-95.
[http://dx.doi.org/10.1016/j.talanta.2012.03.066] [PMID: 22841051]

[26]   Shamsipur, M.; Farzin, L.; Amouzadeh Tabrizi, M.; Sheibani, S. Functionalized Fe_3O_4/graphene oxide nanocomposites with hairpin aptamers for the separation and preconcentration of trace Pb^{2+} from biological samples prior to determination by ICP MS. *Mater. Sci. Eng. C,* **2017**, *77*, 459-469.
        [http://dx.doi.org/10.1016/j.msec.2017.03.262] [PMID: 28532053]

[27]   Qiao, G.; Liu, L.; Hao, X.; Zheng, J.; Liu, W.; Gao, J.; Cheng, C.Z.; Wang, Q. Signal transduction from small particles: Sulfur nanodots featuring mercury sensing, cell entry mechanism and *in vitro* tracking performance. *Chem. Eng. J.,* **2020**, *382*, 122907.
        [http://dx.doi.org/10.1016/j.cej.2019.122907]

[28]   Munawar, A.; Ong, Y.; Schirhagl, R.; Tahir, M.A.; Khan, W.S.; Bajwa, S.Z. Nanosensors for diagnosis with optical, electric and mechanical transducers. *RSC Advances,* **2019**, *2019*(12), 6793-6803.
        [http://dx.doi.org/10.1039/C8RA10144B]

[29]   Vikesland, P.J. Nanosensors for water quality monitoring. *Nat. Nanotechnol.,* **2018**, *13*(8), 651-660.
        [http://dx.doi.org/10.1038/s41565-018-0209-9] [PMID: 30082808]

[30]   Hwang, B-U.; Zabeeb, A.; Trung, T.Q.; Wen, L.; Lee, J.D.; Choi, Y-I.; Lee, H-B.; Kim, J.H.; Han, J.G.; Lee, N-E. A transparent stretchable sensor for distinguishable detection of touch and pressure by capacitive and piezoresistive signal transduction. *NPG Asia Mater.,* **2019**, *11*(1), 1-12.
        [http://dx.doi.org/10.1038/s41427-019-0126-x]

[31]   Nikoleli, G-P.; Nikolelis, D.; Siontorou, C.G.; Karapetis, S. Lipid membrane nanosensors for environmental monitoring: The art, the opportunities, and the challenges. *Sensors (Basel),* **2018**, *18*(1), 284.
        [http://dx.doi.org/10.3390/s18010284] [PMID: 29346326]

[32]   Gu, J.; Li, X.; Zhou, Z.; Liu, W.; Li, K.; Gao, J.; Zhao, Y.; Wang, Q. 2D MnO_2 nanosheets generated signal transduction with 0D carbon quantum dots: synthesis strategy, dual-mode behavior and glucose detection. *Nanoscale,* **2019**, *11*(27), 13058-13068.
        [http://dx.doi.org/10.1039/C9NR03583D] [PMID: 31265041]

[33]   Rong, G.; Tuttle, E.E.; Neal Reilly, A.; Clark, H.A. Recent developments in nanosensors for imaging applications in biological systems. *Annu. Rev. Anal. Chem. (Palo Alto, Calif.),* **2019**, *12*(1), 109-128.
        [http://dx.doi.org/10.1146/annurev-anchem-061417-125747] [PMID: 30857408]

[34]   Wu, J.; Lu, Y.; Wu, Z.; Li, S.; Zhang, Q.; Chen, Z.; Jiang, J.; Lin, S.; Zhu, L.; Li, C.; Liu, Q. Two-dimensional molybdenum disulfide MoS2 with gold nanoparticles for biosensing of explosives by optical spectroscopy. *Sens. Actuators B Chem.,* **2018**, *261*, 279-287.
        [http://dx.doi.org/10.1016/j.snb.2018.01.166]

[35]   Senesac, L.; Thundat, T.G. Nanosensors for trace explosive detection. *Mater. Today,* **2008**, *11*(3), 28-36.
        [http://dx.doi.org/10.1016/S1369-7021(08)70017-8]

[36]   Lee, K.; Yoo, Y.K.; Chae, M-S.; Hwang, K.S.; Lee, J.; Kim, H. Don Hur, and Jeong Hoon Lee. Highly selective reduced graphene oxide rGO sensor based on a peptide aptamer receptor for detecting explosives. *Sci. Rep.,* **2019**, *9*, 1-9.

[37]   Vikrant, K.; Tsang, D.C.W.; Raza, N.; Giri, B.S.; Kukkar, D.; Kim, K.H. Potential Utility of Metal-Organic Framework-Based Platform for Sensing Pesticides. *ACS Appl. Mater. Interfaces,* **2018**, *10*(10), 8797-8817.
        [http://dx.doi.org/10.1021/acsami.8b00664] [PMID: 29465977]

[38]   Wilson, A.D. Application of electronic-nose technologies and VOC-biomarkers for the noninvasive early diagnosis of gastrointestinal diseases. *Sensors (Basel),* **2018**, *18*(8), 2613.
        [http://dx.doi.org/10.3390/s18082613] [PMID: 30096939]

[39]   Ezhilan, M.; Nesakumar, N.; Jayanth Babu, K.; Srinandan, C.S.; Rayappan, J.B.B. An Electronic Nose for Royal Delicious Apple Quality Assessment - A Tri-layer Approach. *Food Res. Int.,* **2018**, *109*, 44-

51.
[http://dx.doi.org/10.1016/j.foodres.2018.04.009] [PMID: 29803469]

[40]   Malegori, C.; Buratti, S.; Benedetti, S.; Oliveri, P.; Ratti, S.; Cappa, C.; Lucisano, M. A modified mid-level data fusion approach on electronic nose and FT-NIR data for evaluating the effect of different storage conditions on rice germ shelf life. *Talanta*, **2020**, *206*, 120208.
[http://dx.doi.org/10.1016/j.talanta.2019.120208] [PMID: 31514827]

[41]   Aheto, J.H.; Huang, X.; Tian, X.; Ren, Y.; Ernest, B.; Alenyorege, E.A.; Dai, C.; Hongyang, T.; Xiaorui, Z.; Wang, P. Multi-sensor integration approach based on hyperspectral imaging and electronic nose for quantitation of fat and peroxide value of pork meat. *Anal. Bioanal. Chem.*, **2020**, *412*(5), 1169-1179.
[http://dx.doi.org/10.1007/s00216-019-02345-5] [PMID: 31912184]

[42]   Paknahad, M.; Ahmadi, A.; Rousseau, J.; Nejad, H.R.; Hoorfar, M. On-chip electronic nose for wine tasting: A digital microfluidic approach. *IEEE Sens. J.*, **2017**, *17*(14), 4322-4329.
[http://dx.doi.org/10.1109/JSEN.2017.2707525]

[43]   Mussi, V.; Ledda, M.; Polese, D.; Maiolo, L.; Paria, D.; Barman, I.; Lolli, M.G.; Lisi, A.; Convertino, A. Silver-coated silicon nanowire platform discriminates genomic DNA from normal and malignant human epithelial cells using label-free Raman spectroscopy. *Mater. Sci. Eng. C*, **2021**, *122*, 111951.
[http://dx.doi.org/10.1016/j.msec.2021.111951] [PMID: 33641882]

[44]   Thiha, A.; Ibrahim, F.; Muniandy, S.; Dinshaw, I.J.; Teh, S.J.; Thong, K.L.; Leo, B.F.; Madou, M. All-carbon suspended nanowire sensors as a rapid highly-sensitive label-free chemiresistive biosensing platform. *Biosens. Bioelectron.*, **2018**, *107*, 145-152.
[http://dx.doi.org/10.1016/j.bios.2018.02.024] [PMID: 29455024]

[45]   Cao, Z.; Duan, F.; Huang, X.; Liu, Y.; Zhou, N.; Xia, L.; Zhang, Z.; Du, M. A multiple aptasensor for ultrasensitive detection of miRNAs by using covalent-organic framework nanowire as platform and shell-encoded gold nanoparticles as signal labels. *Anal. Chim. Acta*, **2019**, *1082*, 176-185.
[http://dx.doi.org/10.1016/j.aca.2019.07.062] [PMID: 31472706]

[46]   Khosla, K.E.; Vanner, M.R.; Ares, N.; Laird, E.A. Displacemon electromechanics: how to detect quantum interference in a nanomechanical resonator. *Phys. Rev. X*, **2018**, *8*(2), 021052.
[http://dx.doi.org/10.1103/PhysRevX.8.021052]

[47]   Lang, H.P.; Hegner, M.; Gerber, C. Nanomechanical cantilever array sensors. In: *Springer handbook of nanotechnology*; Springer: Berlin, Heidelberg, **2017**; pp. 457-485.
[http://dx.doi.org/10.1007/978-3-662-54357-3_15]

[48]   Adil, L.R.; Gopikrishna, P.; Krishnan Iyer, P. Receptor-free detection of picric acid: a new structural approach for designing aggregation-induced emission probes. *ACS Appl. Mater. Interfaces*, **2018**, *10*(32), 27260-27268.
[http://dx.doi.org/10.1021/acsami.8b07019] [PMID: 30022660]

[49]   Mauricio, F.G.M.; José, Y.R.S.; Talhavini, M.; Júnior, S.A.; Weber, I.T. Luminescent sensors for nitroaromatic compound detection: Investigation of mechanism and evaluation of suitability of using in screening test in forensics. *Microchem. J.*, **2019**, *150*, 104037.
[http://dx.doi.org/10.1016/j.microc.2019.104037]

[50]   Huang, Y.; Skripka, A.; Labrador-Páez, L.; Sanz-Rodríguez, F.; Haro-González, P.; Jaque, D.; Rosei, F.; Vetrone, F. Upconverting nanocomposites with combined photothermal and photodynamic effects. *Nanoscale*, **2018**, *10*(2), 791-799.
[http://dx.doi.org/10.1039/C7NR05499H] [PMID: 29256568]

CHAPTER 10

Nanocomposite Materials Interface for Heavy Metal Ions Detection

Prashanth Shivappa Adarakatti[1] and S Ashoka[2,*]

[1] *Department of Chemistry, SVM Arts, Science and Commerce College, ILKAL – 587125, India*

[2] *School of Science, Dayananda Sagar University, Kudlu Gate, Bengaluru – 560068, India*

Abstract: The present chapter provides an overview of the sources, consequences, and quantification of heavy metal ions (HMIs) present in the water sample. Heavy metal ions are recognized as one of the major water pollutants. Long-term consumption of HMIs causes serious health hazards and is also a threat to the ecosystem. In this regard, the synthesis and use of nanocomposites for the selective quantification of HMIs have been discussed in detail.

Keywords: Electrochemical sensor, Heavy metal ions, Modified electrodes, Nanocomposite.

INTRODUCTION

The effluent released by several anthropogenic activities majorly contaminate natural resources [1, 2]. Long-term exposure to these HMIs causes adverse effects on living organisms. Consequently, even exposure to trace level concentrations of toxic heavy metal ions can lead to long-term disorders [3]. Therefore, the accurate and fast detection of HMIs is becoming a challenging issue for analytical chemists. Numerous detection methods have been proposed in the literature to monitor HMIs present in the water and food samples accurately [4, 4b, 5 - 7, 7b].

Chemically modified electrodes (CMEs) have been recognized as potential candidates in developing reliable electrodes to quantify HMIs. In the year 1970, Royce Murray first introduced the concept of CMEs where the SnO_2 electrode has been modified with amines [8].

The CMEs are formed when the modifier is covalently or non-covalently anchored to the surface of the substrate material with the help of a binder. Anchoring of the modifier molecule on the surface of the carbon substrate by

---
* **Corresponding author Ashoka S:** School of Science, Dayananda Sagar University, Kudlu Gate, Bengaluru, India; E-mail: ashok022@gmail.com

specific functional groups provides an excellent pathway for the interaction with the target analyte, such as HMIs.

IMPORTANCE OF CHEMICAL MODIFICATION

The electrochemical reactions occur at the interface of electrode and electrolyte solution. Hence, the surface structure of the electrode at the interface plays a vital role in promoting electrode kinetics [9]. The commonly used electrodes are carbon-based substrate materials. The carbon-based electrodes possess low background current, broad potential range, and chemically inertness and have low cost. These features made them a suitable candidate for various sensing applications [10]. However, these electrodes show some limitations over modified carbon electrodes in terms of sensitivity and selectivity. Hence, the modification of carbon-based electrodes with suitable modifiers having functional groups indeed enhances the electro-analytical signal intensity [11]. CMEs have been extensively employed to quantify HMIs owing to their tailor-made properties, such as selectivity and sensitivity, towards the target analytes. It can be illustrated with an example involving the electrochemical quantification of lead ions using graphene oxide (GO) modified electrode [12]. The presence of GO on the electrode surface will reduce the overpotential required for the redox process of lead ions. Similarly, mercury quantification can be achieved by modifying the carbon-based electrodes with thiol functional groups [13]. The modification of the substrate with suitable modifier molecules with specific functional groups can be achieved using several approaches like physical adsorption, chemical adsorption, covalent attachment, electrochemical, ball milling, and microwave-assisted covalent modification procedures [14, 15]. The carbon substrates like graphite, glassy carbon, carbon nanotubes, and screen-printed electrodes (SPEs) have been modified with specific modifier molecules.

TYPES OF ELECTRODES

Glassy Carbon Electrode (GCE)

The glassy carbon electrode is fabricated by using controlled charring of polymeric resin (phenol/formaldehyde or polyacrylonitrile) at a high temperature of about 1000 to 3000 °C. Glassy carbon has a ribbon-like structure with graphitic sheets that are cross linked (Fig. **1**). Hence, it is stronger and more robust than graphite [16].

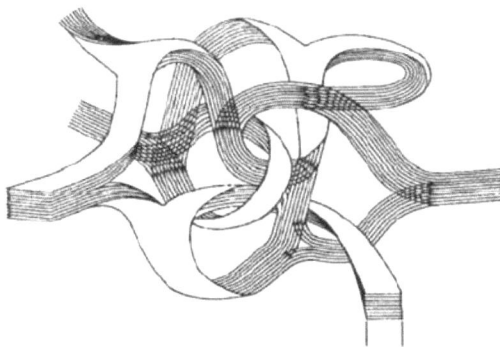

Fig. (1). Structure of glassy carbon [17].

The GCE has been extensively used in electroanalytical chemistry owing to its outstanding mechanical and conducting properties together with broad working potential. Generally, the surface of GCE is pretreated with alumina slurry of different particle sizes to get improved analytical response by means of enhanced electron transfer on the surface of GCE [18, 19]. This GCE is modified with a suitable modifier to enhance analytical signals.

Carbon Paste Electrode (CPE)

Carbon paste electrode is generally used as a working electrode for the measurement of target analytes in electro-analytical chemistry. It mainly consists of graphite powder and a binder like paraffin oil, Nujol, bromonaphthalene, and silicone grease. These electrodes offer easy surface renewability, low cost, and low background current. The fabrication of working electrodes can be made by mixing modifier molecules along with graphite powder in the presence of a binder. The potential drawback of this electrode includes continuous leaching of the modifier molecule from the surface of CPE along with the binder, which hinders electrocatalytic reaction.

Screen-printed Electrodes (SPEs)

Screen-printed electrodes (SPEs) have gained considerable attention in the electrochemical community in recent years [5]. The SPE contains all three electrodes (working, counter, and reference electrodes) in one strip and offers easy fabrication, on-site detection, low cost, broad potential range, and low sample volume. Being a disposable sensor, it requires no polishing or smoothening, but it can often be seen in the conventional solid electrodes [20]. The analytical signal could be improved by modifying suitable functional groups upon the carbon-based SPEs.

Pencil Graphite Electrodes (PGEs)

The graphite rods are known as pencil graphite electrodes (PGEs), which serve as working electrodes in electrochemical studies [21, 22]. PGEs, like other traditional electrodes, have a readily regenerable surface, need less time to clean, and are inexpensive. Utilizing this electrode, well-defined voltammetric peaks can be observed, followed by reproducible results. These electrodes are considered inexpensive and exhibit improved analytical response.

Basal and Edge Plane Pyrolytic Graphite Electrodes

The basal plane and edge plane surfaces can be found in the pyrolytic graphite. Further, the basal plane consists of layers of graphite parallel to the surface, and the spacing between the two layers of graphite is approximately 3.35 Å [23]. Moreover, the basal plane and edge plane of the two types exhibit different electrochemical properties as this occurs because of the chemical bonding nature of the graphite [24]. The impressive electrocatalytic activity in terms of low background current occurred when edge plane electrodes were employed in the electrochemical measurement studies rather than basal plane electrodes [22, 24].

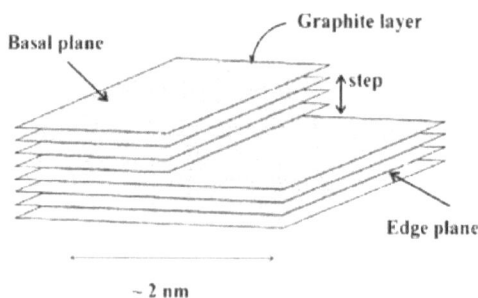

Fig. (2). Schematic representation of a basal and edge plane graphite [22].

TOXIC METAL IONS, THEIR DISTRIBUTION, AND HEALTH HAZARDS

Lead

Among heavy metal ions, lead is toxic and considered to be a metallic pollutant to both mankind and aquatic systems. The best-known organolead compounds are the two simplest additives used, *i.e.,* tetramethyl lead (TML) and tetraethyl lead (TEL), as anti-knocking agents in gasoline. It is a soft, malleable, and non-degradable metal ion, which has received the second rank in the list recommended by the United States Agency for Toxic Substances and Disease Registry (USATSDR) [25]. Lead as a metal/ion is one of the most toxic substances that has been associated with environmental pollution and shows severe problems because of its long-term stability and complex biological toxicity. It can cause adverse effects on the nervous system, kidney failure, hypertension, abdominal pain, nausea, vomiting, and the immune system. It also causes renal blood non-functioning, irreversible brain damage, and cardiovascular effects [26]. The sources of lead include electroplating industries, batteries, gasoline pipes, paints, and industrial alloys. Lead is naturally occurring in the form of ores like galena (PbS) and is found in the soil since it is a strong accumulator of lead. Galena is mostly found with other minerals, mostly zinc ores. It may also be released into the environment as a byproduct of mining and the industrial processing of other metals, such as silver, gold, and bismuth. Additionally, cosmetics like kajal and sindoor contain lead-based materials. The various sources of lead are shown in Fig. (3).

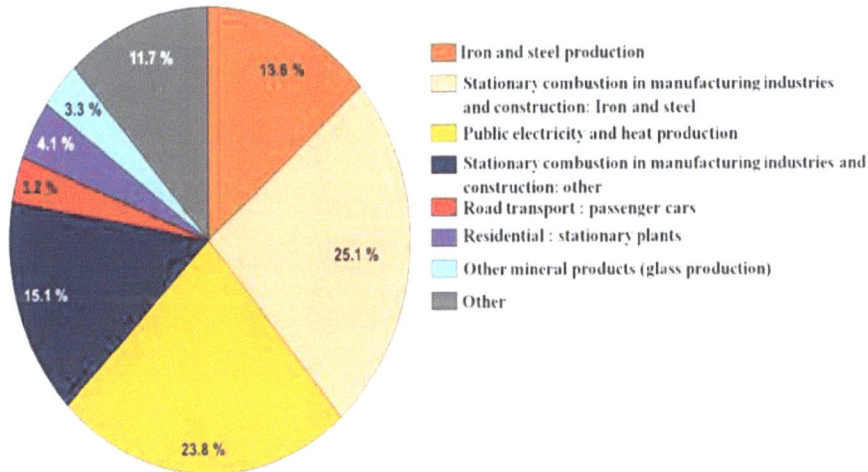

Fig. (3). Various sources of lead.

Cadmium

It is one of the most toxic trace heavy metal ions found in food and the environment [27]. Long-term exposure to cadmium ion, even at trace level, may cause pancreatic cancer, enhanced growth of tumor, anemia, hypertension and decalcification of bones [28]. Exposure to cadmium may result in symptoms like chill, fever, and muscle ache [29].

The main natural source of cadmium in the atmosphere is volcanic action, and this volcanic action emits large quantities of particulate matter and aerosols. It is also found in the environment, both naturally and as a result of human activities. The weathering of crustal materials releases cadmium into soils and aquatic systems.

In addition to this, it is released into the environment as a byproduct of extraction processes, smelting, and refining of zinc, lead and copper ores. Cadmium shows superior resistance to high stress, temperature and it has good dispersion ability in polymers and produces strong color. Hence, cadmium-containing compounds are extensively used as pigments, coatings, and as stabilizers in the case of plastics, glass, and enamel materials. Phosphate fertilizers also contain cadmium (100 mg/kg), and the use of these fertilizers in the agricultural sector increases the cadmium concentration in soil. The other source of cadmium is batteries made of Ni and Cd [30]. The sources of cadmium are depicted in Fig. (**4**).

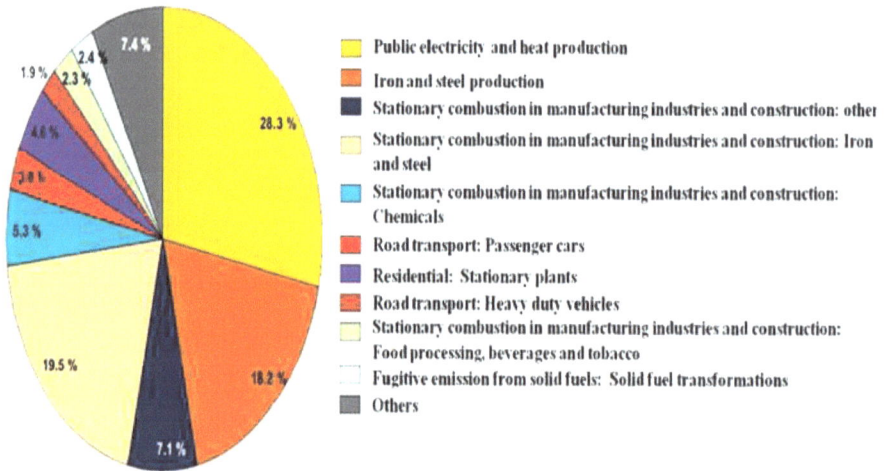

Fig. (**4**). Cadmium sources and its distribution [30].

Mercury (Hg)

Mercury (Hg) is a naturally occurring element and it is present throughout the environment [31]. Mercury exists in three forms, mainly elemental, ionic and organic, in the environment [32]. It is considered a global contaminant since it can travel large distances in the atmosphere and in the environment [33]. It is a highly reactive molecule that produces toxic effects by binding strongly to sulfhydryl and to a lesser degree to hydroxyl, carboxyl, and phosphoryl groups. It is widely distributed as an environmental and industrial pollutant. The earth's crust contains 0.5 parts per million of mercury. The prominent sources of human exposure to mercury nevertheless come from dental amalgams, pharmaceuticals, cosmetics, and food, primarily contaminated fish [34]. Acute mercury poisoning was known only after the Minamata bay disease in Japan in the year 1952 [35]. The sources of mercury are presented in Fig. (**5**).

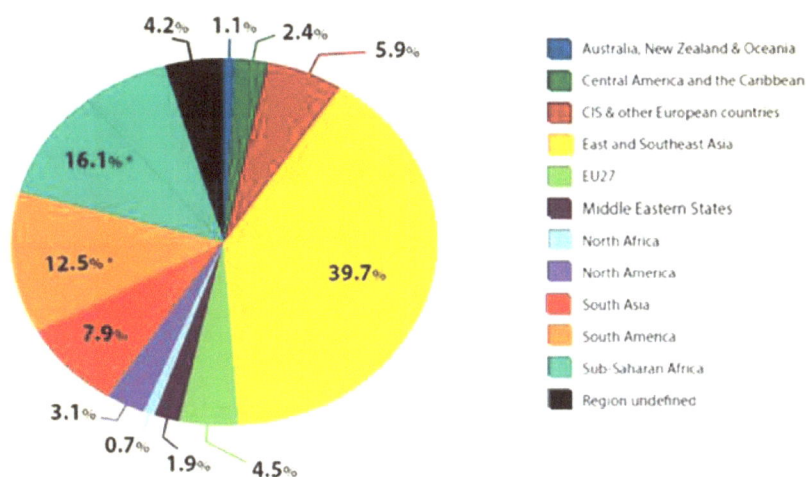

Legend:
- Australia, New Zealand & Oceania
- Central America and the Caribbean
- CIS & other European countries
- East and Southeast Asia
- EU27
- Middle Eastern States
- North Africa
- North America
- South Asia
- South America
- Sub-Saharan Africa
- Region undefined

Fig. (5). Various sources of mercury.

Copper

Copper comes from the Latin word "cuprum", which is a corruption of the Greek word "cyprium", the source of which traces back to Egyptian and Roman languages [36]. Soft, malleable, ductile copper has been considered as one of the toxic elements. Copper is found in two forms in the bloodstream, bound form and free form. Free copper eliminates the reactive oxygen species and thereby reduces oxidative stress. However, excessive free copper increases oxidative stress. Regular excess intake of copper leads to zinc deficiency. The organic copper present in the food is an essential micronutrient to maintain good health, whereas Metallic copper, inorganic or inorganic, is neurotoxic and causes physical and psychological symptoms similar to mercury and lead [37, 38]. Copper deficiency results in anemia in the vertebrates [39]. Excess copper accumulation causes irreversible harm [40].

Generally, copper poisoning occurs from vending machines and copper/brass vessels [41]. Further, anthropogenic activities release copper into the global biosphere. Copper is also used in agricultural fertilisers, as well as in the veterinary, medical, and food industries. The naturally occurring copper sources are minerals, such as Cu_2S, $CuFeS_2$, and $CuCO/Cu(OH)_2$ [42]. The sources of copper are presented in Fig. (**6**).

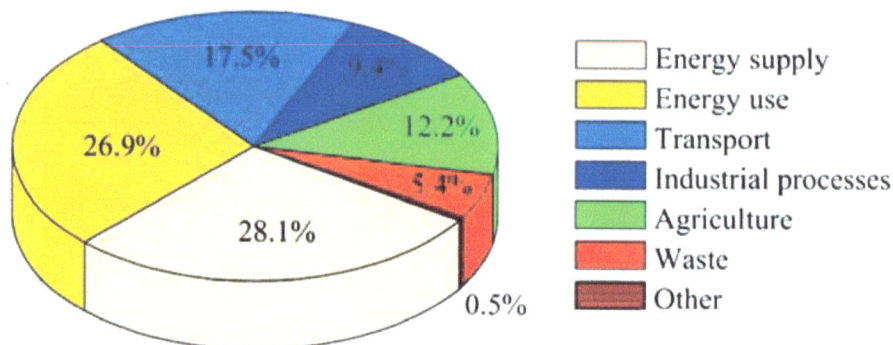

Fig. (6). Various sources of copper.

THRESHOLD LIMIT VALUES (TLV) OF TOXIC METAL IONS IN DRINKING WATER

The Threshold Limit Value (TLV) is the maximum average floating concentration of toxic metals to which workers can be exposed for eight hours per day and 40 hours per week without experiencing significant adverse health effects over the course of their working lives [43].

Table 1. TLV's of various toxic metal ions in drinking water.

Agency	Concentration (ppb)				Ref
	Lead	Cadmium	Mercury	Copper	
WHO	10	3	2	3	[44]
USEPA	50	25	10	15	[45]
CPCB	100	50	10	35	[46]
KSPCB	50	10	3	5	[47]

Abbreviations: USEPA: United States Environmental Protection Agency, WHO: World Health Organization, CPCB: Central Pollution Control Board, New Delhi, India, KSPCB: Karnataka State Pollution Control Board

ELECTROCHEMICAL SENSING OF HEAVY METAL IONS

Toxic metal ions can be quantified, at trace level, using CMEs. The modified CMEs exhibit high sensitivity and selectivity towards target analytes owing to the presence of potential functional groups on the modifier molecule [48]. For instance, Pikna *et al.* used carbon nanotubes modified graphite electrodes for the electrochemical measurement of Cd(II), Pb(II), Cu(II) and Hg(II) simultaneously at trace level [49]. Similarly, Wang *et al.* described the use of polyaniline and polyaniline/multi-walled modified GCE for the detection of Pb(II) and Cd(II) ions using square wave anodic stripping voltammetry (SWASV) [50]. Afkhami *et al.*

have prepared multi-walled carbon nanotubes (MWCNTs) and 3-(4-methoxybenzylideneamino)-2-thioxothiazolodin-4-one modified carbon paste electrodes to detect Hg(II) and Pb(II) simultaneously using SWASV [51]. Mafa *et al*. have fabricated the bismuth-modified exfoliated graphite electrode and used it in the measurement of As(III), Hg(II) and Pb(II) ions in real samples using SWASV [52]. Wang *et al*. have proposed a new membrane-based visualization protocol with good selectivity and high sensitivity for the detection of Pb(II) and Cd(II) ions at ultra-trace levels [53]. Rajkumar *et al*. reported the preparation of bismuth nanoribbons and the use of prepared bismuth nanoribbons in the quantification of lead and cadmium ions [54]. Mandil *et al*. have fabricated a disposable gold-coated screen-printed electrode (SPE) to quantify Pb(II)and Hg(II) simultaneously in tap water [55]. Salih *et al*. used Poly(1,8-diaminonaphthalene) and bismuth film modified CPS to detect Pb(II) in real water samples [56]. Gaurav *et al*. have fabricated SnO_2 based electrochemical sensor to detect Cd(II) [57]. Armas *et al*. have developed an electrochemical sensor using polystyrene sulfonate, NiO, and carbon nano-powder modified SPE to detect Hg(II) [58]. Similarly, Lu *et al*. have constructed the modified electrode using SnS nanosheets supported on graphene as an electrochemical sensor to detect Cu(II) using GCE, where it has shown good anti-interference performance [59].

NANOCOMPOSITE BASED ELECTRODES FOR HEAVY METAL ION SENSOR

The fabrication of eco-friendly and potential electrode materials with good selectivity and selectivity is a key parameter to improve the electrochemical performance to monitor HMIs even at trace levels. Nanocomposites are hybrid materials resulting from the blending of two or more constituents and expected to deliver new and deficient properties of each individual constituent. These new hybrid materials with new properties are always advantageous in terms of improving the sensing performance [60, 61]. Nanocomposites have been used as modifiers to modify the carbon-based electrodes to enhance the selectivity and sensitivity towards HMIs. Recently, bismuth-based electrochemical sensors attained significant interest in the detection of HMIs, and this may be due to the fact that bismuth can form an alloy with a wide range of heavy metals [62]. Carbon-based compounds containing bismuth are attracted in large quantities. Graphene is a two-dimensional material with sp^2 hybridised bonded carbon atoms that is one atom thick. This graphene possesses excellent and exceptional thermal, mechanical, electrochemical properties with high surface area as compared to the other forms of carbon-based materials.

The integration of nanoparticles and graphene provides enhanced electrochemical performance owing to its higher electrical conductivity and high surface-to-volume ratio. In this direction, solventless thermal decomposition method has been reported to prepare Fe_2O_3/graphene nanocomposite, where mixtures of iron oleate and graphene were subjected to the thermal treatment. The reported method is able to produce fine Fe_2O_3 nanoparticles of size 30 nm with uniform distribution on graphene, presented in Fig. (**7a** and **7b**).

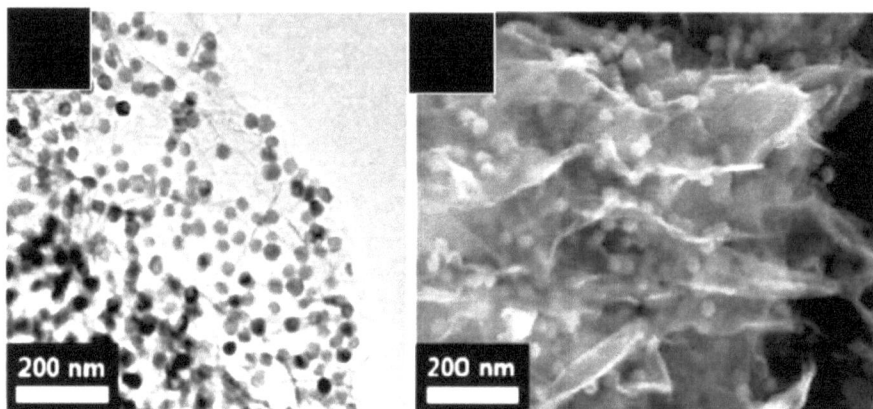

Fig. (7). (a) TEM and **(b)** SEM images of Fe_2O_3/graphene nanocomposite [62].

This proposed method has the advantage of being simple, cost-effective, and suitable for bulk production. The cyclic voltammetric technique was used to find the electrochemically effective surface area of Fe_2O_3/graphene nanocomposite and bare glassy carbon electrode (GCE) using $K_3[Fe(CN)_6]$, according to the following Randles–Sevcik equation:

$$i_p = 2.69 \times 10^5 \eta A C_0 D^{1/2} v^{1/2}$$

where D is diffusion coefficient, C_0 is $K_3[Fe(CN)_6]$ concentration, n is the number of electrons, v is the scan rate, and A is the electrochemically active surface area. The electrochemical active surface area of the Fe_2O_3/graphene nanocomposite modified GCE is found to be 0.082 cm^2, which is ~1.86 times higher than bare GCE (0.044 cm^2).

The prepared nanocomposite of Fe_2O_3/graphene has been coated on GCE and then modified with bismuth (Fe_2O_3/graphene/Bi). After that, it was used to measure Zinc(II), Cadmium(II), and Lead(II) simultaneously. The sensing performance of Fe_2O_3/graphene/Bi has been compared with bare GCE/Bi and GCE/graphene/Bi (Fig. (**8a** and **8b**). As shown in Fig. (**8a**), the stripping signal of Fe_2O_3/graphene/Bi was enhanced as compared to the stripping signal exhibited with bare GCE/Bi and GCE/graphene/Bi. The low response of GCE/Bi towards Zn^{2+}, Cd^{2+}, and Pb^{2+} is attributed to the poor conductivity at the interface.

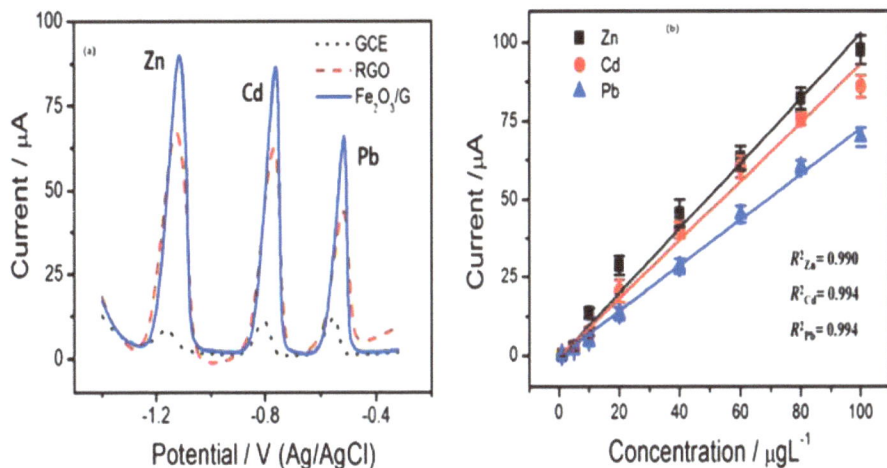

Fig. (8). a) Sensing response at GCE, RGO/GCE, and Fe$_2$O$_3$/G/GCE electrode and **b)** the calibration graphs of Zn^{2+}, Cd^{2+}, and Pb^{2+} [62].

The GCE/graphene/Bi electrode has a better sensing response than the GCE/Bi electrode, which is due to the addition of graphene, which boosts conductivity and electrochemical surface area, resulting in a faster rate of electron transfer and increased metal loading amounts. On the other hand, Fe$_2$O$_3$/graphene/Bi nanocomposite exhibits the highest stripping response towards Zn^{2+}, Cd^{2+}, and Pb^{2+}. This high response is attributed to the combined effects of graphene and the Fe$_2$O$_3$ nanoparticles, where the synergetic effect of Fe$_2$O$_3$ and graphene endows the essential conduction path to enhance the electrochemical performance. Also, the combination of graphene and Fe$_2$O$_3$ nanoparticles enhances the conductivity and surface area. The developed Fe$_2$O$_3$/graphene/Bi sensor was applied for the simultaneous detection of Zn^{2+}, Cd^{2+}, and Pb^{2+} in the concentration range of 1 to 100 mg L^{-1}, where it exhibits detection limits of 0.11, 0.08, and 0.07, respectively, for Zn^{2+}, Cd$^{2+,}$ and Pb^{2+}. The different bismuth-based electrochemical sensors used for the detection of heavy metal ions are summarized in Table **2**.

Table 2. Comparision of the bismuth modified electrodes with previously reported electrodes.

Electrode substrate	Linear range (µg L^{-1})	Limits of detection (µgL^{-1})			Ref.
-		Zn^{2+}	Cd^{2+}	Pb^{2+}	
G/AlOOH/GCE	200-800	-	0.03	0.09	[63]
Bi/G/MWCNT/GCE	5 – 30	-	0.10	0.20	[64]
DMSA/Fe$_3$O$_4$	1-50	-	-	0.5	[65]
Bi/NA/G/PANI	1-300	1.00	0.10	0.10	[66]
Bi/Au/G/Cys/GCE	0.5-40	-	0.1	0.05	[67]

(Table 2) cont.....

NA/MgO/GCE	1-6.7	-	7.21	0.43	[68]
G/Fe₃O₄/GCE	20 – 40	-	2.68	-	[69]
Fe₃O₄/MWCNT/GCE	1-130	-	-	0.5	[70]
Bi/NA/MWCNT/SPE	0.5-100 (Cd²⁺; 0.5 - 80)	0.3	0.1	0.07	[71]
Bi/NA/IL/G/SPCE	0.1-100	0.09 (ngL⁻¹) 0.06 (ngL⁻¹) 0.08 (ngL⁻¹)			[72]
Bi/Fe₂O₃/G/GCE	1-100	0.11 0.08 0.07			[62]

Abbreviations: Bi: bismuth, Cys: cysteine, DMSA: dimercaptosuccinic acid, G: graphene, GCE: glassy carbon electrode, IL: ionic liquid, MWCNT: Multi-walled carbon nanotube, NA: Nafion, PANI: polyaniline, SPE: screen-printed electrode, SPCE: screen-printed carbon electrode

Table 3. Comparision of the screen-printed electrodes with previously reported electrodes.

Electrode substrate	Linear range (µgL⁻¹)		Limits of detection (µgL⁻¹)		Ref.
-	Pb²⁺	Cd²⁺	Pb²⁺	Cd²⁺	
In-situ Bi film modified SPCPE	0.05–30	1–30	0.03	0.34	[73]
Bi NPs modified SPCE	--	--	1.3	1.7	[74]
Bi₂O₃ modified SPCE	0–12	0–12	0.2	0.2	[75]
Bi₂O₃ powder bulk modified SPCE	20–300	20–300	8	16	[76]
Bi citrate bulk modified SPCE	10–80	5–40	0.9	1.1	[77]
Bi₂O₃ SPE	20–100	20–100	2.3	1.5	[78]
Bi NPs bulk modified SPCPE	5–100	5–100	3.9	2.1	[79]
Bi NPs bulk modified SPCPE	1–50	1–50	2.3	1.5	[79]

Abbreviations: SPCPE: screen-printed porous carbon electrode, SPCE: screen-printed carbon electrode, SPE: screen-printed electrode, NPs: nanoparticles.

A simple one-pot hydrothermal method was adopted to synthesize γ-AlOOH/graphene nanocomposite using graphene oxide, aluminium nitrate and urea as precursors (Fig. **9**) [63].

Initially, Al^{3+} adsorb on the surface of the graphene oxide through electrostatic force of attraction between positively charged Al^{3+} ions and electronegative oxygen-containing functional groups (-COOH, -OH, and C-O-C) present in graphene. Then, the adsorbed Al^{3+} ions react with hydroxide ions, resulting from the decomposition of urea, which lead to the formation of $Al(OH)_3$ on graphene oxide nanosheets. Further, $Al(OH)_3$ is converted into AlOOH nanoplates on graphene nano-sheets, under hydrothermal reaction at 180 °C, together with the reduction of pristine graphene oxide to graphene. The chemical reaction involved in the formation of AlOOH-reduced graphene nanocomposite is shown below.

$$CO(NH_2)_2 + 3H_2O \rightarrow 2NH_4^+ + CO_2 + 2OH^-$$
$$Al^{3+} + 3OH^- \rightarrow Al(OH)_3$$
$$Al(OH)_3 \rightarrow AlOOH + H_2O$$

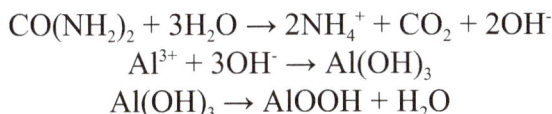

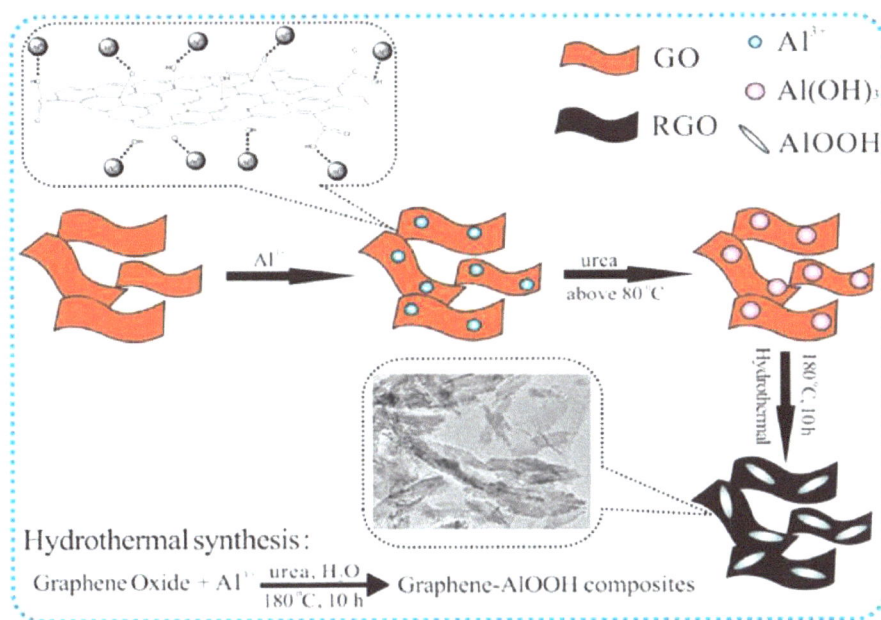

Fig. (9). Schematic representation of AlOOH/graphene nanocomposites synthesis [63].

The intercalation of AlOOH between the graphene nanosheets avoids the restacking of the reduced graphene nanosheets, which eventually leads to the formation of AlOOH-reduced graphene oxide nanocomposite where the graphene nanosheets exhibit enhanced surface area and conductivity.

Raman spectroscopy has been used to distinguish ordered and disordered carbon in graphene. Generally, graphene is characterized using G and D bands. The G band, observed at 1600 cm^{-1}, is attributed to sp^2 hybridised carbon, while the D band, observed at 1355 cm^{-1}, is attributed to A_{1g} symmetry. The ratio of D-band/G-band intensity demonstrates the disorder level in the graphene. The ratio of D-band intensity/G-band intensity for graphene oxide is ~1.2 while the same ratio increases to ~1.5 for AlOOH- graphene oxide and ~1.7 for reduced graphene oxide. The increased D-band/G-band intensity reveals the decrease in the size of the sp^2 carbon domains in the graphene oxide nanosheets and the reestablishment of the conjugated graphene network (sp^2 carbon).

The prepared AlOOH-reduced graphene oxide nanocomposite was used to modify the GCE, and this modified GCE has been used to quantify Cd(II) and Pb(II) using SWASV. The interface properties of the AlOOH-reduced graphene oxide modified GCE are characterized using electrochemical impedance spectrum (EIS).

Charge transfer resistance of AlOOH-reduced graphene oxide modified GCE is drastically reduced compared to the bare GCE and AlOOH coated GCE, demonstrating the fast electron transfer process at the interface. The ESCA is calculated using the Randles-Sevcik equation, where the surface of AlOOH-RGO (5.77×10^{-2}) and RGO $(6.28 \times 10^{-2}$ cm$^{-2})$ are almost the same, but the response towards Pb^{2+} and Cd^{2+} is higher for AlOOH-reduced graphene oxide. The high response is due to AlOOH accumulating more Cd^{2+} and Pb^{2+} ions. The AlOOH-reduced graphene oxide modified GCE works for the quantification of Cd^{2+} between the concentration 0.1 and 0.8 μM with LOD 4.46×10^{-11} M. Pb^{2+} was observed in concentrations ranging from 0.3 to 1.1 M, with a LOD of 7.60 1011 M. Pure NiO has a conductivity of ~10^{-13} S cm$^{-1,}$ and it is considered as a Mott–Hubbard insulator [80]. This low conductivity significantly diminishes the electrochemical performance towards the electrochemical detection of HMIs. The addition of tungsten atoms into NiO to form NiWO$_4$ significantly enhances the conductivity; therefore, NiWO$_4$ exhibits high electrical conductivity in the range 10^{-7} to 10^{-2} S cm^{-1}, which is higher than that of many simple and binary oxides. Therefore, this binary transition metal oxide-reduced graphene, NiWO$_4$/RGO nanocomposite, has been used as an effective electrode interface for the simultaneous quantification of Cd^{2+}, Pb^{2+}, Cu$^{2+,}$ and Hg^{2+}. The hydrothermal method has been adopted to synthesize NiWO$_4$/RGO nanocomposite using graphene oxide, sodium tungstate and nickel chloride at 180 °C for 12 h. The hydrothermally derived product was subjected to thermal treatment (400 °C/2 h) in the air followed by pyrolysis (600 °C/2 h) in an inert atmosphere to prepare NiWO$_4$/RGO nanocomposite, as shown in Scheme. (**1**).

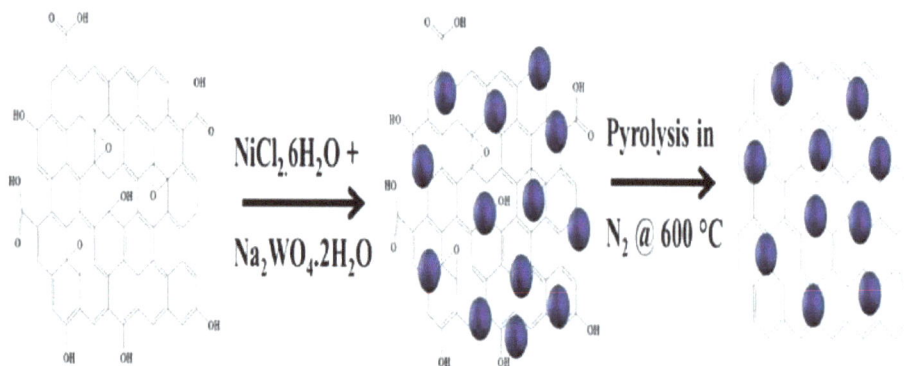

Scheme. (1). Schematic representation of the preparation of RGO/NiWO$_4$ nanocomposite [80].

The formation of $NiWO_4$ during hydrothermal treatment is shown according to the following equation:

$$2Na^+ + WO_4^{2-} + 8H_2O + Ni^{2+} + 2Cl^- \rightarrow NiWO_4 + 2Na^+ + 2Cl^- + 8H_2O$$

This proposed method allows the homogenous distribution of $NiWO_4$ nanoparticles of sizes 20-50 nm on RGO sheets, thereby demonstrating that RGO sheets provide 3D conductive support and a basal plane for the attachment of $NiWO_4$ nanoparticles.

The prepared $NiWO_4$/RGO was mixed with acetylene black and poly(vinylidenefluoride) in the weight ratio of 80:10:10, and then the resulting mixture was coated on carbon paper. The simultaneous determination of three heavy metal ions, using $NiWO_4$/RGO nanocomposite, was performed using DPASV. The peak currents of the Cd^{2+}, $Cu^{2+,}$ and Hg^{2+} increase with an increase in concentrations. The sensitivity and LOD were found to be 0.0622, 0.1411 and 0.1652 mA mM^{-1} cm^{-2} and 1.2 x 10^{-10}, 1.16 x10^{-10} and 1.36 x 10^{-10} M for Cd^{2+}, Cu^{2+} and Hg^{2+}, respectively. Similarly, sensitivity and LOD for Pb^{2+}, Cu^{2+} and Hg^{2+} were found to be 0.0857, 0.07432, 0.0809 mA mM^{-1} cm^{-2} and 2.04 x10^{-1}0, 1.11 x 10^{-10} and 1.92 x 10^{-10} M for Pb^{2+}, Cd^{2+} and Hg^{2+} ions, respectively. The stripping signal of four heavy metal ions, namely, Cd^{2+}, Pb^{2+}, Cu^{2+}, and $Hg^{2+,}$ and their corresponding calibration plots are presented in Fig. (**10a & 10b**).

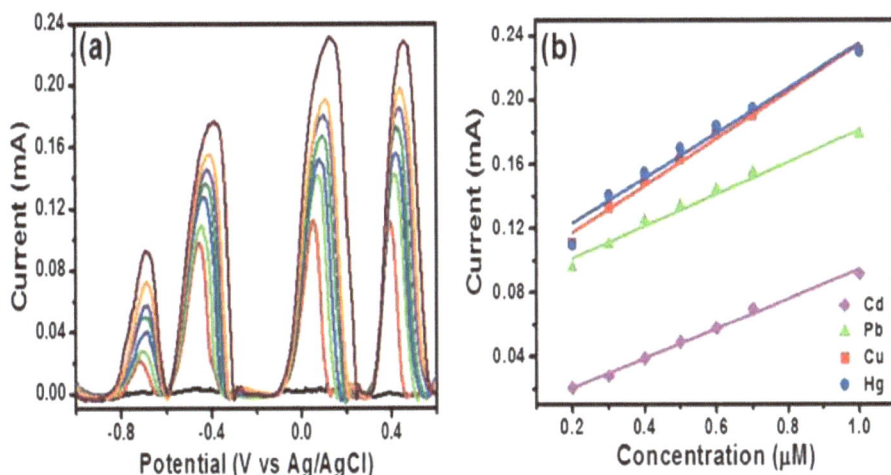

Fig.(10). DPASV response of $NiWO_4$/RGO nanocomposite for (a) the simultaneous detection of Cd^{2+}, Pb^{2+}, Cu^{2+}, and Hg^{2+} and (b) calibration plot [80].

The peak currents of Cd^{2+}, Pb^{2+}, Cu^{2+}, and Hg^{2+} increase with an increase in respective concentrations. The LOD values for Cd^{2+}, Pb^{2+}, Cu^{2+}, and Hg^{2+} were found to be 1.015 x 10^{-10}, 1.839 x 10^{-10}, 2.269 x 10^{-10} and 2.789 x 10^{-10} M, respectively, and the observed values were below the values prescribed by World Health Organization. This proposed $NiWO_4$/RGO electrochemical sensor was used for the quantification of Cd^{2+}, Pb^{2+}, Cu^{2+}, and Hg^{2+} in carbonated drinks, milk, and orange juice. The observed results for these real samples are presented in Figs. (**11a-f**).

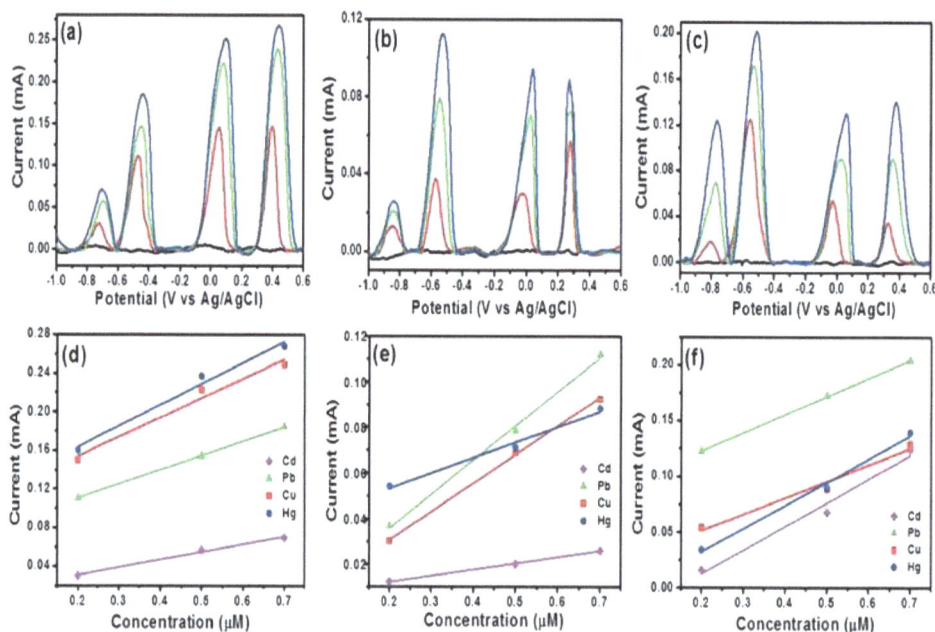

Fig. (11). DPASV response of NiWO4/RGO nanocomposite for the quantification of Cd^{2+}, Pb^{2+}, Cu^{2+}, and Hg^{2+} ions in (**a**) carbonated drinks (**b**) milk (**c**) orange juice and (**d–f**) their respective calibration [80].

Polymer composites with graphene have gained significant interest owing to their vast applications including transport, electronics, telecommunications, biomedicine, sensors, and energy industries. Graphene oxide/methylaniline nanocomposite has been developed for the detection of Pb^{2+}. Graphene oxide/methylaniline nanocomposite has been synthesized *via* inverse microemulsion technique [81]. In the preparation of nanocomposite, graphene oxide was mixed with sodium dodecylbenzenesulfonate (SDBS) solution, and then methylaniline monomer was added. Black green coloured nanocomposite was obtained after polymerization using ammonium persulfate. The SDBS is used to control the morphology of the nanocomposite. The prepared graphene oxide/methylaniline was coated on GCE and then used for the detection of Pb(II). The proposed sensor demonstrated a quick response to detect Pb(II) in just 16 seconds. The sensitivity and LOD of the proposed sensor are found to be 20.56 μAm M^{-1} cm^{-2} and 50.0 ± 2.5 pM, respectively.

Ethylenediaminetetraacetic acid-modified Polyaniline-SWNTs (EDTA modified PANI/SWNTs) nanocomposite has been reported for the sensing of Cu^{2+} using DPASV [82]. The nanocomposite has been prepared on stainless steel (SS) substrate through electrodeposition, where the mixture of aniline monomer was mixed with single-walled carbon nanotube in an acidic medium. Then, the electrodepositing was carried out using clean SS as a working electrode, Pt as a counter electrode, and Ag/AgCl as a reference electrode between the potential +0.1 and +1.0 V. Finally, the electrodeposited PANI/SWNT was functionalized with EDTA. The PANI was formed on the surface of the SWNTs where the π-π^* stacking between PANI and SWNTs enhanced the charge carrying capacity, which led to the faster signal and thereby fast sensor response towards Cu(II). The proposed sensor exhibits LOD and sensitivity of 1.4 μM and 189 $mA/\mu M$, respectively, towards Cu^{2+}.

The silver nanoparticles-reduced graphene oxide (AgNPs/RGO) has been reported for the sensing of Pb^{2+}, Cd^{2+}, Cu^{2+}, and Hg^{2+} simultaneously [83]. AgNPs/RGO nanocomposite has been prepared using RGO and silver nitrate dispersed in dimethyl formamide and then ascorbic acid was used for the reduction of silver ions directly on RGO. AgNPs/RGO nanocomposite was coated on a magnetic glassy carbon electrode (MGCE) to fabricate an electrochemical sensor. The proposed method endows the uniform distribution of AgNPs on the surface RGO and thereby ensures the strong interaction between AgNPs and RGO nanosheets. This synergic effect ensures a high electrochemical response. The prepared AgNPs/RGO nanocomposite modified MGCE exhibits the sensitivities of 48.69, 40.06, 15.66 and 43.18 mA mM^{-1} and the LODs of 0.141, 0.254, 0.178 and 0.285 mM towards Pb^{2+}, Cd^{2+}, Cu^{2+} and Hg^{2+}, respectively.

Table 4. Comparision of the current sensitivity with previously reported electrodes.

Electrode Substrate	Sensitivity (μA μM^{-1})	Limits of Detection (μM)	Ref
RGO/GCE	31.98	0.533	[83]
AgNPs/RGO/GCE (RGO:Ag = 1: 0.25)	42.15	0.141	[83]
AgNPs/RGO/GCE (RGO:Ag = 1: 0.75)	48.69	0.155	[83]
AuNPs/RGO/GCE	47.76	12.69 nM	[84]
Band Fe_3O_4/rGO GCE	13.6	0.169	[85]
Fe_3O_4/MWCNTs GCE	11.4	6.0 pM	[70]
MnO_2/GCE	22.4		[86]

(Table 4) cont.....

Electrode Substrate	Sensitivity (µA µM⁻¹)	Limits of Detection (µM)	Ref
RGO/MWCNTs GCE	42.1	0.052	[87]
SnO₂/graphene/GCE	18.6	0.0018	[88]
Porous Co₃O₄/GCE	71.5	0.018	[89]

Abbreviations: RGO: Reduced graphene oxide, GCE: Glassy carbon electrode, AgNPs: Silver nanoparticles, AuNPs: Gold nanoparticles, Fe_3O_4: Iron oxide, MWCNTs: Multi-walled carbon nanotubes, MnO_2: Manganese oxide, SnO_2: Tin oxide, Co_3O_4: Cobalt oxide.

CONCLUSION

This chapter discussed the preparation of graphene-based nanocomposites and their use for the quantification of heavy metal ions, particularly the advantage of graphene addition into the Fe_2O_3 and bismuth to prepare nanocomposite and its use in the simultaneous detection of Zn^{2+}, Cd^{2+}, and Pb^{2+} ions. Further, the mechanism involved in the formation of AlOOH-reduced graphene oxide nanocomposite and its application towards the sensing of Cd^{2+} and Pb^{2+} ions were discussed. Additionally, the advancement of $NiWO_4$/RGO nanocomposite towards the simultaneous sensing of Cd^{2+}, Pb^{2+}, Cu^{2+}, and Hg^{2+} ions over pristine nickel oxide and tungsten oxide was discussed. Lastly, the preparation of AgNPs/RGO and its application towards the sensing of Pb^{2+}, Cd^{2+}, Cu^{2+} and Hg^{2+} ions were discussed.

CONSENT FOR PUBLICATION

Not applicable.

CONFLICT OF INTEREST

The author declares no conflict of interest, financial or otherwise.

ACKNOWLEDGEMENTS

Dr. Prashanth S. A. acknowledges Shri. M. V. Patil, Shri. R. R. Sudi, Shri. Dilip N. Devagirikar, Shri. D. S. Kardyal and Dr. P. S. Kandagal SVM Arts, Science and Commerce College, Ilkal, for their support and encouragement. Dr. Ashoka S acknowledges the Dayananda Sagar University, Bengaluru, for their support and encouragement.

REFERENCES

[1]   Hou, H.; Zeinu, K.M.; Gao, S.; Liu, B.; Yang, J.; Hu, J. Recent advances and perspective on design and synthesis of electrode materials for electrochemical sensing of heavy metals. *Energy & Environmental Materials*, **2018**, *1*(3), 113-131.
       [http://dx.doi.org/10.1002/eem2.12011]

[2]   Zhang, P.; Dong, S.; Gu, G.; Huang, T. Simultaneous determination of Cd^{2+}, Pb^{2+}, Cu^{2+} and Hg^{2+} at a carbon paste electrode modified with ionic liquid-functionalized ordered mesoporous silica. *Bull. Korean Chem. Soc.,* **2010**, *31*(10), 2949-2954.
        [http://dx.doi.org/10.5012/bkcs.2010.31.10.2949]

[3]   Oztekin, Y.; Tok, M.; Nalvuran, H.; Kiyak, S.; Gover, T.; Yazicigil, Z.; Ramanaviciene, A.; Ramanavicius, A. Electrochemical modification of glassy carbon electrode by poly-4-nitroaniline and its application for determination of copper (II). *Electrochim. Acta,* **2010**, *56*(1), 387-395.
        [http://dx.doi.org/10.1016/j.electacta.2010.08.064]

[4]   aOztekin, Y.; Ramanaviciene, A.; Ramanavicius, A. Electrochemical Determination of Cu (II) Ions by 4-Formylphenylboronic Acid Modified Gold Electrode. *Electroanalysis,* **2011**, *23*(7), 1645-1653.
        [http://dx.doi.org/10.1002/elan.201100121] bNagaraju, G.; Prashanth, S.; Shastri, M.; Yathish, K.; Anupama, C.; Rangappa, D. Electrochemical heavy metal detection, photocatalytic, photoluminescence, biodiesel production and antibacterial activities of Ag–ZnO nanomaterial. *Mater. Res. Bull.,* **2017**, *94*, 54-63.
        [http://dx.doi.org/10.1016/j.materresbull.2017.05.043]

[5]   Adarakatti, P.S.; Banks, C.E.; Malingappa, P. Amino-thiacalix [4] arene modified screen-printed electrodes as a novel electrochemical interface for Hg (II) quantification at a pico-molar level. *Anal. Methods,* **2017**, *9*(48), 6747-6753.
        [http://dx.doi.org/10.1039/C7AY02468A]

[6]   Adarakatti, P.S.; Siddaramanna, A.; Malingappa, P. Fabrication of a new calix [4] arene-functionalized Mn 3 O 4 nanoparticle-based modified glassy carbon electrode as a fast responding sensor towards Pb 2+ and Cd 2+ ions. *Anal. Methods,* **2019**, *11*(6), 813-820.
        [http://dx.doi.org/10.1039/C8AY02648C]

[7]   aAdarakatti, P.S.; Ashoka S, ; Malingappa, P. A. K.; Adarakatti, P. S.; Ashoka, S.; Malingappa, P., CeO 2 nanoparticle-modified electrode as a novel electrochemical interface in the quantification of Zn 2+ ions at trace level: application to real sample analysis. *J. Solid State Electrochem.,* **2018**, *22*(6), 1711-1719.
        [http://dx.doi.org/10.1007/s10008-017-3872-0] bKempahanumakkagari, S.K.; Adarakatti, P.S.; Malingappa, P. Covalent modification of graphitic carbon through *in-situ* diazonium ion formation and its reduction: A robust pellet electrode as a novel electrochemical interface for divalent lead quantification. *J. Environ. Chem. Eng.,* **2018**, *6*(2), 2674-2683.
        [http://dx.doi.org/10.1016/j.jece.2018.04.006]

[8]   Murray, R.W. Chemically modified electrodes. *Acc. Chem. Res.,* **1980**, *13*(5), 135-141.
        [http://dx.doi.org/10.1021/ar50149a002]

[9]   McCreery, R.L.; Cline, K.K.; McDermott, C.A.; McDermott, M.T. Control of reactivity at carbon electrode surfaces. *Colloids Surf. A Physicochem. Eng. Asp.,* **1994**, *93*, 211-219.
        [http://dx.doi.org/10.1016/0927-7757(94)02899-0]

[10]   Wang, J. *Analytical Electrochemistry*; John Wiley & Sons, **2006**.
        [http://dx.doi.org/10.1002/0471790303]

[11]   Bard, A.J. *Chemical modification of electrodes*; ACS Publications, **1983**.
        [http://dx.doi.org/10.1021/ed060p302]

[12]   McCreery, R.L. Advanced carbon electrode materials for molecular electrochemistry. *Chem. Rev.,* **2008**, *108*(7), 2646-2687.
        [http://dx.doi.org/10.1021/cr068076m] [PMID: 18557655]

[13]   Vieira, E.F.; Simoni, J.A.; Airoldi, C. Interaction of cations with SH-modified silica gel: thermochemical study through calorimetric titration and direct extent of reaction determination. *J. Mater. Chem.,* **1997**, *7*(11), 2249-2252.
        [http://dx.doi.org/10.1039/a704286h]

[14]   Downard, A.J. Electrochemically assisted covalent modification of carbon electrodes. Electroanalysis.

An *International Journal Devoted to Fundamental and Practical Aspects of Electroanalysis,* **2000**, *12*(14), 1085-1096.

[15]   Pandurangappa, M.; Raghu, G.K. Chemically Modified Carbon nanotubes: derivatization and their applications. In: *Carbon nanotubes applications on electron devices*; IntechOpen, **2011**.
[http://dx.doi.org/10.5772/16635]

[16]   Jenkins, G.M.; Kawamura, K. Structure of glassy carbon. *Nature,* **1971**, *231*(5299), 175-176.
[http://dx.doi.org/10.1038/231175a0] [PMID: 16062610]

[17]   Szeluga, U.; Pusz, S.; Kumanek, B.; Olszowska, K.; Czajkowska, S.; Myalski, J.; Kubacki, J.; Trzebicka, B.; Borowski, A.F. Influence of unique structure of glassy carbon on morphology and properties of its epoxy-based binary composites and hybrid composites with carbon nanotubes. *Compos. Sci. Technol.,* **2016**, *134*, 72-80.
[http://dx.doi.org/10.1016/j.compscitech.2016.08.004]

[18]   Fagan, D.T.; Hu, I.F.; Kuwana, T. Vacuum heat-treatment for activation of glassy carbon electrodes. *Anal. Chem.,* **1985**, *57*(14), 2759-2763.
[http://dx.doi.org/10.1021/ac00291a006]

[19]   Weisshaar, D.E.; Kuwana, T. Considerations for polishing glassy carbon to a scratch-free finish. *Anal. Chem.,* **1985**, *57*(1), 378-379.
[http://dx.doi.org/10.1021/ac00279a090]

[20]   Adarakatti, P.S.; Foster, C.W.; Banks, C.E. NS, A. K.; Malingappa, P., Calixarene bulk modified screen-printed electrodes (SPCCEs) as a one-shot disposable sensor for the simultaneous detection of lead (II), copper (II) and mercury (II) ions: Application to environmental samples. *Sens. Actuators A Phys.,* **2017**, *267*, 517-525.
[http://dx.doi.org/10.1016/j.sna.2017.10.059]

[21]   Kawde, A-N.; Baig, N.; Sajid, M. Graphite pencil electrodes as electrochemical sensors for environmental analysis: a review of features, developments, and applications. *RSC Advances,* **2016**, *6*(94), 91325-91340.
[http://dx.doi.org/10.1039/C6RA17466C]

[22]   Banks, C.E.; Compton, R.G. Edge plane pyrolytic graphite electrodes in electroanalysis: an overview. *Anal. Sci.,* **2005**, *21*(11), 1263-1268.
[http://dx.doi.org/10.2116/analsci.21.1263] [PMID: 16317891]

[23]   Velický, M.; Toth, P.S.; Woods, C.R.; Novoselov, K.S.; Dryfe, R.A. Electrochemistry of the Basal Plane *versus* Edge Plane of Graphite Revisited. *J. Phys. Chem. C,* **2019**, *123*(18), 11677-11685.
[http://dx.doi.org/10.1021/acs.jpcc.9b01010]

[24]   Moore, R.R.; Banks, C.E.; Compton, R.G. Electrocatalytic detection of thiols using an edge plane pyrolytic graphite electrode. *Analyst (Lond.),* **2004**, *129*(8), 755-758.
[http://dx.doi.org/10.1039/b406276k] [PMID: 15284921]

[25]   Moreira, F.R.; Moreira, J.C. Effects of lead exposure on the human body and health implications. Revista panamericana de salud publica Pan. *Am. J. Public Health,* **2004**, *15*(2), 119-129.

[26]   Xiao-bo, J.; Yun-hua, W.; Jun-jie, F.; Sheng-shui, H. Electrochemical determination of trace amounts of lead (II) and cadmium (II) at a calix [6] arene modified carbon paste electrode. *Wuhan Univ. J. Nat. Sci.,* **2004**, *9*(6), 943-948.
[http://dx.doi.org/10.1007/BF02850805]

[27]   Friberg, L. *Cadmium in the Environment*; CRC press, **2018**.

[28]   Waalkes, M.P. Cadmium carcinogenesis in review. *J. Inorg. Biochem.,* **2000**, *79*(1-4), 241-244.
[http://dx.doi.org/10.1016/S0162-0134(00)00009-X] [PMID: 10830873]

[29]   Satarug, S.; Garrett, S.H.; Sens, M.A.; Sens, D.A. Cadmium, environmental exposure, and health outcomes. *Environ. Health Perspect.,* **2010**, *118*(2), 182-190.
[http://dx.doi.org/10.1289/ehp.0901234] [PMID: 20123617]

[30]   Pinot, F.; Kreps, S.E.; Bachelet, M.; Hainaut, P.; Bakonyi, M.; Polla, B.S. Cadmium in the environment: sources, mechanisms of biotoxicity, and biomarkers. *Rev. Environ. Health,* **2000**, *15*(3), 299-323.
[http://dx.doi.org/10.1515/REVEH.2000.15.3.299] [PMID: 11048333]

[31]   Schroeder, W.H.; Munthe, J. Atmospheric mercury—an overview. *Atmos. Environ.,* **1998**, *32*(5), 809-822.
[http://dx.doi.org/10.1016/S1352-2310(97)00293-8]

[32]   Clarkson, T.W. The three modern faces of mercury. *Environ. Health Perspect.,* **2002**, *110* Suppl. 1, 11-23.
[http://dx.doi.org/10.1289/ehp.02110s111] [PMID: 11834460]

[33]   Morel, F.M.; Kraepiel, A.M.; Amyot, M. The chemical cycle and bioaccumulation of mercury. *Annu. Rev. Ecol. Syst.,* **1998**, *29*(1), 543-566.
[http://dx.doi.org/10.1146/annurev.ecolsys.29.1.543]

[34]   Eley, B.M. The future of dental amalgam: a review of the literature. Part 5: Mercury in the urine, blood and body organs from amalgam fillings. *Br. Dent. J.,* **1997**, *182*(11), 413-417.
[http://dx.doi.org/10.1038/sj.bdj.4809401] [PMID: 9217337]

[35]   Liu, G.; Cai, Y.; O'Driscoll, N. *Environmental chemistry and toxicology of mercury.,* **2011**.
[http://dx.doi.org/10.1002/9781118146644]

[36]   Davis, J.R. *Copper and copper alloys*; ASM international, **2001**.
[http://dx.doi.org/10.31399/asm.tb.aub.t61170457]

[37]   Tabelin, C.B.; Igarashi, T.; Villacorte-Tabelin, M.; Park, I.; Opiso, E.M.; Ito, M.; Hiroyoshi, N. Arsenic, selenium, boron, lead, cadmium, copper, and zinc in naturally contaminated rocks: A review of their sources, modes of enrichment, mechanisms of release, and mitigation strategies. *Sci. Total Environ.,* **2018**, *645*, 1522-1553.
[http://dx.doi.org/10.1016/j.scitotenv.2018.07.103] [PMID: 30248873]

[38]   Eisler, R. *Copper hazards to fish, wildlife, and invertebrates: a synoptic review*; US Department of the Interior, US Geological Survey, **1998**.

[39]   Nordberg, G.F.; Fowler, B.A.; Nordberg, M. *Handbook on the Toxicology of Metals*; Academic press, **2014**.

[40]   Hall, W.S.; Bushong, S.J.; Hall, L.W., Jr; Lenkevich, M.J.; Pinkney, A.E. Monitoring dissolved copper concentrations in Chesapeake Bay, U.S.A. *Environ. Monit. Assess.,* **1988**, *11*(1), 33-42.
[http://dx.doi.org/10.1007/BF00394510] [PMID: 24248797]

[41]   Hosler, D.; Macfarlane, A. Copper sources, metal production, and metals trade in Late Postclassic Mesoamerica. *Science,* **1996**, *273*(5283), 1819-1824.
[http://dx.doi.org/10.1126/science.273.5283.1819]

[42]   Health, U. D. O.; Services, H. *Toxicological profile for copper.,* **1990**.

[43]   Galer, D.M.; Leung, H.W.; Sussman, R.G.; Trzos, R.J. Scientific and practical considerations for the development of occupational exposure limits (OELs) for chemical substances. *Regul. Toxicol. Pharmacol.,* **1992**, *15*(3), 291-306.
[http://dx.doi.org/10.1016/0273-2300(92)90040-G] [PMID: 1509122]

[44]   Utriainen, M.; Kärpänoja, E.; Paakkanen, H. Combining miniaturized ion mobility spectrometer and metal oxide gas sensor for the fast detection of toxic chemical vapors. *Sens. Actuators B Chem.,* **2003**, *93*(1-3), 17-24.
[http://dx.doi.org/10.1016/S0925-4005(03)00337-X]

[45]   Regulatory impact analysis of the clean air mercury rule: EPA-452; R-05-003: **2005**.

[46]   Srinivasa Gowd, S.; Govil, P.K. Distribution of heavy metals in surface water of Ranipet industrial area in Tamil Nadu, India. *Environ. Monit. Assess.,* **2008**, *136*(1-3), 197-207.

[http://dx.doi.org/10.1007/s10661-007-9675-5] [PMID: 17457685]

[47]   Mishra, P.R.; Nadda, R. Water resource pollution and impacts on the local livelihood: a case study of Beas river in Kullu district, India. , **2014**.

[48]   Adarakatti, P.S.; Kempahanumakkagari, S.K. *Modified electrodes for sensing*; Int Electrochemistry, **2018**, pp. 58-95.

[49]   Pikna, L.; Heželová, M.; Kováčová, Z. Optimization of simultaneous electrochemical determination of Cd(II), Pb(II), Cu(II) and Hg(II) at carbon nanotube-modified graphite electrodes. *J. Environ. Sci. Health Part A Tox. Hazard. Subst. Environ. Eng.,* **2015**, *50*(8), 874-881.
[http://dx.doi.org/10.1080/10934529.2015.1019810] [PMID: 26030694]

[50]   Wang, Z.; Liu, E.; Zhao, X. Glassy carbon electrode modified by conductive polyaniline coating for determination of trace lead and cadmium ions in acetate buffer solution. *Thin Solid Films,* **2011**, *519*(15), 5285-5289.
[http://dx.doi.org/10.1016/j.tsf.2011.01.176]

[51]   Afkhami, A.; Bagheri, H.; Khoshsafar, H.; Saber-Tehrani, M.; Tabatabaee, M.; Shirzadmehr, A. Simultaneous trace-levels determination of Hg(II) and Pb(II) ions in various samples using a modified carbon paste electrode based on multi-walled carbon nanotubes and a new synthesized Schiff base. *Anal. Chim. Acta,* **2012**, *746*, 98-106.
[http://dx.doi.org/10.1016/j.aca.2012.08.024] [PMID: 22975186]

[52]   Mafa, P.J.; Idris, A.O.; Mabuba, N.; Arotiba, O.A. Electrochemical co-detection of As(III), Hg(II) and Pb(II) on a bismuth modified exfoliated graphite electrode. *Talanta,* **2016**, *153*, 99-106.
[http://dx.doi.org/10.1016/j.talanta.2016.03.003] [PMID: 27130095]

[53]   Wang, C-Y.; Fang, B-Y.; Yao, M-H.; Zhao, Y-D. Visualization detection of ultratrace lead and cadmium ions using cellulose acetate membrane based on silver stain. *Sens. Actuators B Chem.,* **2016**, *228*, 643-648.
[http://dx.doi.org/10.1016/j.snb.2016.01.080]

[54]   Devasenathipathy, R.; Karthik, R.; Chen, S.M.; Mani, V.; Vasantha, V.; Ali, M.A.; Elshikh, M.S.; Lou, B.S.; Al-Hemaid, F.M. Potentiostatic electrochemical preparation of bismuth nanoribbons and its application in biologically poisoning lead and cadmium heavy metal ions detection. *Electroanalysis,* **2015**, *27*(10), 2341-2346.
[http://dx.doi.org/10.1002/elan.201500255]

[55]   Mandil, A.; Idrissi, L.; Amine, A. Stripping voltammetric determination of mercury (II) and lead (II) using screen-printed electrodes modified with gold films, and metal ion preconcentration with thiol-modified magnetic particles. *Mikrochim. Acta,* **2010**, *170*(3-4), 299-305.
[http://dx.doi.org/10.1007/s00604-010-0329-x]

[56]   Salih, F.E.; Ouarzane, A.; El Rhazi, M. Electrochemical detection of lead (II) at bismuth/poly (1, 8-diaminonaphthalene) modified carbon paste electrode. *Arab. J. Chem.,* **2017**, *10*(5), 596-603.
[http://dx.doi.org/10.1016/j.arabjc.2015.08.021]

[57]   Bhanjana, G.; Dilbaghi, N.; Kumar, R.; Umar, A.; Kumar, S. SnO2 quantum dots as novel platform for electrochemical sensing of cadmium. *Electrochim. Acta,* **2015**, *169*, 97-102.
[http://dx.doi.org/10.1016/j.electacta.2015.04.045]

[58]   Armas, M.A.; María-Hormigos, R.; Cantalapiedra, A.; Gismera, M.J.; Sevilla, M.T.; Procopio, J.R. Multiparametric optimization of a new high-sensitive and disposable mercury (II) electrochemical sensor. *Anal. Chim. Acta,* **2016**, *904*, 76-82.
[http://dx.doi.org/10.1016/j.aca.2015.11.016] [PMID: 26724765]

[59]   Lu, J.; Zhang, X.; Liu, N.; Zhang, X.; Yu, Z.; Duan, T. Electrochemical detection of Cu^{2+} using graphene–SnS nanocomposite modified electrode. *J. Electroanal. Chem. (Lausanne),* **2016**, *769*, 21-27.
[http://dx.doi.org/10.1016/j.jelechem.2016.03.009]

[60]   Xiao, Y.; Li, C.M. Nanocomposites: from fabrications to electrochemical bioapplications. Electroanalysis. *An International Journal Devoted to Fundamental and Practical Aspects of Electroanalysis,* **2008**, *20*(6), 648-662.

[61]   Dutta, D.; Chandra, S.; Swain, A.K.; Bahadur, D. SnO(2) quantum dots-reduced graphene oxide composite for enzyme-free ultrasensitive electrochemical detection of urea. *Anal. Chem.,* **2014**, *86*(12), 5914-5921.
       [http://dx.doi.org/10.1021/ac5007365] [PMID: 24830909]

[62]   Lee, S.; Oh, J.; Kim, D.; Piao, Y. A sensitive electrochemical sensor using an iron oxide/graphene composite for the simultaneous detection of heavy metal ions. *Talanta,* **2016**, *160*, 528-536.
       [http://dx.doi.org/10.1016/j.talanta.2016.07.034] [PMID: 27591647]

[63]   Gao, C.; Yu, X-Y.; Xu, R-X.; Liu, J-H.; Huang, X-J. AlOOH-reduced graphene oxide nanocomposites: one-pot hydrothermal synthesis and their enhanced electrochemical activity for heavy metal ions. *ACS Appl. Mater. Interfaces,* **2012**, *4*(9), 4672-4682.
       [http://dx.doi.org/10.1021/am3010434] [PMID: 22924704]

[64]   Huang, H.; Chen, T.; Liu, X.; Ma, H. Ultrasensitive and simultaneous detection of heavy metal ions based on three-dimensional graphene-carbon nanotubes hybrid electrode materials. *Anal. Chim. Acta,* **2014**, *852*, 45-54.
       [http://dx.doi.org/10.1016/j.aca.2014.09.010] [PMID: 25441878]

[65]   Yantasee, W.; Hongsirikarn, K.; Warner, C.L.; Choi, D.; Sangvanich, T.; Toloczko, M.B.; Warner, M.G.; Fryxell, G.E.; Addleman, R.S.; Timchalk, C. Direct detection of Pb in urine and Cd, Pb, Cu, and Ag in natural waters using electrochemical sensors immobilized with DMSA functionalized magnetic nanoparticles. *Analyst (Lond.),* **2008**, *133*(3), 348-355.
       [http://dx.doi.org/10.1039/b711199a] [PMID: 18299749]

[66]   Ruecha, N.; Rodthongkum, N.; Cate, D.M.; Volckens, J.; Chailapakul, O.; Henry, C.S. Sensitive electrochemical sensor using a graphene-polyaniline nanocomposite for simultaneous detection of Zn(II), Cd(II), and Pb(II). *Anal. Chim. Acta,* **2015**, *874*, 40-48.
       [http://dx.doi.org/10.1016/j.aca.2015.02.064] [PMID: 25910444]

[67]   Zhu, L.; Xu, L.; Huang, B.; Jia, N.; Tan, L.; Yao, S. Simultaneous determination of Cd (II) and Pb (II) using square wave anodic stripping voltammetry at a gold nanoparticle-graphene-cysteine composite modified bismuth film electrode. *Electrochim. Acta,* **2014**, *115*, 471-477.
       [http://dx.doi.org/10.1016/j.electacta.2013.10.209]

[68]   Wei, Y.; Yang, R.; Yu, X-Y.; Wang, L.; Liu, J-H.; Huang, X-J. Stripping voltammetry study of ultra-trace toxic metal ions on highly selectively adsorptive porous magnesium oxide nanoflowers. *Analyst (Lond.),* **2012**, *137*(9), 2183-2191.
       [http://dx.doi.org/10.1039/c2an15939b] [PMID: 22421740]

[69]   Sun, Y-F.; Chen, W-K.; Li, W-J.; Jiang, T-J.; Liu, J-H.; Liu, Z-G. Selective detection toward Cd^{2-} using Fe_3O_4/RGO nanoparticle modified glassy carbon electrode. *J. Electroanal. Chem. (Lausanne),* **2014**, *714*, 97-102.
       [http://dx.doi.org/10.1016/j.jelechem.2013.12.030]

[70]   Yang, Y.; You, Y.; Liu, Y.; Yang, Z. A lead (II) sensor based on a glassy carbon electrode modified with Fe_3O_4 nanospheres and carbon nanotubes. *Mikrochim. Acta,* **2013**, *180*(5-6), 379-385.
       [http://dx.doi.org/10.1007/s00604-013-0940-8]

[71]   Fu, L.; Li, X.; Yu, J.; Ye, J. Facile and Simultaneous Stripping Determination of Zinc, Cadmium and Lead on Disposable Multiwalled Carbon Nanotubes Modified Screen-Printed Electrode. *Electroanalysis,* **2013**, *25*(2), 567-572.
       [http://dx.doi.org/10.1002/elan.201200248]

[72]   Chaiyo, S.; Mehmeti, E.; Žagar, K.; Siangproh, W.; Chailapakul, O.; Kalcher, K. Electrochemical sensors for the simultaneous determination of zinc, cadmium and lead using a Nafion/ionic liquid/graphene composite modified screen-printed carbon electrode. *Anal. Chim. Acta,* **2016**, *918*, 26-

34.
[http://dx.doi.org/10.1016/j.aca.2016.03.026] [PMID: 27046207]

[73]   Chen, C.; Niu, X.; Chai, Y.; Zhao, H.; Lan, M. Bismuth-based porous screen-printed carbon electrode with enhanced sensitivity for trace heavy metal detection by stripping voltammetry. *Sens. Actuators B Chem.,* **2013,** *178,* 339-342.
[http://dx.doi.org/10.1016/j.snb.2012.12.109]

[74]   Rico, M.Á.G.; Olivares-Marín, M.; Gil, E.P. Modification of carbon screen-printed electrodes by adsorption of chemically synthesized Bi nanoparticles for the voltammetric stripping detection of Zn(II), Cd(II) and Pb(II). *Talanta,* **2009,** *80*(2), 631-635.
[http://dx.doi.org/10.1016/j.talanta.2009.07.039] [PMID: 19836530]

[75]   Riman, D.; Jirovsky, D.; Hrbac, J.; Prodromidis, M.I. Green and facile electrode modification by spark discharge: Bismuth oxide-screen printed electrodes for the screening of ultra-trace Cd (II) and Pb (II). *Electrochem. Commun.,* **2015,** *50,* 20-23.
[http://dx.doi.org/10.1016/j.elecom.2014.11.003]

[76]   Kadara, R.O.; Tothill, I.E. Development of disposable bulk-modified screen-printed electrode based on bismuth oxide for stripping chronopotentiometric analysis of lead (II) and cadmium (II) in soil and water samples. *Anal. Chim. Acta,* **2008,** *623*(1), 76-81.
[http://dx.doi.org/10.1016/j.aca.2008.06.010] [PMID: 18611460]

[77]   Lezi, N.; Economou, A.; Dimovasilis, P.A.; Trikalitis, P.N.; Prodromidis, M.I. Disposable screen-printed sensors modified with bismuth precursor compounds for the rapid voltammetric screening of trace Pb(II) and Cd(II). *Anal. Chim. Acta,* **2012,** *728,* 1-8.
[http://dx.doi.org/10.1016/j.aca.2012.03.036] [PMID: 22560274]

[78]   Hwang, G-H.; Han, W-K.; Park, J-S.; Kang, S-G. An electrochemical sensor based on the reduction of screen-printed bismuth oxide for the determination of trace lead and cadmium. *Sens. Actuators B Chem.,* **2008,** *135*(1), 309-316.
[http://dx.doi.org/10.1016/j.snb.2008.08.039]

[79]   Niu, P.; Fernández-Sánchez, C.; Gich, M.; Navarro-Hernández, C.; Fanjul-Bolado, P.; Roig, A. Screen-printed electrodes made of a bismuth nanoparticle porous carbon nanocomposite applied to the determination of heavy metal ions. *Mikrochim. Acta,* **2016,** *183*(2), 617-623.
[http://dx.doi.org/10.1007/s00604-015-1684-4]

[80]   Kumar, R.; Bhuvana, T.; Sharma, A. Nickel tungstate–graphene nanocomposite for simultaneous electrochemical detection of heavy metal ions with application to complex aqueous media. *RSC Advances,* **2017,** *7*(67), 42146-42158.
[http://dx.doi.org/10.1039/C7RA08047F]

[81]   Khan, A.; Khan, A.; Asiri, A.; Alam, M.; Rahman, M.; Shaban, M. Surfactant-assisted graphene oxide/methylaniline nanocomposites for lead ionic sensor development for the environmental remediation in real sample matrices. *Int. J. Environ. Sci. Technol.,* **2019,** *16*(12), 8461-8470.
[http://dx.doi.org/10.1007/s13762-019-02447-8]

[82]   Deshmukh, M.A.; Patil, H.K.; Bodkhe, G.A.; Yasuzawa, M.; Koinkar, P.; Ramanaviciene, A.; Shirsat, M.D.; Ramanavicius, A. EDTA-modified PANI/SWNTs nanocomposite for differential pulse voltammetry based determination of Cu (II) ions. *Sens. Actuators B Chem.,* **2018,** *260,* 331-338.
[http://dx.doi.org/10.1016/j.snb.2017.12.160]

[83]   Sang, S.; Li, D.; Zhang, H.; Sun, Y.; Jian, A.; Zhang, Q.; Zhang, W. Facile synthesis of AgNPs on reduced graphene oxide for highly sensitive simultaneous detection of heavy metal ions. *RSC Advances,* **2017,** *7*(35), 21618-21624.
[http://dx.doi.org/10.1039/C7RA02267K]

[84]   Gnanaprakasam, P.; Jeena, S.E.; Premnath, D.; Selvaraju, T. Simple and robust green synthesis of Au NPs on reduced graphene oxide for the simultaneous detection of toxic heavy metal ions and bioremediation using bacterium as the scavenger. *Electroanalysis,* **2016,** *28*(8), 1885-1893.

[http://dx.doi.org/10.1002/elan.201600002]

[85]   Sun, Y.; Zhang, W.; Yu, H.; Hou, C.; Li, D.; Zhang, Y.; Liu, Y. Controlled synthesis various shapes
       Fe₃O₄ decorated reduced graphene oxide applied in the electrochemical detection. *J. Alloys Compd.,*
       2015, *638*, 182-187.
       [http://dx.doi.org/10.1016/j.jallcom.2015.03.061]

[86]   Zhang, Q-X.; Wen, H.; Peng, D.; Fu, Q.; Huang, X-J. Interesting interference evidences of
       electrochemical detection of Zn (II), Cd (II) and Pb (II) on three different morphologies of MnO2
       nanocrystals. *J. Electroanal. Chem. (Lausanne),* **2015**, *739*, 89-96.
       [http://dx.doi.org/10.1016/j.jelechem.2014.12.023]

[87]   Zhang, J-T.; Jin, Z-Y.; Li, W-C.; Dong, W.; Lu, A-H. Graphene modified carbon nanosheets for
       electrochemical detection of Pb (II) in water. *J. Mater. Chem. A Mater. Energy Sustain.,* **2013**, *1*(42),
       13139-13145.
       [http://dx.doi.org/10.1039/c3ta12612a]

[88]   Wei, Y.; Gao, C.; Meng, F-L.; Li, H-H.; Wang, L.; Liu, J-H.; Huang, X-J. SnO2/reduced graphene
       oxide nanocomposite for the simultaneous electrochemical detection of cadmium (II), lead (II), copper
       (II), and mercury (II): an interesting favorable mutual interference. *J. Phys. Chem. C,* **2012**, *116*(1),
       1034-1041.
       [http://dx.doi.org/10.1021/jp209805c]

[89]   Liu, Z-G.; Chen, X.; Liu, J-H.; Huang, X-J. Well-arranged porous Co3O4 microsheets for
       electrochemistry of Pb (II) revealed by stripping voltammetry. *Electrochem. Commun.,* **2013**, *30*, 59-
       62.
       [http://dx.doi.org/10.1016/j.elecom.2013.02.002]

CHAPTER 11

Nanocomposites as Electrochemical Sensing Platforms for Glucose Detection

Prashanth Shivappa Adarakatti[1,*], Suma B. Patri[2] and S Ashoka[3]

[1] *Department of Chemistry, SVM Arts, Science, and Commerce College, ILKAL – 587125, India*

[2] *Department of Chemistry, Bangalore University, Central College Campus, Bengaluru – 560056, India*

[3] *Department of Chemistry, School of Engineering, Dayananda Sagar University, Bengaluru –560068, India*

Abstract: This chapter covers the advances in the development of a variety of composite materials ranging from noble metal/metal oxide nanoparticles, carbon composites, polymer composites and metal-organic framework-based composite materials specific to glucose. The advantages of nanocomposites as 'electrode materials' have been highlighted. The utilization of above-mentioned nanocomposites in non-enzymatic glucose sensors and their mechanism has been discussed. Further, special attention has been given to the MOF-based nanocomposites, which elaborates the applications of MOF-based materials in biosensing in recent years. This chapter gives an overall view of various nanocomposites used as electrochemical glucose sensors and opens up a new trend in materials science research to engineer advanced functional materials with tailor-made properties to suit relevant real-time applications within electroanalysis.

Keywords: Diabetes , Electrochemical Sensor , Glucose sensor, Nanocomposite , ModifiedElectrodes .

INTRODUCTION

Diabetes mellitus is a group of metabolic diseases characterized by variation in the blood glucose level, which results from defects in insulin secretion and its action. This increased glucose level in blood is associated with severe health effects like long-term damage to tissues, failure, and dysfunction of various organs, especially the nerves, kidneys, eyes, heart, and blood vessels [1]. Hence, regular monitoring of glucose levels in the blood has become mandatory in a large sector of people.

* **Corresponding Author P. S. Adarakatti:** Department of Chemistry, SVM Arts, Science and Commerce College, ILKAL – 587125, India; Email: prashanthsa143@gmail.com

Manorama Singh, Vijai K Rai and Ankita Rai (Eds.)

Since the discovery of one-shot disposable biosensors, the development of glucose sensors have grown enormously in the past few decades. The non-enzymatic and enzymatic-based probes are two important protocols [2], which are used for glucose quantification. Fig. **1**. shows the schematic representation of the classification of various types of glucose sensors. The development of the first enzyme-based glucose sensor led to the evolution of various glucose sensors to date.

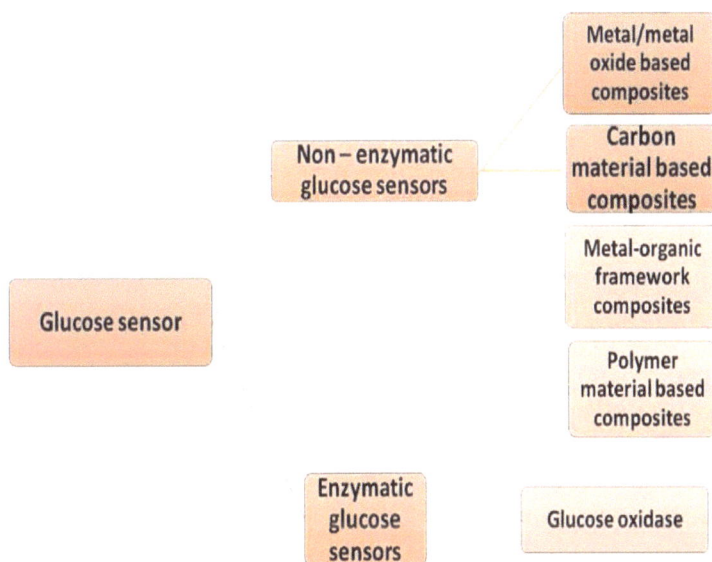

Fig. (1). Schematic representation of the classification of various types of glucose sensors.

NANOCOMPOSITES FOR ENZYMATIC GLUCOSE SENSORS

Enzymatic glucose sensors have gained much attention ever since the beginning of the first generation of glucose sensors by Clark and Lyons in 1962 [3]. The most important component of any blood glucose monitoring device is the detection technology or the sensing mechanism that quantifies the concentration of glucose selectively and sensitively. The detection mechanism evolved from first-generation blood glucose monitoring systems employs oxygen as the electron acceptor and determines the glucose concentration either by consuming the oxygen or by the liberation of hydrogen peroxide, whereas redox dyes and mediators have been utilized in the second generation glucose sensors. Furthermore, sensitive glucose sensors were commercialized, where the direct electron transfer has taken place upon the electrode interface. These types of sensors demonstrated better selectivity and anti-interfering properties in blood glucose monitoring in real samples [4].

Due to the high selectivity and sensitivity of enzymes towards glucose, glucose oxidases (GOxs) and glucose dehydrogenases (GDHs) are the two types of oxidoreductases utilized for glucose monitoring. Other types of enzymes immobilized are Pyrroloquinoline quinine (PQQ), Flavin-Adenine-Dinucleotide (FAD)-dependent Glucose dehydrogenases, Nicotine Adenine Dinucleotide (NADP) (Phosphate)-dependent Glucose Dehydrogenase, which varies immensely in terms of structure, substrate specificity, origin, primary electron acceptor and acceptable final electron acceptor. However, blood glucose monitors using GDH-PQQ can be affected by maltose or galactose, xylose, which is another form of sugar with similar activity, and interferes in glucose measurement at GDH-FAD modified electrodes [5].

The native GOx is incapable of transferring electrons to the electrode interface. Hence, huge works have been done to focus on connecting these enzyme's redox center to the electrode surface and non-biological electrode materials have been attracting significant attention as a substitute to biosensors. Most importantly, non-enzymatic sensors are free from oxygen limitations. Most of the enzyme electrodes with low or high oxygen concentrations may modify the sensor signals, causing divergence from analytical signals measured in the normal range of oxygen concentration. The main approach is to make use of various conducting molecular wires such as metal complex-linked polymers and macromolecules, synthetic co-factor derivatives and carbon nanomaterials. Alternatively, compositing carbon nanomaterials and metal oxide NPs can prevent agglomerating and restacking of metal NPs by steric hindrance and electrostatic attractions. Likewise, these composites can increase the electron transport rates between the electrolyte and modified electrode materials fabricated in an electrochemical device. In recent years, graphene, carbon nanotubes, conducting polymers, metal oxides/sulphides and noble nanoparticles and nanoparticles of transition metals were utilized for the measurement of the enzymatic glucose sensor. For instance, the immobilization of glucose oxidase on gold nanoparticles (AuNPs)-functionalized ZnO nanostructure on a glassy carbon electrode (GCE) is a facile approach to construct third-generation glucose sensors [6]. The non-enzymatic glucose sensing approach is a fascinating area that offers a number of opportunities to material chemists.

NANOCOMPOSITES FOR NON-ENZYMATIC GLUCOSE SENSORS

In recent times, enzymatic glucose sensors have been incorporated with nanomaterials to enhance conductivity, surface area, and electron transfer rates. These nanomaterials include the nanoparticles of noble and transition metals, the nanostructured metal-oxides or metal-sulfides, conductive polymers, carbon nanotubes, and graphene.

Metal/Metal Oxide Based Nanocomposites as Glucose Sensors

Nanostructures based on Pt, Au, Ni, Cu, Zn and their alloys/composites have shown high catalytic activity for the oxidation of glucose at physiological pH values. This is due to its low toxic nature, abundance, excellent electrical conductivity and electrocatalytic activity.

An efficient electrochemical sensor for the measurement of the non-enzymatic glucose sensor is essential, as it reduces the cost of diabetes monitoring. In this view, Imran *et al* [7]. have developed an electrochemical sensing tool by using noble metals such as platinum (Pt) and zinc oxide nanoparticles (ZnO NPs), which further decorated upon graphitic carbon nitride (g-C$_3$N$_4$). The modified electrode was further introduced in the analysis of glucose at physiological conditions (Fig. 2). In this method, the gold electrode has been used to modify indicator molecules *via* the drop coating method. The proposed sensor has shown good stability and selectivity towards the target analyte and hence improved the analytical response, which is ascribed to the presence of ZnO NPs along with Pt providing more hydroxyl groups, which in turn enhances the electrocatalysis. For completeness, the applicability of the developed sensor was checked by measuring glucose levels in human blood, urine and serum samples.

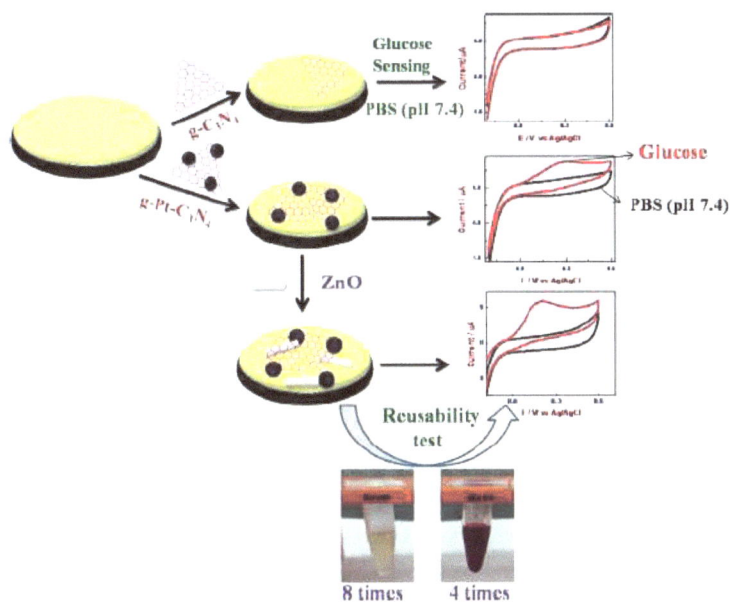

Fig. (2). Reusable non-enzymatic glucose sensing at ZnO-Pt–g-C$_3$N$_4$ nanozyme [7].

Similarly, sandwich type electrochemical interface has been established by Chen *et al*. [8]. In this, p-mercaptobenzoic acid (p-MBA) self-assembled monolayers

(SAMs) and 3-aminophenyl boronic acid (m-APBA) molecules have been used. In the next step, in between ferroceneboronic acid (FcBA) and 3-aminophenyl boronic acid (m-APBA) molecules, glucose molecules were immobilized. The oxidation peak current of the FcBA, which occurs by using this method, is directly proportional to the number of glucose molecules by which glucose can be quantified (Fig. 3). The proposed electrode exhibited good reversibility, sensitivity, and selectivity in the presence of the target analyte. To check the analytical applicability of the modified electrode, it has been introduced in human urine samples.

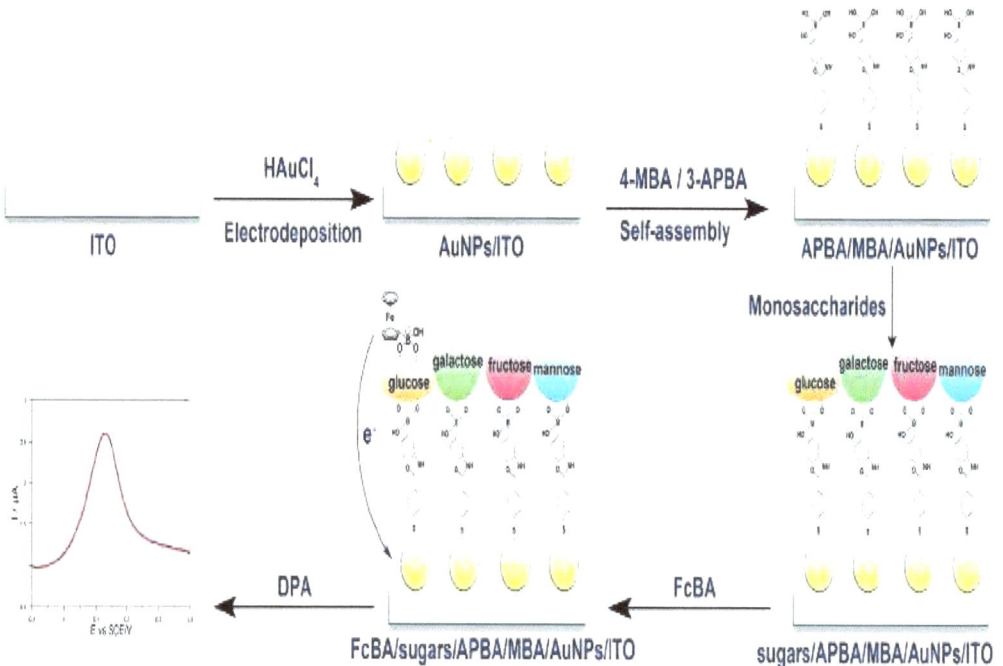

Fig. (3). Preparation and schematic illustration of the sandwich-type sensor for selective detection of glucose [8].

A one-shot disposable sensor such as a screen-printed electrode was used to modify the surface with the aid of reduced graphene oxide decorated with Au NPs towards the detection of glucose in a neutral medium [9]. The modifier was synthesized by the wet chemical method by reducing the gold ions and graphene oxide. Moreover, the presence of AuNPs upon the GO has been confirmed by analyzing the morphology, elemental and spectroscopic characterization studies. Further, AuNPs upon the GO sheet were found to be 50 nm in size. The prepared modified electrode has shown a wide working linear range with good sensitivity.

In a similar way, Ekin *et al.* [10] have developed an electrode using molecularly imprinted polymers (MIPs) decorated with AuNPs towards glucose determination. In this method, an amperometric technique was used to synthesize the AuNP-MIPs. Further, spectroscopic techniques such as atomic force microscopy (AFM) and scanning electron microscopy (SEM) followed by electrochemical tools have been utilized to evaluate the material properties and the electrochemical performance of the sensor. The modified electrode has been checked with its selectivity in the presence of various potential interfering biomolecules such as dopamine, starch, sucrose and bovine serum albumin. The obtained voltammetric signature shows negligible reactivity except for sucrose. The imprinted polymers' behavior in the presence of fluid solvents was tested with AFM. Finally, the sensor was applied to measure glucose from the human serum sample.

A. Mahmoud *et al* [11]. prepared copper-doped zinc oxide nanoparticles (Cu-ZnO) for the electrochemical detection of glucose. The undoped and Cu-ZnO nanoparticles were synthesized *via* the sol-gel process. The synthesized material was characterized, and TEM analysis depicted the spherical shape of the as-prepared samples. The average crystallite size was about 44 nm measured using XRD. It exhibited a large linear detection range from 10^{-9} M to 10^{-5} M and a low detection limit of 10^{-9} M. The Cu doping provided a larger surface area that assisted the interaction between the electrode surface and electrolyte and facilitated the accommodation of glucose towards the active sites. Finally, the electrode was successfully applied to assess the glucose level in human serum samples [11]. Additionally, fluorine-doped tin oxide coated glass was used to modify AuNPs, and poly-aniline blue nanocomposite (AuNP/PAB/FTO) has been explored as an efficient electrochemical interface towards the measurement of glucose [12]. In this method, FTO electrodes were functionalized with PAB *via* the cyclic voltammetric sweeping method. Further, the seed-assisted growth method was utilized to deposit the AuNPs onto the PAB/FTO electrodes. The proposed sensor has shown an improvised analytical response towards the oxidation of glucose with two working linear responses with sensitivities 1.30 and 0.12 $\mu A\ cm^{-2}\mu M^{-1}$, respectively. The developed sensor exhibited the least interference, high reproducibility, long-term stability, and repeatability, followed by good recovery results.

Another electrochemical sensor based on copper (Cu)-nanoflower modified with gold nanoparticles (AuNPs)-graphene oxide (GO) nanofiber (NF) has been established *via* an electrospinning approach for glucose monitoring [13]. In this method, a nanofiber precursor was prepared by mixing the GO with polyvinyl alcohol (PVA), and the results obtained with this composite showed an excellent electrochemical property. The electrode was fabricated by using the dispersed composite solution and coating it onto the surface of the gold chip to form the GO

NFs layer. The schematic illustration is given in Fig. (**4**). In the next step, GO NFs surface was decorated with AuNPs and then combined with organic-inorganic hybrid nanoflower [Cu nanoflower-glucose oxidase (GOx) and horseradish peroxidase (HRP)]. This modified electrode has shown a good analytical response and was found to be selective towards the target analyte. The improved electrocatalytic response is ascribed to the presence of organic-inorganic nanostructured materials on the surfaces of Au chip along with graphene oxide nanofiber.

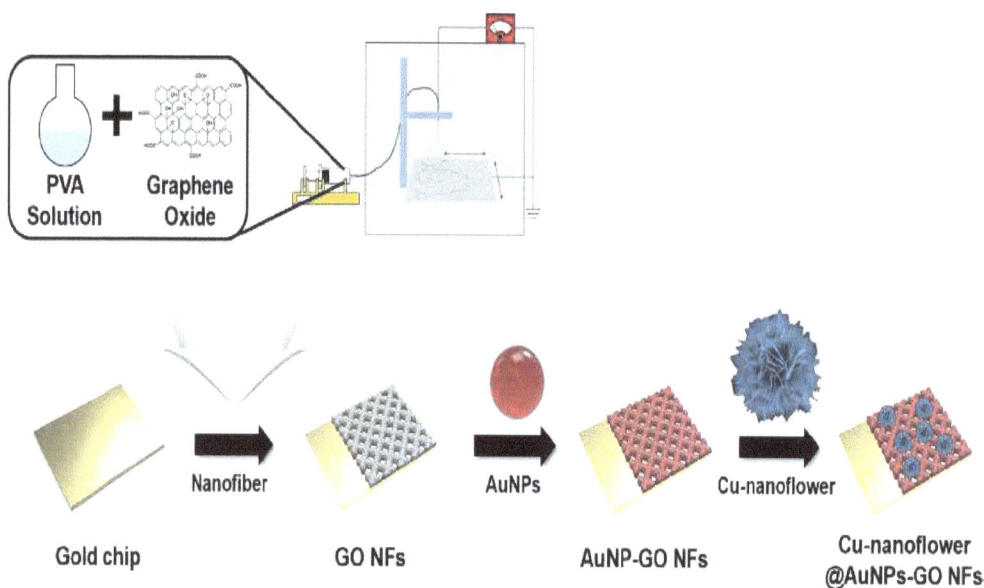

Fig. (4).   Schematic illustration for the fabrication of Cu-nanoflower@AuNPs-GO NFs-based electrochemical glucose nano-biosensor [13].

Fan *et al*. [14] have developed an enzymatic glucose sensor by using ITO electrodes modified with ZnO nanorods and graphene oxide. In this method, ZnO was prepared by hydrothermal method, and AuNPs were electrodeposited on these ZnO nanorods *via* electrostatic interaction. In the proposed method, AuNPs have improved the transfer rate of electrons from redox centers of ZnO nanorods to the GOx. Further, the modified electrode has shown good stability and selectivity and also it was successfully employed in the determination of glucose in human serum samples. The schematic diagram of the redox reaction and electron transfer on this enzymatic electrode is given in (Fig. **5**).

Meng *et al* [15]. have successfully worked on the detection of glucose using palladium Np on a single-walled carbon nanotube (Pd-SWNTs) as a composite. Pd-SWCNTs hybrid nanostructures were developed by depositing palladium Np's of diameter 4-5 nm on the surface of single-walled carbon nanotubes (SWNTs) through chemical reduction of the Pd. The characterization reveals that the Pd-SWNTs electrode possesses a rapid response, a low detection limit (0.2-0.05μM) with acceptable stability and repeatability. Due to the high electrocatalytic activity of the palladium nanostructures and the electrical network formed through their direct binding with the SWNTs, the composite has high sensitivity, which implies that the catalyst has a good electrocatalytic activity towards glucose oxidation in neutral phosphate buffer solution (pH 7.4). Moreover, the well-distributed Pd nanoparticles on the SWNTs surface increase the surface area of the catalyst resulting in the catalyst being easily accessed by substrates, thereby enhancing catalytic activity. Furthermore, unlike most other non-enzymatic sensors developed, which operate under higher pH conditions, the present sensor system demonstrates good performance without surface contamination in neutral PBS even in the presence of excess chloride ions, which makes it an electrode material with great potential for practical use.

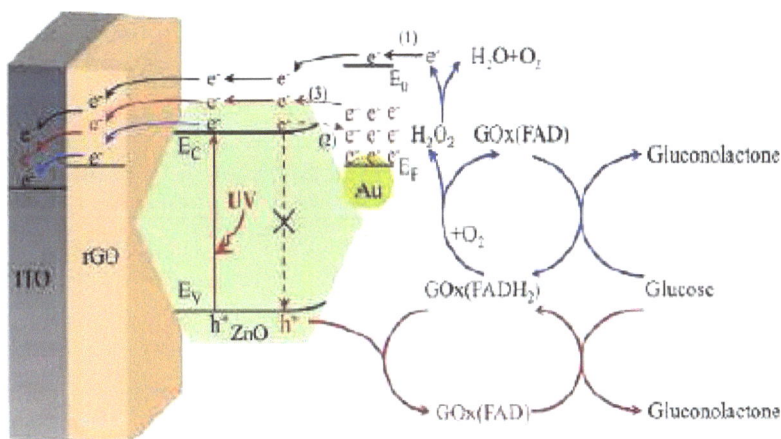

Fig. (5). The schematic diagram of the redox reaction and electron transfer on the enzymatic electrode [14].

Li *et al* . [16] have worked on electrochemical detection of glucose using porous gold. The porous Au film was prepared by the hydrogen bubble template approach followed by a galvanic replacement reaction. The resulting porous Au film has been characterized by XRD, XPS, SEM, EDAX and electrochemical methods. The resulting porous Au film with a large effective surface area, interconnected macropores, and NPs showed excellent electrocatalytic activity, selectivity, sensitivity (11.8 μAcm^2mM^{-1}) and a low detection limit of 5 μM towards the oxidation of glucose. Thus, the porous gold film is promising for the construction of a non-enzymatic glucose biosensor.

Branagan *et al.* [17] have synthesized gold nanoparticles embedded on functionalized multi-walled carbon-nanotubes as a composite for electrochemical detection of glucose at physiological pH. The sensor was fabricated by decorating carbon nanotubes with gold nanoparticles and dispersed and coated onto a glass carbon electrode. The composite was characterized using FEI, energy dispersive X-ray analysis and constant potential amperometry in the detection of glucose. The optimum potential corresponds well with the potential where the reduction of gold oxides occurs, giving an oxide-free electrode surface for the direct oxidation of glucose. A linear calibration curve was obtained giving a sensitivity of 2.77 ± 0.14 μA mm^{-1} with a detection limit of 4.1 μA and a linear region extending beyond 20.0 mM. Good stability and selectivity were observed in the presence of ascorbic acid, fructose and galactose. The interference observed in the presence of uric acid was eliminated successfully by coating the composites with a Nafion film [17]. While this added conductive film reduced the sensitivity, a linear range extending from 0.1 mM-25 mM was observed. This permiselective barrier is quite essential when the composite electrode is operated at physiological pH.

Scandurra *et al* [18]. have worked on multianalyte detection using nanostructured gold onto graphene paper. The electrodes have been synthesized by furnace and laser dewetting of 8 nm thick gold layers onto the graphene paper. These electrodes have been characterized by Micro Raman spectroscopy, SEM, Rutherford backscattering spectroscopy, XRD and cyclic voltammetry. The experimental results showed that both the thermal and laser deserting processes have exhibited a variation in the size and shape of the resulting nanostructures. The particles obtained from the laser dewetting process are of the size ranging from 10-150 nm and are almost spherical, while the nanostructures originated by furnace dewetting are of the size ranging from 200-400 nm and present a characteristic faceting. Both types of electrodes have shown a very low detection limit of 2.5 μm, and wide working linear range of 15 μM to 10^{-8} mM, and high sensitivity of 1240 μA mM^{-1} cm^{-2}. The glucose was detected at a potential of 0.17 V (laser dewetting) or 0.19 V (thermal dewetting) *vs.* SCE. At the potential of 0.4 V *vs.* SCE, fructose was detected. The obtained result revealed the high electrochemical sensitivity of the proposed sensors toward the detection of glucose and fructose in comparison to other glucose sensors.

A simple work on glucose detection using mesoporous platinum has been reported by Park *et al.* [19]. Here, the platinum was deposited on a polished Pt-rod electrode at constant potential (-0.06 V *vs.* Ag/AgCl). The resulting mesoporous Pt electrode was characterized by XRD, cyclic voltammetry and electrochemical analyzer. As synthesized mesoporous Pt electrode surface offers a number of attractive features that have not been reported previously. First, it provides non-enzymatic selectivity over representative interfering species. The mesoporous surface retains sufficient sensitivity in the presence of chloride ions, even at a PBS level of 0.15 M. The Pt-film showed high surface areas and roughness factors. The measured roughness of the mesoporous Pt surface was repeatedly 20 ± 5nm over a 1 mm^2 area of a 300 nm thick film. Highly enhanced electrode areas in nanoscale terms boost the faradaic currents of the sluggish reaction exclusively.

In view of increasing synergistic effects, the use of bimetallic nanocomposites has been described by Mishra *et al.* [20]. In this gold (Au) nanoparticles (NPs) modified copper oxide (CuO) nanowires (NWs) electrode on the copper coil has been used for the detection of glucose. Gold NPs (AuNPs) have been deposited on CuO NWs electrode using *in situ* chemical reactions. The fabricated electrode has been characterized by XRD, SEM, high-resolution transmission electron microscopy (HRTEM), scanning transmission electron microscopy (STEM) and elemental mapping. As a result, the electrode showed a broad linearity range from 0.5 mM to 5.9 mM, a sensitivity of 4398.8 $mAmM^{1}cm^2$, a lower detection limit of 0.5 mM and a very fast response time of 5s. The AuNPs on the surface of CuO NWs increase the effective surface to volume ratio of the electrode, which in turn increases the catalytic capability of the AuNPs modified CuO NWs significantly. Hence, the proposed glucose sensor is used for measuring the amount of glucose in human blood in a practical setup. The combination of higher sensitivity and low detection limit along with the extremely low fabrication cost of this sensor makes it suitable for non-invasive detection of glucose through saliva and urine [20].

Transition metals and noble metals are advantageous materials for the construction of non-enzymatic glucose sensors; however, their applications are limited to a large extent due to their high cost. Hence, metal oxides of these metals offer a cost-effective alternative. They are mostly used in combination with conductive support materials to improve their electronic conductivity and form hierarchical nanostructures with large surface areas and excess active sites. These material properties help in attaining lower detection limits with high sensitivity. In this view, Maghsoudi *et al.* [21] have used cobalt oxide (Co_3O_4) nanoparticles and reduced graphene oxide nanosheets for the electrochemical detection of glucose. They have designed a non-enzymatic, rapid, accurate and sensitive biosensor based on its electrocatalytic reaction on rGO/Co_3O_4 nanoparticlesmodified glassy carbon electrode. As designed, nanocomposite has been characterized by XRD, FTIR, UV-Vis spectroscopy, field emission scanning electron microscopy (FE-SEM) with elemental analysis and dynamic light scattering (DLS). The sensor showed an excellent sensitivity towards glucose with the limit of detection (LOD) as lower as 3.6 nm in low response time 55s. rGO/Co_3O_4 modified electrode provides more active sites on electrode surface using the porous and 1D structure of rGO/Co_3O_4 nanocomposite, which improves the catalytic activity and also increased mobility of electrolyte through the rGO porous structure makes rGO/Co_3O_4 nanocomposite a good candidate for the construction of non-enzymatic glucose sensor in the routine analysis [21]. It can be used for the detection of glucose in real samples, *i.e.*, blood serum.

Carbon Nanocomposites in Glucose Sensing

Carbon and its analogues have gained much interest for biosensors owing to its high thermal conductivity, superior electron mobility, large specific surface area, good biocompatibility, and excellent mechanical flexibility result in the formation of excellent conductive foam with high specific surface areas, robust mechanical strength, and fast electron transport kinetics to replace conventional electrodes. Carbon material-based nanocomposites have been established as effective electrocatalysts in various sensing applications due to their excellent conducting and electrochemical properties. Nanocomposite materials consisting of graphene oxide (GO), reduced graphene oxide (RGO), carbon nanotubes (CNTs), and conducting polymers have shown enhanced stability and improved electrochemical performances. With respect to this, Urban *et al.* [22] have worked on non-enzymatic glucose detection using Ni nanoparticles loaded on electrochemically reduced graphene oxide as a composite electrode. Graphene sheets were coated by the uniform decorated spherical Ni-nanoparticles and have been characterized by X-ray photoelectron spectroscopy, XRD, field-emission scanning electron microscopy and transmission electron microscopy. The nanocomposite (NiNps/ERGO) was fabricated on an Indium-tin oxide electrode using a one-pot electrochemical approach. The morphological and structural analysis results demonstrated that the NiNps were uniformly dispersed on the ERGO sheets. Because of the synergistic effects of NiNps and ERGO, the NiNps/ERGO-ITO electrode exhibited excellent electrocatalytic activity, such as low oxidation potential, wide linear range, short response time, low detection limit, and high sensitivity towards glucose detection. In the NiNps/ERGO-ITO electrode, ERGO nanosheets serve mainly as highly conductive support providing a large surface area for the deposition of NiNps. The excellent interfacial contact and increased bonding interaction between NiNps and ERGO can significantly promote fast ion and electron transportation, resulting in a high electrical conductivity of the electrode. Since the electrical conductivity of most metal nanoparticles is poor, using the graphene-metal nanocomposites is a promising method to achieve easy accessibility to the composite interface by the electrolyte and improved electrochemical utilization of Ni. The results showed that the NiNps/ERGO nanocomposite has potential application in the electrochemical oxidation of biomolecules as well as in non-enzymatic glucose [22] quantification studies.

In a similar way, Parashuram *et al.* [23] have developed a controlled co-precipitation method under dilute conditions for the non-enzymatic sensor. In this method, the authors used a ZrO_2-Cu(I) based carbon paste electrode towards the electrochemical detection of glucose in raw citrus *Aurantium varsinensis*. The synthesized electrode has been characterized by different spectroscopic and electrochemical analysis. The modified electrode showed a very low detection limit, high sensitivity, wide working linear range and excellent anti-interference. The presence of spherical morphology of the active material, alkaline environment and copper (+1) have attributed to the electro-catalytic oxidation of glucose on the carbon paste platform. From the TEM image, it is clearly seen that the copper is embedded in the zirconia nanostructures indicating the core-shell model. Zirconia nanomaterials not only provide stability but also improve conduction due to the presence of oxygen vacancy defects. These defects significantly influence the formation and stabilization of +1 oxidation of Cu. This particular construction would result in a stable, reproducible and high-performance electrochemical sensor [23]. Also, this sensor was used to detect the glucose in real orange juice. All these features showed that Gr/Zr-Cu electrode is attractive and promising for the enzyme-less electrochemical detection of glucose.

Among the various carbon materials recommended, carbon nanotubes (CNTs) have wide applications due to their inherent material chemistry. They have a larger specific surface area than other carbon structures, are more stable, have superior conductivity, are thermally and chemically stable, and have a particularly good adhesion property. In this regard, CuO *et al.* [24] have used platinum-lead alloy NPs/carbon nanotube nanocomposites for the measurement of glucose. PtPbNP/MWCNT was prepared by the electrodeposition of Pt-PbNPs to multi-walled carbon nanotubes (MWCNTs). The prepared nanocomposite has been characterized by SEM and cyclic voltammetry. This electrode possessed high electrocatalytic activity towards glucose oxidation in both alkaline and neutral media. Further, the sensitive electrochemical amperometric response was shown by the modified electrode in neutral phosphate buffer solution. Submonolayer electrodeposition of PtPbNPs on the Nafion coated PtPbNP/MWCNT nanocomposite exhibited a low detection limit with high selectivity for glucose at -0.15 V. In neutral solution, the electrode exhibited linearity of up to 11 mM of glucose with a sensitivity of $17.8 Acm^{-2}mM^{-1}$ and detection limit of 1.8 μM (S/N=3) at 0.30 V. The Nafion coating lowered the detection limit by reducing the background noise, while the second layer of PtPbNPs reduced the sensitivity to the level before Nafion coating.

Polymer Nanocomposites for Glucose Detection

Conductive Polymers (CPs) are the electroactive polymeric materials used in the preparation of a variety of nanocomposites in glucose sensing. CPs possess high redox activity and electron affinity. These materials have electronic properties which complement the metals. They support selectivity and specificity through hydrophilic and hydrophobic affinities, ion-exchange abilities, and electrostatic interaction. Generally, CPs are prepared by the oxidation of monomer using oxidizing agents or applying constant potential or current *via* the electropolymerization method. Using carbon materials alone has a significant drawback, especially when used in sensor devices, as they lack selectivity for target molecules. The incorporation of polymers favors the formation of interconnecting conductive bridges between metal, carbon substrate and electrified interface. Thus, designing and constructing functionalized carbon composites are, therefore, crucial for use in precise applications. Hence, selectivity is an important parameter to be used in electrochemical studies, and preparation of composites with CPs could be done to improve the performance of the sensor. The various strategies used in the preparation of polymer nanocomposites have been shown in (Fig. **6**).

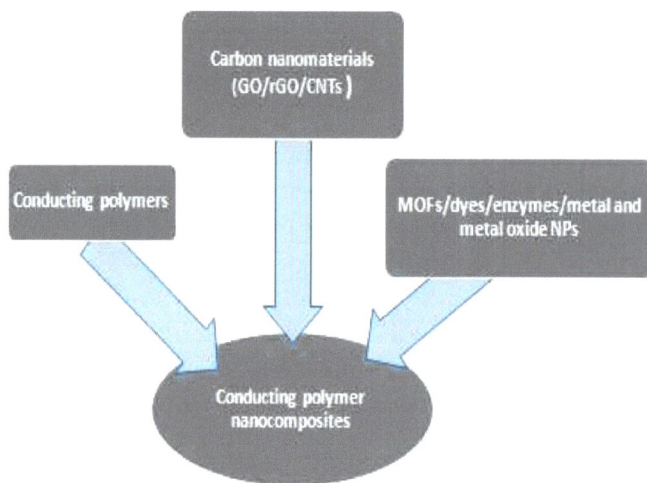

Fig. (6). Schematic representation of the preparation of conducting polymer nanocomposites.

Wang *et al* . [25] have worked on the detection of glucose by using a gold electrode modified with thin-film PET (PGE). PGE was synthesized by using ultraviolet mediated chemical plating technique, which was simple, and cheap and minimum equipment was required. As synthesized electrode exhibited excellent sensitivity (22.05 μA mM^{-1}cm^{-2}), linear range (0.02 to 1.11 mM) with a low detection limit (2.7 μm S/N=3), sensitivity, long term stability, reproducibility, thus PGE-glucose sensor could be employed in the measurement of glucose.

Tianjiaoliu *et al* . [26] have fabricated a hollow CuO/polyaniline (PANI) hybrid nanofibres for non-enzymatic electrochemical detection of glucose and hydrogen peroxide. The hollow CuO/PANI nano-hybrid fibres have been prepared to promote the detection performance for H_2O_2 and glucose sensing. Further, PANI facilitates the large surface area and many reactive sites followed by rich in imine and amine groups, and also it provides strong adsorption ability. The as-prepared nanofibers have been characterized by XRD, SEM and FTIR. As a result, the as-prepared electrode exhibited a wide dynamic linear range, long-term stability, high selectivity, and low detection limit. This confirms that this electrode is suitable for the monitoring of glucose in pharmaceuticals, biomedicine, and food security analysis.

Ma *et al* . have worked on the sensitive detection of glucose using two-dimensional molybdenum disulphide decorated with gold nanoparticles polypyrrole upon glassy carbon electrode (GCE). As the prepared electrode has been characterized by XRD, SEM, FTIR and cyclic voltammetry, the modified electrode was further utilized for the examination of glucose *via* differential pulse voltammetry technique in sodium hydroxide solution. As a result, the fabricated electrode improved the electrode kinetics and enhanced the electrochemical behavior towards the glucose measurement. Cu(II)/Cu(III) redox couple was applied as a catalytic center, which oxidized glucose into glucanolactone at an oxidation potential of +0.45 V. The modified electrode has shown a very low detection limit of 0.08 nM, quantification limit of 0.26 nM, long-term stability, good reproducibility, high selectivity and good recovery in human serum samples towards glucose detection [27].

The analytical parameters of the various modified electrode systems during electrochemical sensing studies are summarized in Table **1**.

Table 1. Comparison of various modified electrodes with existing sensors.

Electrode	Technique	Linear Range	LOD (µM)	Real Sample	Ref
ZnO-Pt-g-C$_3$N$_4$	Chronoamperometry	0.1 to 0.5 µM	0.1	Human blood, urine, and serum	[7]
FcBA/glucose/3APBA/ 4MBA/AuNPs/ITO	DPV / EIS / CV	0.5–30 mM	43	Human urine	[8]
AuNP-MIPs	SWV	1.25 nM–2.56 µM	00125	Human serum	[10]
AuNP/PAB/FTO	CV	2-50 µM and 50–250 µM	0.40	Human serum	[12]
Cu-nanoflower@AuNPs-GO NFs	CV	0.001–0.1 mM	0.018	NA	[13]

(Table 1) cont.....

Electrode	Technique	Linear Range	LOD (µM)	Real Sample	Ref
Nafion/GOx/ZnO/rGO/ ITO	CV / Amperometry	0–6.5 mM	1	Human serum	[14]
Pd-SWNTs	CV / Amperometry	0.5-17 mM	0.2	Blood sample	[15]
rGO/Co₃O₄/ Nafion	CV	25 nM – 2 µM	0.0036	Human serum and urine	[21]
NiNPs/ERGO	Amperometry	0.5–244 µM	0.004	Human serum	[22]

Abbreviations: ZnO-Pt-g-C_3N_4: Zinc oxide-platinum-graphitic carbon nitride, FcBA/glucose/3APBA/ 4MBA/AuNPs/ITO: Ferrocene boronic acid/4-mercapto benzoic acid/gold nanoparticles/indium tin oxide, AuNP-MIPs: Gold nanoparticle-molecularly imprinted polymers, PAB/FTO: Polyaniline blue/Fluorine doped tin oxide, GO: Graphene oxide, NFs: nanofibers, Pd-SWCNTs: Palladium-single walled carbon nanotubes, rGO: reduced graphene oxide, ITO: Indium tin oxide, Co_3O_4: Cobalt oxide, NiNPs/ERGO: Nickel nanoparticles/electrochemically reduced graphene oxide

MOF Based Composites

Metal-organic frameworks are the class of organic-inorganic hybrid materials possessing high stability, porosity, large surface area and chemical inertness. These versatile materials are generally prepared by linking metal atoms with organic linker molecules to form a stable framework structure [28]. Owing to their brilliant properties like high porosity, adjustable structure, and large surface area, they have been widely utilized in the field of catalysis, energy storage and conversions, electrochemical sensing and gas adsorption-desorption studies. The potential of MOF composites has been explored as an ideal host material for diversified applications. Most of the MOFs are poor conductors and unstable in aqueous media. This aspect restricts their use in electrochemical sensing platforms. Hence to overcome these constraints, MOF materials are combined with carbon-based materials such as graphene and its derivatives, carbon nanotubes, carbon dots, *etc* [29]. These carbon materials are the promising substrates to anchor MOF structure owing to their outstanding conductivity, matrix-forming ability, high stability, and large surface area. The electrochemical performance can be further enhanced by forming these carbon composites in comparison to their pristine forms. These composites promote the development of advanced functional materials along with broadening the scope of MOF materials in electrochemical applications [30].

In this regard, Liu *et al.* [31] developed an electrocatalytic glucose sensor by combining MOFs and metal nanoparticles. In this method, the silver nanoparticles were wrapped with Co(II) based three-dimensional porous MOF *via* deposition-reduction method. The schematic representation of the modified electrode and its analytical response is displayed in Fig. (**7**). The modified electrode exhibited much improved electrocatalytic performance. As a result, the fabricated sensor showed a very low detection limit of 1.32 μM and good sensitivity followed by selectivity. Furthermore, the authors have studied the magnetic property, which has great importance for revealing the structure-property relationship based on specific Co(II) cluster MOFs.

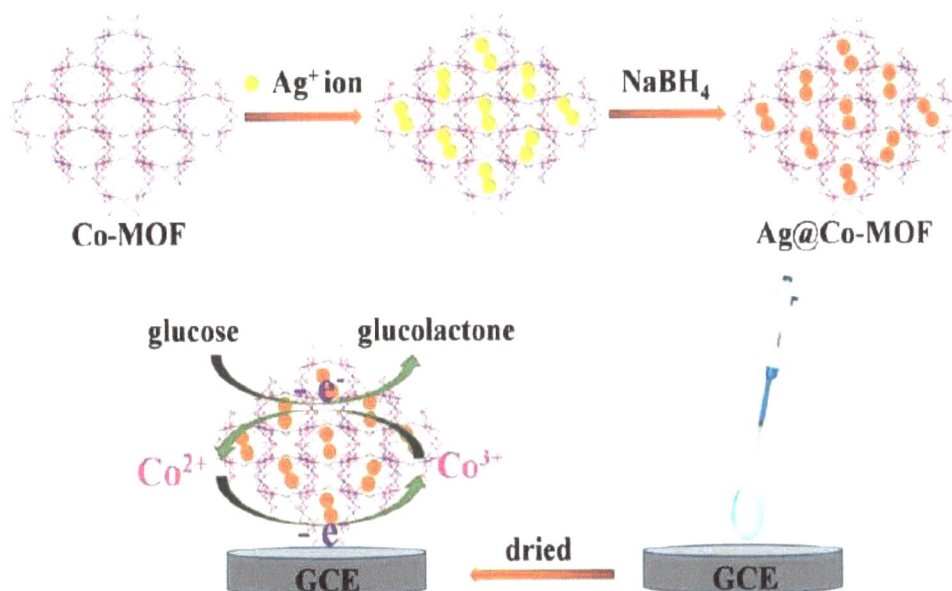

Fig. (7). Schematic Diagram for Preparing the Modified GCE and Application in Electrocatalyst for Glucose Oxidization [31].

The graphene film decorated with nanoparticles would display a superior electrochemical response towards the target analyte with improved electrode kinetics. In this direction, Zhang *et al* [32]. prepared the Ni_2P nanoparticles and which were grown upon graphene film for the measurement of glucose in an alkaline medium. $Ni_2P/G/GCE$ shows excellent electrocatalytic performance for the electro-oxidation of glucose. This enhanced anodic response is due to the electrocatalytic activity of Ni-OOH. The $Ni_2P/G/GCE$ showed excellent sensing performances in a wide detection range of 5 μM to 1.4 mM with a detection limit of 0.44 μM. Furthermore, it also showed good specificity for glucose detection in human serum samples with an excellent linear analytical response.

Lola *et al* . [33] have successfully utilized the metal azolate framework (MAF---Co) modified glassy carbon electrode (GCE) for the measurement of glucose in human serum samples. The synthesized electrode has been characterized by X-Ray Diffraction, FTIR, FE-TEM and High- angle annular dark-field scanning (HAADF). As a result, the prepared composite has shown good electrochemical stability and chemical stability in an alkaline medium. The modified electrode has shown high sensitivity towards glucose, where the formation of MAF- 4- CoII/MAF-4- CoIII redox pair occurs at the interface of the electrode, which enables direct electro-oxidation of the target analyte. The sensor exhibited two linear ranges one from 2-50 µm and another from 100 to 1800 µm with the LOD of 0.6 µm (S/N=3). The developed electrode exhibited good reproducibility and recovery, followed by long-term signal stability. Thus, MAF-4-CoII could be used as a promising candidate for the detection of glucose.

The non-oxidized metal nanoparticle embedded porous carbon composites prepared from MOFs as glucose sensor has been demonstrated by Song *et al.* [34]. They prepared copper nanoparticles embedded porous carbon composites by the simple pyrolysis of a copper-containing MOF (HKUST-1) at various temperatures under an N_2 atmosphere. The as-prepared composite has shown a large surface area, which was prepared at 500 °C. Cu@C-500 was utilized for glucose monitoring *via* colorimetric assay, and it mimicked peroxidase-like catalytic activity. Interestingly, the material exhibits a low detection limit of 3.2 nM in colorimetric glucose sensing. Further, the modified electrode showed high selectivity towards glucose even in the presence of other molecules such as fructose, galactose, mannose, and sucrose. The presence of pure metal nanoparticles and ordered porosity of the material is attributed to its application for glucose sensing.

Peng *et al.* [35] have reported a hydrothermal treatment and agitation work on AgNPs/MOF-74(Ni). The AgNPs/MOF-74(Ni) based composite sensing system for the detection of glucose. The AgNPs/MOF-74(Ni) was characterized by SEM, TEM, XRD, Brunner-Emmet-Teller (BET), Barette-Joyner- Halenda (BJH) and XPS. The electrochemical behavior of the modified electrode has been examined with electrochemical techniques such as cyclic voltammetry and current-time curve with three-electrode systems. The electrochemical sensor based on AgNPs/MOF-74(Ni) exhibited high sensitivity (1.29 mA $mM^{-1}cm^{-2}$) for the determination of glucose with a low detection limit (4.7 µm) and wide linear range (0.01~4mM). In addition, the AgNPs/MOF-74(Ni) sensing system has shown good stability and reproducibility. The good performance of the synthesized electrode is attributed to the high conductivity of Ag nanoparticles and MOF-74(Ni). Therefore, the AgNPs/MOF-74(Ni) composite material is found to have a high potential for the determination of glucose in biomedical applications.

Due to excellent electrochemical features, layered double hydroxides (LDHs) have gained much attention within the electrochemical community. These materials can conduct electricity due to electron hopping between metal centers that are close together and ionic displacement in the material. But still, their conductivity is not highly appreciated. As a result, LDH and their carbon derivatives have gained much attention in recent years to develop electrochemical sensors. In view of this, Gualandi *et al.* developed an electrosynthesis method to deposit Ni/Al LDH on pretreated glassy carbon electrodes. The graphical representation of preparation and fabrication of LDH modified electrodes is shown in (Fig. **8**). These composite materials exhibited a large number of active sites, good chemical and thermal stability, mechanical strength and high electrical conductivity. Cyclic voltammetry technique showed the presence of graphene ensured a large electrochemically active area, which is higher than observed for an LDH deposited alone on a bare glassy carbon. Moreover, carbon nanomaterials have minimized the charge transfer resistance and enhanced the electrical conductivity of the prepared composite, which was examined by the electrochemical impedance spectroscopic technique [36]. Lu *et al.* [37] explored the MOF@MOF composite combination of UiO-67 and Ni-MOF as promising materials in electrochemical sensing of glucose. A core-shell UiO-67@Ni-MOF material that acts as an electrode material to achieve glucose detection was synthesized for the first time. The composites were synthesized through an internal extended growth method under polyvinyl pyrrolidone (PVP) regulation. The pre-prepared UiO-67 served as the core for the growth of shell Ni-MOF. The ability of UiO-67@Ni-MOF to catalyze glucose was elucidated by cyclic voltammetry (CV) through an amperometric study in alkaline media. A good linear relationship between oxidation peak current and glucose concentration has been demonstrated. In the corresponding calibration plot, UiO-67@Ni-MOF/GCE shows two linear response ranges at glucose concentrations of 5 µM-550 µM and 550 µM to 3.9 mM. The limit of detection (LOD) was 0.98 µM. Finally, the practical applicability of the modified electrode was assessed within human serum samples.

Wen *et al.* [38] reported a novel CoII-MOF $\{[Co_2(Dcpp)(Bpe)0.5 (H_2O)(\mu2\text{-}H_2O)]\cdot(Bpe)0.5\}n$ (H_4Dcpp=4,5-bis(4-carboxylphenyl)-phthalic acid, Bpe = 1,2-bis(4-pyridyl)ethane) synthesized *via* a hydrothermal reaction and utilized for the measurement of glucose monitoring. Then, by combining the advantages of the CoII-MOF and acetylene black (Acb), a new composite of CoII-MOF/Acb was constructed. The CoII-MOF/Acb modified electrode has shown a superior response towards the target analyte. The improved analytical response could be due to the synergistic effect of the Acb and CoII-MOF. Furthermore, Acb promotes the charge transfer and improves the electrode kinetics between CoII-MOF and the GCE, and also, the presence of uniform and active metal sites can be found upon the CoII-MOF. The proposed sensor has shown a wide working linear range from 60 to 700 µM with a sensitivity of $0.212 \mu A \, \mu M^{-1} \, cm^{-2}$.

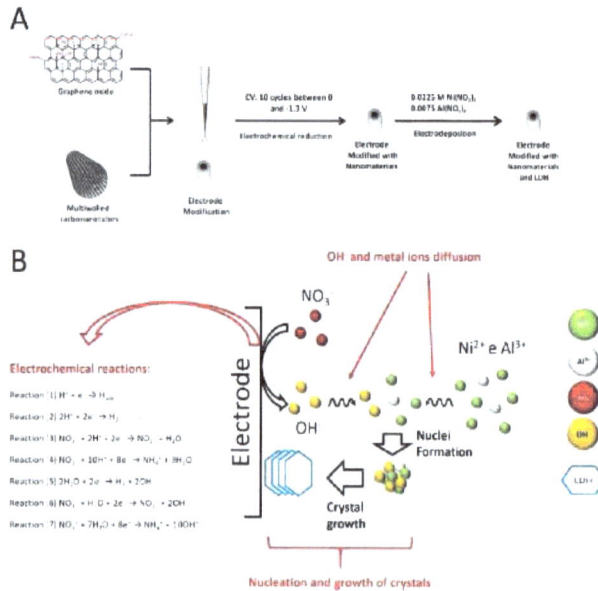

Fig. (8). (A) Schematic representation of the electrode fabrication (B) Electrochemical reactions [36].

Xiao *et al.* [39] developed a facile method to make metal oxide@carbon composites (M/MO@C) resulting from a bimetallic Cu/Ni-based metal-organic framework. Due to the synergistic effect, the composite has shown an improved electrochemical response towards glucose. The current exhibited a linear dependence on the glucose concentration from 0.1 µM to 2.2 mM. The detection limit for glucose was 0.06 µM. The practical applicability of the modified electrode was investigated by the standard addition technique. The measurement of glucose concentration from honey samples was done by using the I-t curves technique.

An electrochemical glucose sensor has been developed by Haihan *et al.* by using ultrathin Ni/Co bimetallic metal-organic-frameworks nanosheets with various metal ratios. The proposed composite has shown low overpotential towards the oxidation of sensing analyte and subsequently improved electrode kinetics. As a result, the developed sensor exhibited high sensitivity of 2086.7 µA mM cm^{-2} with a wide linear range of 0.1 µM - 1.4 mM and a very low detection limit of 0.047 µM [40].

Polyoxometalates (POMs) are the class of organic-inorganic hybrid functional materials and are considered to be the substrates for many composite materials. These POMs act as a pool of electrons and protons and hence finds many electrochemical applications. Ayranci *et al.* [41] have prepared Co-

POM/MWCNT based sensing interface for the electro-catalytic oxidation of glucose. Co-POM/MWCNT sensor platform was developed by using multi-walled carbon nanotube composite and Keggin type $K_7[CoIII(H_2O)W11039].15H_2O$(Co-POM) upon graphite electrode. Further, there would be an increase in the electron transfer rate due to the presence of mixed-valence of Co(III) and Co(II) structures, which is present within the Keggin type (Co-POM) composite. The developed sensor has been characterized by TEM, SEM, Raman spectra, FTIR and AFM. As a result, the proposed sensor displayed a wide dynamic working linear range from 0.1 mM to 10.0 mm (R^2=0.99) with a detection limit of 1.21μM and 6s of quick response time and sensitivity of 256.4 $\mu mM^{-1}cm^{-2}$. The porous nature exhibited by the layers of MWCNT, which is very important within this proposed work. These impressive results indicate that by combining MWCNT and Co-POM, one can construct the electrochemical glucose sensor as a promising tool.

The analytical parameters of the carbon composite and MOF-based modified electrodes are summarized in Table **2**.

Table 2. Comparison of various modified electrodes with existing glucose sensors.

Electrode	Technique	Linear Range	LOD (μM)	Real Sample	Ref
AgNPs/MOF-74(Ni)	CV	0.01-4 mM	4.7	Human blood serum	[35]
Au NPs modified CuO NWs	Amperometry	0.0005–5.9 mM	0.5	Human blood	[20]
ZrO$_2$-Cu(I)	CV	1 mM–10 mM	250	Orange juice	[23]
Co-POM/MWCNT	CV / Amperometry	0.1- 10.0 mM	1.21	Coke and Juice	[41]
Ag@1/GCE	CV / Amperometry	5–550 μM	1.32	NR	[31]
Ni$_2$P/G	Amperometry	5 μM to 1.4 mM	0.44	Human serum	[32]
UiO-67@Ni-MOF/GCE	Amperometry	5 μM-550 μM	0.98	Human serum	[37]
CoII-MOF/Acb-0.2%/GCE	Amperometry	5–1000 μM	1.7	NR	[38]
M/MO@C-800	Amperometry	0.1–2200 μM	0.06	Honey	[39]

Abbreviations: AgNPs/MOF: Silver nanoparticles/metal-organic framework, CuO NWs: Copper oxide nanowires, ZrO$_2$-Cu(I): Zirconium oxide-copper(I), Co-POM/MWCNT: Cobalt-polyoxometallate/multi walled carbon nanotube, GCE: Glassy carbon electrode, Ni$_2$P/G: Nickel phosphorous/graphene, Acb: Acetylene black, M/MO: Metal/metal oxide, NR-not reported.

CONCLUDING REMARKS

As diabetes continues to become ever more prevalent nowadays, the need for improved sensing technology is evident. Although the enzyme-based glucose sensors are highly specific and sensitive, the analytical signal can be altered by various conditions like the presence of oxygen, temperature, pH and activity of the immobilized enzyme, thereby reducing the selectivity. It is important to understand the basic properties and limitations of the enzymes employed in the sensor and the principles behind their catalytic activity to avoid potential dangers in glucose monitoring. More prominently, nanocomposite-based non-enzymatic sensors are free from the above-mentioned limitations. The nanocomposite-based sensors are advantageous as they can operate in neutral pH conditions sensibly well. However, they show lower stability, anti-interference and reproducibility.

All types of glucose sensors have advantages and disadvantages, in turn, demanding intensive research that embark upon the key problems in terms of analytical sensing performance rather than adhering to the material property of newly synthesized materials. Most studies have been strongly biased towards materialistic approaches based on various combinations of substances and structural engineering. The conditions followed for analysis are far from practical applications, which clearly reveal the functional limitations of the proposed systems or materials in the viewpoint of the practical glucose sensors.

For practical use and mass production, however, it requires more effort to look into specificity, selectivity, sensing mechanism and miniaturization. The ideal sensor for glucose remains elusive, but there is a significant advancement towards the development of novel glucose sensors, and the nanocomposites in glucose sensing need exploration.

CONSENT FOR PUBLICATION

Not applicable.

CONFLICT OF INTEREST

The author declares no conflict of interest, financial or otherwise.

ACKNOWLEDGEMENTS

Prashanth S. A. acknowledges Shri. M. V. Patil, Shri. R. R. Sudi, Shri. Dilip N. Devagirikar, Shri. D. S. Kardyal and Dr. P. S. Kandagal SVM Arts, Science and Commerce College, Ilkal, for their support and encouragement.

REFERENCES

[1]   Mellitus, D. Diagnosis and classification of diabetes mellitus. *Diabetes Care,* **2005**, *28*(S37) Suppl. 1, S37-S42.
        [PMID: 15618111]

[2]   Oliver, N.S.; Toumazou, C.; Cass, A.E.; Johnston, D.G. Glucose sensors: a review of current and emerging technology. *Diabet. Med.,* **2009**, *26*(3), 197-210.
        [http://dx.doi.org/10.1111/j.1464-5491.2008.02642.x] [PMID: 19317813]

[3]   Clark, L.C., Jr; Lyons, C. Electrode systems for continuous monitoring in cardiovascular surgery. *Ann. N. Y. Acad. Sci.,* **1962**, *102*(1), 29-45.

[http://dx.doi.org/10.1111/j.1749-6632.1962.tb13623.x] [PMID: 14021529]

[4]   Bruen, D.; Delaney, C.; Florea, L.; Diamond, D. Glucose sensing for diabetes monitoring: recent developments. *Sensors (Basel),* **2017**, *17*(8), 1866.
[http://dx.doi.org/10.3390/s17081866] [PMID: 28805693]

[5]   Fleischer, J. Diabetic autonomic imbalance and glycemic variability. *J. Diabetes Sci. Technol.,* **2012**, *6*(5), 1207-1215.
[http://dx.doi.org/10.1177/193229681200600526] [PMID: 23063048]

[6]   Hwang, D-W.; Lee, S.; Seo, M.; Chung, T.D. Recent advances in electrochemical non-enzymatic glucose sensors - A review. *Anal. Chim. Acta,* **2018**, *1033*, 1-34.
[http://dx.doi.org/10.1016/j.aca.2018.05.051] [PMID: 30172314]

[7]   Imran, H.; Vaishali, K.; Antony Francy, S.; Manikandan, P.N.; Dharuman, V. Platinum and zinc oxide modified carbon nitride electrode as non-enzymatic highly selective and reusable electrochemical diabetic sensor in human blood. *Bioelectrochemistry,* **2021**, *137*, 107645.
[http://dx.doi.org/10.1016/j.bioelechem.2020.107645] [PMID: 32916428]

[8]   Chen, M.; Cao, X.; Chang, K.; Xiang, H.; Wang, R. A novel electrochemical non-enzymatic glucose sensor based on Au nanoparticle-modified indium tin oxide electrode and boronate affinity. *Electrochim. Acta,* **2021**, *368*, 137603.
[http://dx.doi.org/10.1016/j.electacta.2020.137603]

[9]   Chaandini, J.; Suneesh, P.V.; Babu, T.S. Gold nanoparticle decorated reduced graphene oxide for the nonenzymatic electrochemical sensing of glucose in neutral medium. *Mater. Today Proc.,* **2020**, *33*, 2414-2420.
[http://dx.doi.org/10.1016/j.matpr.2020.08.072]

[10]   Sehit, E.; Drzazgowska, J.; Buchenau, D.; Yesildag, C.; Lensen, M.; Altintas, Z. Ultrasensitive nonenzymatic electrochemical glucose sensor based on gold nanoparticles and molecularly imprinted polymers. *Biosens. Bioelectron.,* **2020**, *165*, 112432.
[http://dx.doi.org/10.1016/j.bios.2020.112432] [PMID: 32729546]

[11]   Mahmoud, A.; Echabaane, M.; Omri, K.; El Mir, L.; Chaabane, R.B. Development of an impedimetric non enzymatic sensor based on ZnO and Cu doped ZnO nanoparticles for the detection of glucose. *J. Alloys Compd.,* **2019**, *786*, 960-968.
[http://dx.doi.org/10.1016/j.jallcom.2019.02.060]

[12]   Ansari, S.A.; Ahmed, A.; Ferdousi, F.K.; Salam, M.A.; Shaikh, A.; Barai, H.R.; Lopa, N.S.; Rahman, M.M. Conducting poly (aniline blue)-gold nanoparticles composite modified fluorine-doped tin oxide electrode for sensitive and non-enzymatic electrochemical detection of glucose. *J. Electroanal. Chem. (Lausanne),* **2019**, *850*, 113394.
[http://dx.doi.org/10.1016/j.jelechem.2019.113394]

[13]   Baek, S.H.; Roh, J.; Park, C.Y.; Kim, M.W.; Shi, R.; Kailasa, S.K.; Park, T.J. Cu-nanoflower decorated gold nanoparticles-graphene oxide nanofiber as electrochemical biosensor for glucose detection. *Mater. Sci. Eng. C,* **2020**, *107*, 110273.
[http://dx.doi.org/10.1016/j.msec.2019.110273] [PMID: 31761219]

[14]   Zhou, F.; Jing, W.; Liu, S.; Mao, Q.; Xu, Y.; Han, F.; Wei, Z.; Jiang, Z. Electrodeposition of gold nanoparticles on ZnO nanorods for improved performance of enzymatic glucose sensors. *Mater. Sci. Semicond. Process.,* **2020**, *105*, 104708.
[http://dx.doi.org/10.1016/j.mssp.2019.104708]

[15]   Meng, L.; Jin, J.; Yang, G.; Lu, T.; Zhang, H.; Cai, C. Nonenzymatic electrochemical detection of glucose based on palladium-single-walled carbon nanotube hybrid nanostructures. *Anal. Chem.,* **2009**, *81*(17), 7271-7280.
[http://dx.doi.org/10.1021/ac901005p] [PMID: 19715358]

[16]   Li, Y.; Song, Y-Y.; Yang, C.; Xia, X-H. Hydrogen bubble dynamic template synthesis of porous gold for nonenzymatic electrochemical detection of glucose. *Electrochem. Commun.,* **2007**, *9*(5), 981-988.

[http://dx.doi.org/10.1016/j.elecom.2006.11.035]

[17]   Branagan, D.; Breslin, C.B. Electrochemical detection of glucose at physiological pH using gold nanoparticles deposited on carbon nanotubes. *Sens. Actuators B Chem.*, **2019**, *282*, 490-499.
[http://dx.doi.org/10.1016/j.snb.2018.11.089]

[18]   Scandurra, A.; Ruffino, F.; Sanzaro, S.; Grimaldi, M.G. Laser and thermal dewetting of gold layer onto graphene paper for non-enzymatic electrochemical detection of glucose and fructose. *Sens. Actuators B Chem.*, **2019**, *301*, 127113.
[http://dx.doi.org/10.1016/j.snb.2019.127113]

[19]   Park, S.; Chung, T.D.; Kim, H.C. Nonenzymatic glucose detection using mesoporous platinum. *Anal. Chem.*, **2003**, *75*(13), 3046-3049.
[http://dx.doi.org/10.1021/ac0263465] [PMID: 12964749]

[20]   Mishra, A.K.; Mukherjee, B.; Kumar, A.; Jarwal, D.K.; Ratan, S.; Kumar, C.; Jit, S. Superficial fabrication of gold nanoparticles modified CuO nanowires electrode for non-enzymatic glucose detection. *RSC Advances*, **2019**, *9*(4), 1772-1781.
[http://dx.doi.org/10.1039/C8RA07516F]

[21]   Maghsoudi, S.; Mohammadi, A. Reduced graphene oxide nanosheets decorated with cobalt oxide nanoparticles: A nonenzymatic electrochemical approach for glucose detection. *Synth. Met.*, **2020**, *269*, 116543.
[http://dx.doi.org/10.1016/j.synthmet.2020.116543]

[22]   Urhan, B.K.; Demir, Ü.; Özer, T.Ö.; Doğan, H.Ö. Electrochemical fabrication of Ni nanoparticles-decorated electrochemically reduced graphene oxide composite electrode for non-enzymatic glucose detection. *Thin Solid Films*, **2020**, *693*, 137695.
[http://dx.doi.org/10.1016/j.tsf.2019.137695]

[23]   Parashuram, L.; Sreenivasa, S.; Akshatha, S.; Udayakumar, V.; Sandeep Kumar, S. A non-enzymatic electrochemical sensor based on ZrO_2: Cu(I) nanosphere modified carbon paste electrode for electro-catalytic oxidative detection of glucose in raw Citrus aurantium var. sinensis. *Food Chem.*, **2019**, *300*, 125178.
[http://dx.doi.org/10.1016/j.foodchem.2019.125178] [PMID: 31326677]

[24]   Cui, H-F.; Ye, J-S.; Zhang, W-D.; Li, C-M.; Luong, J.H.; Sheu, F-S. Selective and sensitive electrochemical detection of glucose in neutral solution using platinum-lead alloy nanoparticle/carbon nanotube nanocomposites. *Anal. Chim. Acta*, **2007**, *594*(2), 175-183.
[http://dx.doi.org/10.1016/j.aca.2007.05.047] [PMID: 17586112]

[25]   Wang, Y.; Wang, X.; Lu, W.; Yuan, Q.; Zheng, Y.; Yao, B. A thin film polyethylene terephthalate (PET) electrochemical sensor for detection of glucose in sweat. *Talanta*, **2019**, *198*, 86-92.
[http://dx.doi.org/10.1016/j.talanta.2019.01.104] [PMID: 30876607]

[26]   Liu, T.; Guo, Y.; Zhang, Z.; Miao, Z.; Zhang, X.; Su, Z. Fabrication of hollow CuO/PANI hybrid nanofibers for non-enzymatic electrochemical detection of H2O2 and glucose. *Sens. Actuators B Chem.*, **2019**, *286*, 370-376.
[http://dx.doi.org/10.1016/j.snb.2019.02.006]

[27]   Ma, K.; Sinha, A.; Dang, X.; Zhao, H. Electrochemical preparation of gold nanoparticles-polypyrrole co-decorated 2D MoS2 nanocomposite sensor for sensitive detection of glucose. *J. Electrochem. Soc.*, **2019**, *166*(2), B147-B154.
[http://dx.doi.org/10.1149/2.1231902jes]

[28]   Tang, J.; Yamauchi, Y. Carbon materials: MOF morphologies in control. *Nat. Chem.*, **2016**, *8*(7), 638-639.
[http://dx.doi.org/10.1038/nchem.2548] [PMID: 27325086]

[29]   Ding, M.; Flaig, R.W.; Jiang, H-L.; Yaghi, O.M. Carbon capture and conversion using metal-organic frameworks and MOF-based materials. *Chem. Soc. Rev.*, **2019**, *48*(10), 2783-2828.
[http://dx.doi.org/10.1039/C8CS00829A] [PMID: 31032507]

[30]   Shekhah, O.; Liu, J.; Fischer, R.A.; Wöll, Ch. MOF thin films: existing and future applications. *Chem. Soc. Rev.,* **2011**, *40*(2), 1081-1106.
       [http://dx.doi.org/10.1039/c0cs00147c] [PMID: 21225034]

[31]   Yang, Liu; Wen-Juan, Shi; Yu-Ke, Lu; Ge, Liu; Lei, Hou; Yao-Yu, Wang Nonenzymatic Glucose Sensing and Magnetic Property Based On the Composite Formed by Encapsulating Ag Nanoparticles in Cluster-Based Co-MOF. *Inorg. Chem.,* **2019**, *58*(24), 16743-16751.
       [http://dx.doi.org/10.1021/acs.inorgchem.9b02889] [PMID: 31794201]

[32]   Zhang, Y.; Xu, J.; Xia, J.; Zhang, F.; Wang, Z. MOF-derived porous Ni2P/graphene composites with enhanced electrochemical properties for sensitive nonenzymatic glucose sensing. *ACS Appl. Mater. Interfaces,* **2018**, *10*(45), 39151-39160.
       [http://dx.doi.org/10.1021/acsami.8b11867] [PMID: 30350939]

[33]   Lopa, N.S.; Rahman, M.M.; Ahmed, F.; Ryu, T.; Lei, J.; Choi, I.; Kim, D.H.; Lee, Y.H.; Kim, W. A chemically and electrochemically stable, redox-active and highly sensitive metal azolate framework for non-enzymatic electrochemical detection of glucose. *J. Electroanal. Chem. (Lausanne),* **2019**, *840*, 263-271.
       [http://dx.doi.org/10.1016/j.jelechem.2019.03.081]

[34]   Song, Y.; Cho, D.; Venkateswarlu, S.; Yoon, M. Systematic study on preparation of copper nanoparticle embedded porous carbon by carbonization of metal–organic framework for enzymatic glucose sensor. *RSC Advances,* **2017**, *7*(17), 10592-10600.
       [http://dx.doi.org/10.1039/C7RA00115K]

[35]   Peng, X.; Wan, Y.; Wang, Y.; Liu, T.; Zou, P.; Wang, X.; Zhao, Q.; Ding, F.; Rao, H. Flower□like Ni (II)□based Metal□organic Framework□decorated Ag Nanoparticles: Fabrication, Characterization and Electrochemical Detection of Glucose. *Electroanalysis,* **2019**, *31*(11), 2179-2186.
       [http://dx.doi.org/10.1002/elan.201900259]

[36]   Gualandi, I.; Vlamidis, Y.; Mazzei, L.; Musella, E.; Giorgetti, M.; Christian, M.; Morandi, V.; Scavetta, E.; Tonelli, D. Ni/Al layered double hydroxide and carbon nanomaterial composites for glucose sensing. *ACS Appl. Nano Mater.,* **2018**, *2*(1), 143-155.
       [http://dx.doi.org/10.1021/acsanm.8b01765]

[37]   Lu, M.; Deng, Y.; Li, Y.; Li, T.; Xu, J.; Chen, S-W.; Wang, J.; Core-shell, M.O.F. Core-shell MOF@MOF composites for sensitive nonenzymatic glucose sensing in human serum. *Anal. Chim. Acta,* **2020**, *1110*, 35-43.
       [http://dx.doi.org/10.1016/j.aca.2020.02.023] [PMID: 32278398]

[38]   Wen, Y.; Meng, W.; Li, C.; Dai, L.; He, Z.; Wang, L.; Li, M.; Zhu, J. Enhanced glucose sensing based on a novel composite Co^{II}-MOF/Acb modified electrode. *Dalton Trans.,* **2018**, *47*(11), 3872-3879.
       [http://dx.doi.org/10.1039/C8DT00296G] [PMID: 29451291]

[39]   Xiao, X.; Peng, S.; Wang, C.; Cheng, D.; Li, N.; Dong, Y.; Li, Q.; Wei, D.; Liu, P.; Xie, Z.; Qu, D.; Li, X. Metal/metal oxide@ carbon composites derived from bimetallic Cu/Ni-based MOF and their electrocatalytic performance for glucose sensing. *J. Electroanal. Chem. (Lausanne),* **2019**, *841*, 94-100.
       [http://dx.doi.org/10.1016/j.jelechem.2019.04.038]

[40]   Zou, H.; Tian, D.; Lv, C.; Wu, S.; Lu, G.; Guo, Y.; Liu, Y.; Yu, Y.; Ding, K. The synergistic effect of Co/Ni in ultrathin metal-organic framework nanosheets for the prominent optimization of non-enzymatic electrochemical glucose detection. *J. Mater. Chem. B Mater. Biol. Med.,* **2020**, *8*(5), 1008-1016.
       [http://dx.doi.org/10.1039/C9TB02382H] [PMID: 31930260]

[41]   Ayranci, R.; Torlak, Y.; Ak, M. Non-Enzymatic Electrochemical detection of glucose by mixed-valence cobalt containing keggin polyoxometalate/multi-walled carbon nanotube composite. *J. Electrochem. Soc.,* **2019**, *166*(4), B205-B211.
       [http://dx.doi.org/10.1149/2.0581904jes]

CHAPTER 12

Tailored Nanocomposites for Hydrazine Electrochemical Sensors

A. Rebekah[1], **G. Srividhya**[1] and **N. Ponpandian**[1,*]

[1] *Department of Nanoscience and Technology, Bharathiar University, Coimbatore 641046, India*

Abstract: In the field of nanotechnology, nanocomposites have gained increasing and significant attention owing to their unique physico-chemical characteristics. This outstanding characteristic makes them a suitable candidate in various fields of application, such as electronics, sensors, biotechnology and catalysis. The development of nanocomposites has proven to be a basis for the development of accurate electrochemical sensors with low limit of detection, high sensitivity and selectivity. The high performance electrochemical sensors have found their way in various application fields, such as biomedical, analysis of food products and other environmental pollutants present in the atmosphere. In this chapter, we present a survey of the application of various tailored nanocomposites as sensing platforms for hydrazine. Particularly, electrochemical sensors based on carbon-based nanomaterials, metallic nanomaterials, and related nanocomposites are given special attention.

Keywords: Composites, Hydrazine, Hybrid materials, Sensors, Tailored nanocomposites.

INTRODUCTION

The improved technological development in the past few decades has led to a colossal growth of industries.

This was combined with the economical growth of urban and semi-urban populations that has caused an upsurge in the productivity and consumption of industrial products in a huge scale. Most industrial products like automobiles, hardware, paints, domestic appliances, clothing and cosmetics are produced at the expense of many inorganic and organic chemical, some of which are highly dangerous for human health and environmental sustainability. Hence, screening of industrial surroundings, effluent and wastes are highly important to protect our environment by preventing hazardous chemicals from contaminating the biosys-

---

* **Corresponding author N. Ponpandian:** Department of Nanoscience and Technology, Bharathiar University, Coimbatore 641 046, India; E-mail: ponpandian@buc.edu.in

Manorama Singh, Vijai K Rai and Ankita Rai (Eds.)
All rights reserved-© 2022 Bentham Science Publishers

tems. Owing to the aforementioned reason, the development of gas and vapour sensors has received a massive research interest from the past two decades. A sensor is a device that can detect a physical component and produce a user understandable output signal corresponding to the quantity of the target component. There are different types of sensors that detect different physical entities by using different analytical methods. For example, based on the sensor target, they are classified into gas sensors, vapour sensors, temperature sensors, stress sensors, biosensors and so on. Similarly, based on the principle of operation, they are classified into chemi-resistive sensors, electrochemical sensors, optical sensors, piezoelectric sensors, chemi-luminescence sensors and so on. Among them, gas and vapour sensors functioning on the basis of chemi-resistive principle, optical and electrochemical techniques are widely developed to protect the working environment from various types of toxic and flammable gases like carbon monoxide, carbon di-oxide, methane and vapours of volatile organic compounds like ethanol, methanol, ammonia, chloroform, di- and trimethylamine, *etc*.

Hydrazine (N_2H_4) is an organic compound with two amino groups and is also dubbed as diamidogen. At normal conditions, it is a colourless liquid with a faint ammonia smell. It has strong basic nature and is a powerful reducing agent and so often used as an anti-oxidant, oxygen scavenger and corrosion inhibitor. It has high enthalpy of combustion and hence is used as a propellant in space vehicles. Apart from the aforementioned applications, hydrazine is frequently used as a chemical reactant in different industries like textile dying, pharmaceuticals, pesticides and fertilizers. Despite its numerous applications, hydrazine is highly toxic and inhaling hydrazine may cause a range of health problems. The health risk of exposure to hydrazine fumes, include mild symptoms like irritation in eyes, nose, and throat, dizziness, headache, nausea and also severe health deterioration like the damage of internal organs like liver, kidney and the central nervous system. Moreover, hydrazine easily dissolves in water and there is a high risk of drinking water contamination with hydrazine in industrial surroundings. The US Environmental Protection Agency identified hydrazine as a potential carcinogen and has defined a threshold value below 10 ppb in drinking water to prevent health complications.

Hence, it is a matter of urgency and necessary to design an efficient sensor that can detect even the small traces of hydrazine in drinking water and other consumable industrial products. Considering this, many researchers have aimed to fabricate sensors that can effectively detect hydrazine. Techniques like chromatography, chemi-luminescence, fluorescence and voltammetry have been employed to fabricate hydrazine sensors. Among them, the electrochemical sensors are highly desirable due to low cost, ease of fabrication, lower limit of detection (LOD) and prospective commercialization. Nanocomposites possess the advantage and synergistic effect of the constituent nanomaterials and hence, nanocomposites based devices show better performance than those based on simple devices. Tailoring of composites for designing the hybrid materials of carbon derivates and metal oxides is very important in the view of electrochemical based sensors. In this chapter, electrochemical sensors based on tailored nanocomposites for hydrazine detection are discussed.

COPPER BASED NANOCOMPOSITES FOR HYDRAZINE SENSOR

Zhao *et al.*, synthesized cupric oxide functionalized with carbon nanotube and reduced graphene oxide (CuO/CNTs-rGO) nanocomposites for sensitive hydrazine sensors. The nanocomposite exhibited a sensitivity of 4.28 $\mu A \cdot \mu M^{-1} \cdot cm^{-2}$, a linear range of 1.2–430 μM, and detection limit of 0.2 μM towards oxidation of hydrazine. The efficiency of the sensor was attributed to the higher electron conductivity and large specific surface area of the hybrid structure, which would have enabled a promising nanocomposite for the effective determination of hydrazine. The fabricated CuO/CNTs-rGO/GCE sensor also retained about 90.5% of initial sensitivity even after 20 days of duration, demonstrating that the sensor possesses good stability [1]. Guo *et al.* demonstrated the electrochemical performance towards oxidation of hydrazine using hollow CuO nanospheres prepared using reduction reaction of copper ions on porous Si nanowires followed by calcination for uniformly anchoring on their surfaces. The sensor showed a better electrochemical performance with a rapid response time of below 3 s, a linear range from 1 to 5 mM, and a detection limit down of 0.25 mM [2]. Ramachandran *et al.*, fabricated graphene supported CuO nanorods for sensitive determination of hydrazine and the nanocomposite revealed a detection limit of 9.8 nM, linear response range from 0.1 to 400 μM and a sensitivity of 3.87 $\mu A/\mu M/cm^2$. Moreover, the sensor disclosed good selectivity, reproducibility and long durability towards oxidation of hydrazine [3]. Teymoori *et al.* investigated the electrochemical oxidation of hydrazine using copper oxide nanoparticles/ionic liquid/carbon paste electrode (CuO NPs/IL/CPE) nanocomposite. The modified electrode exhibited sensitivity for hydrazine oxidation which has two linear ranges 0.1–15 and 15–150 μM, found to be 1.4741 and 0.4562 $\mu A/\mu M$. The stability of the modified electrode was investigated after it was stored at room temperature for 3 weeks. It was observed that the oxidation currents showed less than 2.35% decrease relative to the initial responses for hydrazine [4]. Zang *et al.* demonstrated hydrazine electrochemical sensor by painting Cu/Cu$_2$O@carbon on a glassy carbon electrode and immobilized using Nafion. The Cu/Cu$_2$O@carbon modified GCE was evaluated using cyclic voltammograms and amperometry technique, which exhibited high sensitivity of 2.37 $\mu A \cdot \mu M^{-1} \cdot cm^{-2}$, linear concentration ranges from 0.25 to 800 μM, and a low detection limit of 0.022 µM. Moreover, the Cu/Cu$_2$O@carbon based sensor has excellent stability for hydrazine sensing in interference conditions and sensitive response after the addition of hydrazine in which the oxidation current attained 95% steady state within 3s. This result clearly displays that Cu/Cu$_2$O@carbon nanocomposite shows better electrochemical performance in oxidizing hydrazine [5]. L. Wang *et al.*, prepared leaf-like copper oxide (CuO) which is anchored over worm like ordered mesoporous carbon (OMC) composite was employed as an electrochemical sensing probe for hydrazine. OMC serves as an effective binder for holding leaf shaped CuO to form novel nanocomposites. The increased electrocatalytic performance was attributed to the synergetic catalytic effect and unique structural properties. The nanocomposite exhibited better performance towards oxidation of hydrazine with a catalytic rate constant (k_{cat}) of 1.28 x 10^{-5} M^{-1}s^{-1}. It also exhibited a wide linear range of 1 µM to 2.11 mM with a sensitivity of 0.00487 µA and good stability [6]. Yin *et al.*, prepared nano-copper oxide modified GCE by electrodepositing CuCl$_2$ solution at −0.4 V on the surface of GCE and its electrochemical performance towards hydrazine oxidation was evaluated in 0.01 M NaOH solution. The hydrazine was oxidized on nano-copper oxide modified GCE at a low potential of 0.27 V and the oxidative current increased linearly with increasing concentration of hydrazine with a sensitivity of 94.21 µA /mM [7]. Rostami *et al.* fabricated a novel electrochemical sensor using CuO doped in ZSM-5 nanoparticles (CuO/ZSM-5 NPs/CPE) for simultaneous determination of hydrazine and hydroxylamine using DPV and amperometry technique. From amperometric results, it was observed that the current response and the hydrazine concentration have linear relationships in the ranges of 25 mM to 0.9 mM and 0.9 mM to 4.5 mM, respectively. Also, there observed two linear current–time responses in the ranges of 20 mM to 0.9 mM and 0.9 mM to 7.0 mM and revealed a low detection limit of 3.6 mM for hydrazine. The high porous nature of ZSM-5 NPs is the main reason for the enhanced activity, as the improved surface area provides enough substrates for CuO particles [8]. Y. Ma *et al.*, synthesized ultrathin willow-like CuO nanoflakes by one-step process, which displayed elevated electrocatalytic activity toward oxidation of hydrazine at low onset potential of 0.1 V and higher oxidation peak current of 0.9 mA. The result demonstrated that low cost and environmentally friendly CuO catalysts can effectively oxidize hydrazine [9]. Karim-Nezhad *et al.*, prepared copper hydroxide modified copper electrode by cyclic voltammetry in 0.1M NaOH solution in the potential range of −300 to 800 mV. The oxidation of hydrazine was evaluated using CV and it was observed that initially, the oxidation of hydrazine takes place on the bare copper electrode, which has high over-potential. However, in the preceding cycles, the oxidation took place on the copper hydroxide layer which is formed on the copper electrode during the first cycle. There was a negative shift of the peak potential of about 120 mV, which denoted the catalytic activity of the hydroxide layer [10]. Yang *et al.*, constructed a novel copper sulfide (CuS) enzyme-free electrode with varying microstructures for sensitive and selective detection of hydrazine. The flower like CuS exhibited enhanced electrocatalytic performance towards hydrazine oxidation which offered superior performances with a wide linear range of 0.5 μM to 4.775 mM, a high sensitivity of 359.3 μA/mM/cm^2 and a low LOD of 0.097 μM [11].

M. Faisal, M.A. Rashed, M.M. Abdullah, *et al.* fabricated 5% polyaniline (PANI)-doped mesoporous SrTiO$_3$ nanocomposite modified GCE by sol-gel technique using F127 structured binding agent to form mesoporous SrTiO$_3$ followed by an ultra-sonication technique to produce the final PANI doped SrTiO$_3$. The prepared PANI/SrTiO$_3$ nanocomposite modified GCE exhibited sensitivity of 0.2128 μA/μM/cm^2 over the concentration range of 0.2–3.56 mM (LSV) and 0.2438 μAμM/cm^2 over the concentration range of 16–58 μM (amperometry) with a rapid response time of < 10 s. The LOD was found to be 1.09 μM and 0.95 μM at (S/N=3) for LSV and amperometric technique, respectively [12].

H. Akhter *et al.* prepared silica-coated Fe$_2$O$_3$ magnetic nanoparticles by a rapid microwave irradiation method and its electrochemical performance towards oxidation of hydrazine was studied. The LOD was determined from the calibration curve with the target analyte hydrazine range of 0.2 nM ~ 0.2 M. The sensitivity value was found to be ~12.658 μA/mM/cm^2 and the LOD of silica-coated Fe$_2$O$_3$ NPs/GCE sensor was determined as 76.0 pM [13].

ZINC BASED NANOCOMPOSITES FOR HYDRAZINE SENSOR

A. M. Ali *et al.*, synthesized amorphous zinc oxide-silica (a-ZnO–SiO$_2$) nanocomposite and it was subjected for the determination of phenyl hydrazine by measuring the current-potential (I–V) characteristics. The calibration curve revealed good linearity over all concentration from 390 μM to 50 mM and the LOD value was found to be 1.42 μM. The sensitivity was derived as 10.8 μA/mM/cm^2. This outstanding performance was ascribed to good adsorption ability and large surface area of the prepared nanocomposite. In addition, the modified ZnO–SiO$_2$ nanocomposite electrode was extremely stable with no significant sensitivity decrease even after it is re-used five times. Moreover, the sensor was highly stable even when stored in ambient conditions, with its I-V characteristics not degrading for a long period of three months [14]. Fang *et al.* fabricated a ZnO-MWCNTs nanocomposite for sensitive detection of hydrazine, which revealed excellent stability with fast response time, wide linear range of 0.6–250 μM and low LOD of 0.18 μM. The amperometric sensor exhibited a rapid and sensitive oxidative current response to the addition of varying concentrations of hydrazine. This finding implies that ZnO-MWCNTs-GCE sensor would be efficient in sensitive analysis of hydrazine [15]. M. Faisal *et al.*, prepared polythiophene (PTh)/ZnO nanocomposite by sol-gel technique using F127 structure directing agent. Cyclic voltammetry studies revealed that the higher sensor response for PTh/ZnO modified GCE, than bare and pure ZnO modified GCE. The sensor showed a significant sensitivity of 1.22 $\mu A/\mu M/cm^2$ with a rapid sensor response of 5 s. The sensor could detect a range of concentrations of hydrazine from 0.5 up to 48 μM, with a very low LOD value of 0.207 μM at (S/N=3) [16]. R. Madhu *et al.* reported the electrochemical sensing performance of GCE/RGO/ZnO-Au modified electrode towards the sensitive evaluation of hydrazine in water samples. The electrochemical detection was performed using CV and chronoamperometry methods. RGO/ZnO-Au modified GCE has a lower over-potential of 0.1 V and improved oxidation peak current of hydrazine of about 4.1 times and 2.4 times greater when compared with RGO-Au and ZnO/Au-modified GCE, respectively. The sensitivity of the sensor was as high as 5.54 μA μM^{-1} cm^{-2} and the LOD was as low as 18 nM [18]. M. Rahman *et al.* fabricated tin-doped zinc oxide nanoparticles (Sn/ZnO NPs) using wet-chemical method. The calibration plot displayed a linear fit over the concentration ranging from 2.0 nM to 20.0 mM. The sensitivity and detection limit is ~5.0108 $\mu A/cm^2/\mu M$ and ~0.01895 $\pm$ 0.0002 nM respectively. It also exhibited lower detection limit of 2.0 nM ~ 0.20 mM, and long-term stability [35]. Zhao *et al.* fabricated sensitive hydrazine electrochemical sensor based on zinc oxide nano-wires (ZnO NWs) synthesized from hydrothermal method, which showed a high sensitivity of 3.8 $\mu A/\mu M/cm^2$, a short response time of 3 s, and a very low LOD of 0.0144 μM with a linear hydrazine concentration response range from 3.0 to 562 μM [17].

NOBLE METALS BASED NANOCOMPOSITES FOR HYDRAZINE SENSOR

Baron *et al.* investigated the electrochemical properties of different metallic nanoparticles (Au, Pd, and Ag) supported on carbon microspheres for hydrazine detection. Among the combinatorial approach, Pd nanoparticles were identified as an exclusive efficient material in oxidizing hydrazine with a limit of detection of 2 μM [19]. L. Chen *et al.* prepared Pd nanoparticles anchored on carbon nanotubes as a catalyst for the determination of hydrazine in which Pd nanoparticles (1-3 nm) were anchored over CNTs by including a cross linker benzyl mercaptan. The electrocatalytic oxidation of hydrazine was evaluated using CV at 0.35 V on Pd/CNT-modified GC electrode and it was found that the peak on the negative potential scan is not detected, whereas it is oxidized completely on the positive potential scan on Pd/CNT electrode. The better electrocatalytic performance was due to the small particle size and high dispersion of Pd nanoparticles on CNT surface [20]. Dutta *et al.* successfully prepared $Au_{core}@Pd_{shell}$ with an average size of ~11.5 nm on rGO support (denoted as GAP) through a surfactant-free synthetic procedure. The developed sensor revealed a limit of detection of 0.08 μM with a linear range of 2–40 μM from the chronoamperometric (CA) result recorded at -0.15 V *vs.* SCE [23]. X. Ji *et al.* produced palladium nanoparticle (below 1 nm) decorated bamboo multi-walled carbon nanotubes and employed it for the electrochemical oxidation of hydrazine. A linear voltammetric response was observed with the addition of 56 μM to 219 μM and acquired a low LOD value of 10 μM, which insists that it was excellent in sensing hydrazine [24]. J. Li, X. Lin *et al.* developed a novel electrochemical sensor by electrodepositing gold nanoparticles on pre-synthesized polypyrrole (PPy) nanowire, forming an Au/PPy composite matrix on glassy carbon electrode (Au/PPy/GCE). From the DPV analysis, there observed two linear segments, with the current sensitivity of 126 μA/mM in the range of 0.0010–0.50 mM and 35.6 μA/mM in the range of 0.50–7.5 mM. The stability of the Au/PPy/GCE sensor was also good as it retained the current response after 500 successive cycling of the modified electrode [26]. Liu *et al.*, designed bimetallic alloyed AuPd nanorod chains (NRCs) by simple wet-chemical method, using 4-aminopyridine (4-AP) as the structure directing and stabilizing probe. An amperometric sensor for detecting hydrazine exhibited a linear range of 0.10–501.00 μM and low LOD (S/N = 3) value of 0.02 μM with ultra-high sensitivity for the determination of hydrazine [27]. Zhao *et al.*, developed a novel sensor based on electrodepositing gold nanoparticles on the single-walled carbon nanohorns modified GCE. It showed two wide linear segments between the catalytic currents with a detection limit of 1.1 μM (S/N=3) among the concentration of hydrazine ranging between 0.005-3.345 mM [25]. Zhao *et al.*, prepared Pd nanoparticles supported multiwall carbon nanotubes (Pd/MWNTs) and employed it for the electrochemical determination of hydrazine. From the electrochemical studies, it was observed that as the pH increases, the oxidation peak current of hydrazine decreases. Pd/MWNT–Nafion-based sensor showed a linear response for a broad concentration range of hydrazine from 2.5 to 700 μM, with a low detection limit of 1.0 μM [28]. Y. Zhang *et al.*, prepared nanosized platinum particles supported on ordered mesoporous carbons (OMCs) and a positively charged poly-(diallydimethylammonium chloride, PDDA) was introduced to wrap OMCs and bound Pt nanoparticles. The sensor exhibited a sensitivity of 237 μA mM/cm² and detection limit of 0.17 μM with S/N = 3. The response time of the signal was found to be less than 0.5 s. The relative standard deviation of the sensitivity of the prepared sensor was less than 4.1% [29]. Yang *et al.* demonstrated the electrochemical performance using Au@Pt-nFs/GO nanocomposite modified GCE. The enhanced electrochemical performance of Au@Pt-nFs/GO was due to large surface area, abundant active sites and open structure, which reduced the overpotential and boosted the kinetics of hydrazine oxidation. The modified sensor exhibited a wide linear range of 0.8 μM to 0.429 mM, a high sensitivity of 1695.3 μA/ mM/cm^2 and a low detection limit of 0.43 μM [30].

OTHER NANOCOMPOSITES FOR HYDRAZINE SENSOR

Devasenathipathy *et al.*, electrodeposited calcium ions cross linked pectin film (CCLP) with gold nanoparticles (GNPs) on the graphene modified GCE and employed it for the efficient and sensitive determination of hydrazine. The electron transfer coefficient (α) and diffusion coefficient (D_o) for oxidation of hydrazine were determined as 0.46 and 2.91×10^{-6} cm^{2}s^{-1}, correspondingly. Two linear ranges were observed (1) 10–600 nM with sensitivity of 47.6 nA/nM/cm^2 and (2) 0.6-197.4 μM with sensitivity of 1.786 μA/μM/cm^2. The sensor exhibited a very low deletion limit of 1.6 nM. The sensor was highly selective to hydrazine even in the presence of a very high concentration (about 500 times) of commonly interfering ions in water and biofluids. The sensor showed excellent results with repeatability, reproducibility and stability. The sensor could also successfully detect hydrazine in practical samples like water and urine [21]. Devasenathipathy *et al.*, developed an amperometric sensor for the determination of hydrazine using bismuth nanoparticles (Bi) decorated graphene nanosheets(GR) composite film modified GCE. The sensor exhibited a linear range from 20 nM to 0.28 mM and a very low LOD of 5 nM, which is the lowest LOD achieved for the determination of hydrazine in neutral pH. The sensor was highly sensitive and could detect hydrazine effectively in the presence of interfering compounds of even 1000 times higher concentration [22]. Y. Fang *et al.* prepared NiFe$_2$O$_4$/MWCNTs nanoparticles synthesized by facile hydrothermal method. The NiFe$_2$O$_4$/MWCNTs modified GCE showed better catalytic behavior in the oxidation of hydrazine. The designed amperometric sensor showed a linear response in the concentration range from 5.0 μM to 2.5 mM and a low LOD value of 1.5 μM, for a signal-t--noise ratio (S/N) of 3. The modified electrode also showed an acceptable reproducibility with an RSD of 4.2% for the current determined at 250 μM hydrazine concentration and good stability [31]. G. Hu *et al.*, synthesized rhodium nanoparticle loaded carbon nanofibers (nano-Rh/CNF) using electrospinning technique. The developed sensor nano-Rh/CNF/Nafion/PGE displayed a linear response ranging from 0.5 to 175 μM and a sensitivity of 527 μA/mM. The nano-Rh/CNF/Nafion/PGE based sensor showed good reproducibility with an RSD of - time parallel determinations for 10 μM hydrazine is 2.6%, as determined from the amperometric method [32]. Kannan *et al.*, reported the electrochemical sensing of hydrazine using multi-layer Graphene nanobelts. The fabricated hydrazine sensor displayed a sensitivity value of 0.08 μA μM/cm^2 with a linear range from 10 μM to 1.36 mM. The interference studies revealed high selectivity to hydrazine in the presence of interfering compounds, such as ascorbic acid, uric acid, glucose and lactic acid [33]. H Peng *et al.* reported the electrochemical performance of polydopamine-reduced graphene oxide (PDA-RGO) nanocomposites. The elevated electrochemical oxidation of hydrazine was attributed to the increased surface area of RGO. An amperometric response was observed in which the sensor exhibited a wide linear range of 0.03 to 100 μM, with the LOD value of 0.01 μM [34]. M. Rahman *et al.* prepared cobalt pyrite (CoS_2) decorated carbon nanotube nanocomposites (CoS_2-CNT NCs) by the simple wet-chemical method. It also displayed excellent sensitivity, low detection limit, long-term stability, and reproducibility. The calibration plot ($r^2 = 0.9992$) revealed a linear fit ranging from 0.1 nM to 1.0 mM, with a sensitivity value of 4.430 μA/nM/cm² and extremely low detection limit (LOD) of 0.1 nM [36]. M. Rahman *et al.* developed strontium oxide nanoparticles decorated carbon nanotube nanocomposites (SrO.CNT NCs) using a wet-chemical technique at low temperatures. The detection limit and sensitivity were calculated as 0.036 nM and ~26.37 μA/mM/cm²⋅ respectively. The detection of hydrazine by current *vs* voltage (I-V) method using SrO.CNT NCs modified GCE electrode displayed higher sensitivity when compared to other nanomaterials [37].

Zhou *et al.*, prepared porous Mn_2O_3 nanorods synthesized using thermal decomposition technique towards hydrazine electrochemical detection. The performance of hydrazine was improved obviously due to the anchoring of Au nanoparticles onto Mn_2O_3 nanorods. The modified sensor revealed a sensitivity of 500 mA/mol L/cm² in the linear range of 2×10^{-6} - 1.3×10^{-3} mol/L, LOD value was found to be 1.1×10^{-6} mol/L [38]. P. Yue *et al.* developed a novel and efficient hydrazine sensor based on nitrogen-doped carbon nanopolyhedra (CNP), Prussian blue (PB) and conductive polymer. The modified electrode PB/CNP/PPy revealed a rapid response of less than 2 s, a sensitivity of 0.22 A/M, a wide linear range of 7.5×10^{-7} M to 1.7×10^{-3} M and a low detection limit of 2.9×10^{-7} M. The sensor also showed remarkable stability and selectivity [39]. Yao *et al.*, fabricated MXene/ZIF-8 nanocomposite and it was employed as a sensitive probe for the detection of hydrazine. The enhanced electrocatalytic performance was ascribed to the large specific surface area and its porous structure provided by ZIF-8. Also, Mxene, which is a highly conductive material greatly enhances the peak current response with a rapid electron transfer rate. This makes the response current to reach a steady state more rapidly, and revealed a linear response over a wide range from 10-7700 μM [40]. Wang *et al.*, reported about the electrochemical sensing behavior of three-dimensional interconnected $Co(OH)_2$ nanosheets electrodeposited onto Ti mesh (TM) surface. The prepared electrode with the electrodeposition time of 10 min exhibited prominent behavior in detecting hydrazine with the linear range from 5 μM to 3.0 mM and LOD of 1.98 μM. It also showed a sensitivity of 2793.9 μA/mM/cm², good selectivity, reproducibility and stability with RSD less than 3.5% [41]. Wang *et al.*, developed Nano-

Au/Porous-TiO_2 composite modified GCE and it was electrochemically determined towards oxidation of hydrazine. The Nano-Au/Porous- modified sensor exhibited a wide linear range of hydrazine from 2.5 to 500 µM, with a detection limit of 0.5 µM at a signal-to-noise ratio of 3 and fast response time [42].

CONCLUDING REMARKS

In this chapter, we discussed the state of the art research and development of different types of nanocomposites, particularly metal based nanocomposites for the electrochemical sensing of hydrazine. Tailored nanocomposites were excellent candidates for fabricating electrochemical sensors as they have lower over-potential value of hydrazine oxidation. Moreover, these sensors are highly sensitive and also selective for the effective quantization of hydrazine, and thus, help in improving environmental and human health. However, more work should be done to develop effective sensors for real sample analysis for the successful commercialization of electrochemical hydrazine sensors.

CONSENT FOR PUBLICATION

Not applicable.

CONFLICT OF INTEREST

The author declares no conflict of interest, financial or otherwise.

ACKNOWLEDGEMENTS

The authors thank the DST-FIST; DST-PURSE and UGC-SAP, Government of India, for financial support to build the instrumental facilities.

REFERENCES

[1]   Zhao, Z.; Wang, W.; Tang, W.; Xie, Y.; Li, Y.; Song, J.; Zhuiykov, S.; Hu, J.; Gong, W. Synthesis and electrochemistry performance of CuO-functionalized CNTs-rGO nanocomposites for highly sensitive hydrazine detection. *Ionics,* **2020**, *26*(5), 2599-2609.
       [http://dx.doi.org/10.1007/s11581-019-03305-w]

[2]   Guo, Z.; Seol, M-L.; Kim, M-S.; Ahn, J-H.; Choi, Y-K.; Liu, J-H.; Huang, X-J. Hollow CuO nanospheres uniformly anchored on porous Si nanowires: preparation and their potential use as electrochemical sensors. *Nanoscale,* **2012**, *4*(23), 7525-7531.
       [http://dx.doi.org/10.1039/c2nr32556j] [PMID: 23099737]

[3]   Ramachandran, K.; Babu, K.; Kumar, G.G.; Kim, A.R.; Yoo, D.J. One-pot synthesis of graphene supported CuO nanorods for the electrochemical hydrazine sensor applications. *Sci. Adv. Mater.,* **2015**, *7*(2), 329-336.
       [http://dx.doi.org/10.1166/sam.2015.2025]

[4]   Teymoori, N; Raoof, JB; Khalilzadeh, MA; Ojani, R An electrochemical sensor based on CuO nanoparticle for simultaneous determination of hydrazine and bisphenol A. *J Iran Chem Soc,* **2018**, *15*(10), 2271-2279.

[5]   Zhao, Z.; Wang, Y.; Li, P.; Sang, S.; Zhang, W.; Hu, J.; Lian, K. A highly sensitive electrochemical sensor based on $Cu/Cu_2O@$ carbon nanocomposite structures for hydrazine detection. *Anal. Methods,* **2015**, *7*(21), 9040-9046.
[http://dx.doi.org/10.1039/C5AY02122G]

[6]   Wang, L.; Meng, T.; Jia, H.; Feng, Y.; Gong, T.; Wang, H.; Zhang, Y. Electrochemical study of hydrazine oxidation by leaf-shaped copper oxide loaded on highly ordered mesoporous carbon composite. *J. Colloid Interface Sci.,* **2019**, *549*, 98-104.
[http://dx.doi.org/10.1016/j.jcis.2019.04.063] [PMID: 31026767]

[7]   Yin, Z.; Liu, L.; Yang, Z. An amperometric sensor for hydrazine based on nano-copper oxide modified electrode. *J. Solid State Electrochem.,* **2011**, *15*(4), 821-827.
[http://dx.doi.org/10.1007/s10008-010-1161-2]

[8]   Rostami, S.; Azizi, S.N.; Ghasemi, S. Simultaneous electrochemical determination of hydrazine and hydroxylamine by CuO doped in ZSM-5 nanoparticles as a new amperometric sensor. *New J. Chem.,* **2017**, *41*(22), 13712-13723.
[http://dx.doi.org/10.1039/C7NJ02685D]

[9]   Ma, Y.; Li, H.; Wang, R.; Wang, H.; Lv, W.; Ji, S. LvW, Ji S, Ultrathin willowlike CuO nanoflakes as an efficient catalyst for electro-oxidation of hydrazine. *J. Power Sources,* **2015**, *289*, 22-25.
[http://dx.doi.org/10.1016/j.jpowsour.2015.04.151]

[10]   Karim-Nezhad, Ghasem; Jafarloo, Roghieh; Dorraji, Parisa Seyed Copper (hydr)oxide modified copper electrode for electrocatalytic oxidation of hydrazine in alkaline media, Electrochimica Acta, 2009 **2009**, *54*, 5721-5726.

[11]   Yang, Z.; Zhang, S.; Zheng, X.; Fu, Y.; Zheng, J. Controllable synthesis of copper sulfide for nonenzymatic hydrazine sensing. *Sens. Actuators B Chem.,* **2018**, *255*, 2643-2651.
[http://dx.doi.org/10.1016/j.snb.2017.09.075]

[12]   Faisal, M.; Rashed, M.A.; Abdullah, M.M.; Harraz, F.A.; Jalalah, M.; Al-Assiri, M.S. Md.A. Rashed, M.M. Abdullah, Farid A. Harraz, Mohammed Jalalah, M.S. Al-Assiri, Efficient hydrazine electrochemical sensor based on PANI doped mesoporous $SrTiO_3$ nanocomposite modified glassy carbon electrode. *J. Electroanal. Chem. (Lausanne),* **2020**, *879*, 114805.
[http://dx.doi.org/10.1016/j.jelechem.2020.114805]

[13]   Akhter, H.; Murshed, J.; Rashed, M.A.; Oshima, Y.; Nagao, Y.; Rahman, M.M.; Asiri, A.M.; Hasnat, M.A.; Uddin, M.N.; Siddiquey, I.A. Md. A. Rashed, Yoshifumi Oshima, Yuki Nagao, Mohammed M. Rahman, Abdullah M. Asiri, M.A. Hasnat, Md. Nizam Uddin, Iqbal Ahmed Siddiquey, Fabrication of hydrazine sensor based on silica-coated Fe_2O_3 magnetic nanoparticles prepared by a rapid microwave irradiation method. *J. Alloys Compd.,* **2016**, *698*, 921-929.
[http://dx.doi.org/10.1016/j.jallcom.2016.12.266]

[14]   Ali, A.M.; Harraz, F.A.; Ismail, A.A.; Al-Sayari, S.A.; Algarni, H.; Al-Sehemi, A.G. S.A. Al-Sayari, H. Algarni, Abdullah G. Al-Sehemi, Synthesis of amorphous $ZnO–SiO_2$ nanocomposite with enhanced chemical sensing properties. *Thin Solid Films,* **2015**, *605*, 277-282.
[http://dx.doi.org/10.1016/j.tsf.2015.11.044]

[15]   Fang, B.; Zhang, C.; Zhang, W.; Wang, G. A novel hydrazine electrochemical sensor based on a carbon nanotube-wired ZnO nanoflower-modified electrode. *Electrochim. Acta,* **2009**, *55*(1), 178-182.
[http://dx.doi.org/10.1016/j.electacta.2009.08.036]

[16]   Faisal, M.; Farid, A. Harraz, A.E. Al-Salami, S.A. Al-Sayari, A. Al-Hajry, M.S. Al-Assiri, Polythiophene/ZnO nanocomposite-modified glassy carbon electrode as efficient electrochemical hydrazine sensor. *Mater. Chem. Phys.,* **2018**, *214*, 126-134.
[http://dx.doi.org/10.1016/j.matchemphys.2018.04.085]

[17]   Zhao, Z.; Sun, Y.; Li, P.; Sang, S.; Zhang, W.; Hu, J.; Lian, K. A Sensitive Hydrazine Electrochemical Sensor Based on Zinc Oxide Nano-Wires. *J. Electrochem. Soc.,* **2014**, *161*(6), B157-B162.
[http://dx.doi.org/10.1149/2.095406jes]

[18] Madhu, R.; Dinesh, B.; Chen, S-M.; Saraswathi, R.; Mani, V. An electrochemical synthesis strategy of composite based ZnO microspheres-Au nanoparticles on reduced graphene oxide for the sensitive detection of hydrazine in water samples. *RSC Advances,* **2015**, *5*(67), 54379-54386.
[http://dx.doi.org/10.1039/C5RA05612H]

[19] Baron, R.; Šljukić, B.; Salter, C.; Crossley, A.; Compton, R.G. Biljana Sˇ ljukic´, Chris Salter, Alison Crossley, Richard G. Compton, Development of an Electrochemical Sensor Nanoarray for Hydrazine Detection Using a Combinatorial Approach. *Electroanalysis,* **2007**, *19*(10), 1062-1068.
[http://dx.doi.org/10.1002/elan.200703822]

[20] Chen, L.; Hu, G.; Zou, G.; Shao, S.; Wang, X. Efficient anchorage of Pd nanoparticles on carbon nanotubes as a catalyst for hydrazine oxidation. *Electrochem. Commun.,* **2009**, *11*(2), 504-507.
[http://dx.doi.org/10.1016/j.elecom.2008.12.047]

[21] Devasenathipathy Veerappan Mani, R.; Shen-Ming Chen Daneial Arulraj, V.S.V. Highly stable and sensitive amperometric sensor for the determination of trace level hydrazine at cross linked pectin stabilized gold nanoparticles decorated graphene nanosheets. *Electrochim. Acta,* **2014**, *135*, 260-269.
[http://dx.doi.org/10.1016/j.electacta.2014.05.002]

[22] Devasenathipathy, R.; Mani, V.; Chen, S-M. Highly selective amperometric sensor for the trace level detection of hydrazine at bismuth nanoparticles decorated graphene nanosheets modified electrode. *Talanta,* **2014**, *124*, 43-51.
[http://dx.doi.org/10.1016/j.talanta.2014.02.031] [PMID: 24767444]

[23] Dutta, S.; Ray, C.; Mallick, S.; Sarkar, S.; Roy, A.; Pal, T. Au@Pd Core-shell Nanoparticles Decorated Reduced Graphene Oxide: A Highly Sensitive and Selective Platform for Electrochemical Detection of Hydrazine. *RSC Advances,* **2015**, *5*(64), 51690-51700.
[http://dx.doi.org/10.1039/C5RA04817F]

[24] Ji, X.; Banks, C.E.; Holloway, A.F.; Jurkschat, K.; Thorogood, C.A.; Wildgoose, G.G.; Compton, R.G.; Palladium, S-N.D.B.M-W.C.N.E.E.M. Voltammetric Sensing in Otherwise Inaccessible pH Ranges. *Electroanalysis,* **2006**, *18*(24), 2481-2485.
[http://dx.doi.org/10.1002/elan.200603681]

[25] Zhao, S.; Wang, L.; Wang, T.; Han, Q.; Xu, S. Liang liang, Wang Tingting, Wang Qinghua, Han Shukun Xu, A high-performance hydrazine electrochemical sensor based on gold nanoparticles/single-walled carbon nanohorns composite film. *Appl. Surf. Sci.,* **2016**, *369*, 36-42.
[http://dx.doi.org/10.1016/j.apsusc.2016.02.013]

[26] Li, J.; Lin, X. Electrocatalytic oxidation of hydrazine and hydroxylamine at gold nanoparticle-polypyrrole nanowire modified glassy carbon electrode. *Sens. Actuators B Chem.,* **2007**, *126*(2), 527-535.
[http://dx.doi.org/10.1016/j.snb.2007.03.044]

[27] Liu, Y.; Chen, S-S.; Wang, A-J.; Feng, J-J.; Wu, X.; Weng, X. An ultra-sensitive electrochemical sensor for hydrazine based on AuPd nanorod alloy nanochains. *Electrochim. Acta,* **2016**, *195*, 68-76.
[http://dx.doi.org/10.1016/j.electacta.2016.01.229]

[28] Zhao, J.; Zhu, M.; Zheng, M.; Tang, Y.; Chen, Y.; Lu, T. Electrocatalytic oxidation and detection of hydrazine at carbon nanotube supported palladium nanoparticles in strong acidic solution conditions. *Electrochim. Acta,* **2011**, *56*(13), 4930-4936.
[http://dx.doi.org/10.1016/j.electacta.2011.03.014]

[29] Zhang, Y.; Bo, X.; Luhana, C.; Guo, L. Preparation and electrocatalytic application of high dispersed Pt nanoparticles/ordered mesoporous carbon composites. *Electrochim. Acta,* **2011**, *56*(17), 5849-5854.
[http://dx.doi.org/10.1016/j.electacta.2011.05.016]

[30] Yang, Z.; Zheng, X.; Zheng, J. Facile synthesis of three-dimensional porous Au@Pt core-shell nanoflowers supported on graphene oxide for highly sensitive and selective detection of hydrazine. *Chem. Eng. J.,* **2017**, *327*, 431-440.
[http://dx.doi.org/10.1016/j.cej.2017.06.120]

[31] Fang, B.; Feng, Y.; Liu, M.; Wang, G.; Zhang, X.; Wang, M. Electrocatalytic oxidation of hydrazine at a glassy carbon electrode modified with nickel ferrite and multi-walled carbon nanotubes. *Mikrochim. Acta,* **2011**, *175*(1-2), 145-150.
[http://dx.doi.org/10.1007/s00604-011-0662-8]

[32] Hu, G.; Zhou, Z.; Guo, Y.; Hou, H.; Shao, S. Electrospun rhodium nanoparticle-loaded carbon nanofibers for highly selective amperometric sensing of hydrazine. *Electrochem. Commun.,* **2010**, *12*(3), 422-426.
[http://dx.doi.org/10.1016/j.elecom.2010.01.009]

[33] Kannan, P.K.; Moshkalev, S.A.; Rout, C.S. Electrochemical Sensing of Hydrazine using Multilayer Graphene Nanobelts. *RSC Advances,* **2016**, *6*(14), 11329-11334.
[http://dx.doi.org/10.1039/C5RA24912K]

[34] Peng, H.; Liang, C. *Electrochemical determination of hydrazine based on polydopamine reduced graphene oxide nanocomposite*; Taylor & Francis, **2016**, pp. 29-33.

[35] Rahman, M.M.; Balkhoyor, H.B.; Asiri, A.M. Ultrasensitive and selective hydrazine sensor development based on Sn/ZnO nanoparticles. *RSC Advances,* **2016**, *6*(35), 29342-29352.
[http://dx.doi.org/10.1039/C6RA02352E]

[36] Rahman, M.M.; Ahmed, J.; Asiri, A.M.; Siddiquey, I.A.; Hasnat, M.A. Development of ultra-sensitive hydrazine sensor based on facile CoS_2-CNT nanocomposites. *RSC Advances,* **2016**, *6*(93), 90470-90479.
[http://dx.doi.org/10.1039/C6RA08772H]

[37] Rahman, M.M.; Hussain, M.M.; Asiri, A.M. A novel approach towards the hydrazine sensor development by SrO.CNT nanocomposites. *RSC Advances,* **2016**, *6*(70), 65338-65348.
[http://dx.doi.org/10.1039/C6RA11582A]

[38] Zhou, B.; Yang, J.; Jiang, X. JiaoYang, Xiaoqing Jiang, Porous Mn_2O_3 nanorods synthesized from thermal decomposition of coordination polymer and used in hydrazine electrochemical sensing. *Mater. Lett.,* **2015**, *159*, 362-365.
[http://dx.doi.org/10.1016/j.matlet.2015.07.031]

[39] Yue, P.; Tricard, S.; Pang, T.; Yang, Y.; Zhao, J.; Fang, J. Prussian blue (PB)/ carbon nanopolyhedra/polypyrrole composite as electrode: a high performance sensor to detect hydrazine with long linear range. *Sens. Actuators B Chem.,* **2017**, *251*, 706-712.
[http://dx.doi.org/10.1016/j.snb.2017.05.042]

[40] Yao, Y.; Han, X.; Yang, X.; Zhao, J.; Chai, C. Detection of Hydrazine at MXene/ZIF-8 Nanocomposite Modified Electrode. *Chin. J. Chem.,* **2020**, •••, 38.

[41] Wang, J.; Xie, T.; Deng, Q.; Wang, Y.; Zhu, Q.; Liu, S. Three-dimensional interconnected $Co(OH)_2$ nanosheets on Ti mesh as a highly sensitive electrochemical sensor for hydrazine detection. *New J. Chem.,* **2019**, *43*(7), 3218-3225.
[http://dx.doi.org/10.1039/C8NJ06008H]

[42] Wang, G.; Zhang, C.; He, X.; Li, Z.; Zhang, X.; Wang, L.; Fang, B. Detection of hydrazine based on Nano-Au deposited on Porous-TiO_2 film. *Electrochim. Acta,* **2010**, *55*(24), 7204-7210.
[http://dx.doi.org/10.1016/j.electacta.2010.07.053]

CHAPTER 13

Optical Detection Of Toxic Cations And Anions By Nanocomposite Materials

D. Amilan Jose[1,*], Nancy Sharma[1] and **Srushti Gadiyaram[1]**

[1] *Department of Chemistry, National Institute of Technology (NIT) Kurukshetra, Kurukshetra-136119, Haryana, India*

Abstract: Nanocomposite materials have appeared as appropriate replacements to overcome the limitations of microcomposites and simple nanomaterials. There is an upsurge of interest in nanocomposite materials due to the significant applications and all the research areas related to chemical, physical and biological sciences. Making composites with the combination of nanomaterials could convert them into superior materials for the sensing analytes, such as anions, cations, biomolecules, explosives, toxic gases, food toxins and organic compounds. This chapter emphasizes only on the recent investigations of nanocomposite materials for the detection of toxic metals and anions by optical detection methods, such as fluorescent and colorimetric methods. Appropriately selected examples are discussed in detail.

Keywords: Anion sensor, Cation sensor, Colorimetric, Cyanide detection, Fluorescent, Mercury detection, Nanocomposites.

INTRODUCTION

Nanocomposite (NC) materials are composed of at least one phase having a nanometer size in scale. Nanocomposite materials are anticipated to show novel properties evolving from the mixture of individual phases. They could be considered as suitable substitutes to overcome the limitations of small molecules, nanomaterials, and microcomposites [1].

Currently, polymers are being used to encapsulate nanomaterials to develop new nanocomposites [2]. They have immense applications due to their extraordinary properties depending upon the combination of polymer and nanomaterials embedded within them. Nanocomposites have shown favorable applications in several fields, such as biomedical, food package, optics, electronics, ion separation, chemical/bio-sensor, catalysis, and protective coatings [1, 3, 4].

---

* **Corresponding Author D. Amilan Jose**: Department of Chemistry, National Institute of Technology (NIT), Kurukshetra-136119, Haryana, India. Email: amilanjosenit@nitkkr.ac.in

Manorama Singh, Vijai K Rai and Ankita Rai (Eds.)

In the conventional methods, an optical sensor for anion and cation is prepared by employing receptors with signaling units, such as organic dyes, metal complexes, and nanomaterials. Choosing suitable materials to combine with the conventional sensors could enhance their sensitivity, selectivity, and water compatibility. Subsequently, several nanocomposite materials composed of inorganic nanoparticles, such as gold nanoparticles (AuNPs), MoS_2 nanosheets, Carbon nanotubes (CNTs), Carbon dots (CDs), and Quantum dots (QDs), have been developed. Nanocomposite materials composed of conjugated polymers (CP) and metal organic framework (MOF) with nanoparticles (NPs) have also been explored for various applications [2, 3, 5 - 10]. This chapter aims to describe the recent developments of the nanocomposite materials as optical (colorimetric and fluorescent) sensors to detect toxic cations and anions. Relevant selected crucial examples are discussed in detail.

DETECTION OF TOXIC HEAVY METALS

Mercury Detection

Mercury ions are toxic to the environment. They exist in the environment as inorganic ions within the variety of Hg(II) cations or monomethyl mercury as an alkylated form. Hg^{2+} is also known to bioaccumulate through the biomagnification process. It causes various health problems and diseases in humans and animals. Nanocomposites exhibit distinctive properties as sensing platforms for mercury detection. They have an advantage of simultaneously detecting and separating Hg^{2+} pollutants without causing secondary pollution.

Dithizone-based porous polymers are known for the elimination of cations from wastewater samples. Roya Sedghi *et al.* described a dithizone-based chromogenic nanocomposite probe to detect trace levels of lead and mercury ions in an aqueous solution [11]. In this sensor system, dithizone was anchored on TiO_2/poly (2-V--MBAm-AA) nanocomposite through π– π stacking interactions. The trace levels of ions exhibit a substantial colorimetric response from grey to violet for Hg^{2+} ions and red for Pb^{2+} ions. This nanocomposite sensor has the advantage of easy preparation, rapid response time, being cheap, highly selective and sensitive with LOD as low as 10 ppb.

Another group explored the polymeric TiO_2/poly(acrylamide-co-methylene bisacrylamide) nanocomposites as an ultra-trace level sensor to detect two heavy metals, lead and mercury ions, in the aqueous solution [12]. The LOD of the polymeric nanocomposites for the mercury and lead ions was found to be 1 and 10 μgL^{-1}, respectively. Dithizone is used as an indicator for sensitive sensing of heavy metal ions.

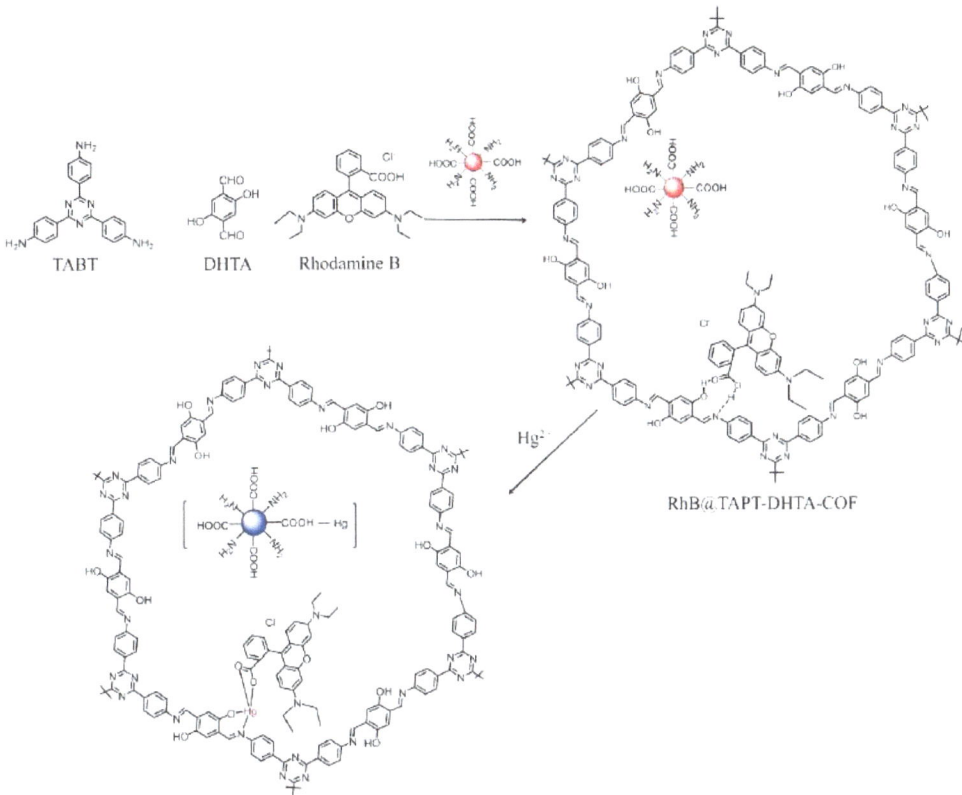

Fig. (1). Hg^{2+} detection mechanism and schematic representation for the synthesis of composite NCDs-RhB@COF.

Acrylamide is used as a monomer and methylene bisacrylamide as the cross-linker to arrange as a polymer on the TiO$_2$ nanoparticles' surface. The secondary interaction of dithizone with adsorbed target ions plays a significant role in detecting heavy metals on the hydrogel. At pH 6, the extreme adsorption of Pb^{2+} ion, and at pH 9, adsorption of Hg^{2+}, were realized. The response time for both metal ions is less than 10 min. A prominent color change from violet to pink was observed for mercury ions and from red to light red for lead ions. The reversibility of the complex was verified by the addition of HNO$_3$. The selectivity of the metal ions can be derived from the HSAB principle. The adsorption of the target Pb^{2+} and Hg^{2+} is not interfered by other ions at optimum pH and sorption time. However, by changing the pH of the hydrogel, other ions could react with heteroatoms (O and N). The adsorption capacity of the new nanocomposite for Pb^{2+} and Hg^{2+} ions was determined as 96 mg/g and 87 mg/g, respectively. Up to 200 mL of sample volume can be used for the simultaneous recovery of mercury and lead ions by using the hydrogel.

Graphene-based silver nanocomposite (G-Ag-NPs) has been prepared and used as a selective sensor for Hg^{2+}. Ascorbic acid has been used as a reducing agent for the preparation of G-Ag NPs through an easy single-pot redox reaction [13]. The peak at 420 nm decreases in response to Hg^{2+}, but other metal ions show insignificant change at 420 nm. The LOD was determined as 3.1×10^{-8} mol/L. Ag NPs are toxic, but the cytotoxicity studies confirm that the silver nanocomposite G-Ag NPs are non-toxic to the environment. Importantly, the G-Ag NPs system is also useful to estimate trace amounts of Hg^{2+} in various water samples collected from different water bodies with a recovery rate of >98%.

Agarose-based hydrogel films composed of N-doped CDs and Rhodamine B covalent organic frameworks (NCDs-RhB@COF) as nanocomposites were prepared by Wang *et al.* [14]. This nanocomposite is used for the semi-quantitative naked eye recognition of Hg(II) by using a UV lamp. The double emission peaks were exhibited by NCDs-RhB@COF at 440 and 570 nm for both NCDs and RhB, respectively. The fluorescence emission at 440 nm decreased and at 570 nm increased with the addition of Hg^{2+}. This ratiometric sensor was applied to detect the trace of Hg^{2+} with a linear range of 0.048–10 µM and LOD of 15.9 nM.

Zhang *et al.* [15] prepared a PVDF-g-PAA-CDs composite sensor membrane through an amidation reaction by grafting CDs onto a poly(vinylidene fluoride) (PVDF)-g-PAA composite membrane, and the composite sensor membrane showed stable fluorescence. The complexation of Cu^{2+}, Hg^{2+} and Fe^{3+} quenched the fluorescence of CDs on the surface of the composite sensor. The complexation of these metals with $-NH_2$ and -COOH groups of CDs surface led to electron transfer, and thereby, fluorescence quenching was observed. The LOD of 1 nmol/L and 1 µmol/L was obtained for Cu^{2+}/Hg^{2+} and Fe^{3+}, respectively.

Colorimetric non-toxic Graphene-Gold nanocomposite (G-AuNPs) was reported for the separation and naked-eye detection of Hg^{2+}. The color change of purple-red to light brown was observed [16]. G-AuNPs and Hg^{2+} react together to form gold amalgam as they have high affinity and high specific surface area. Hg^{2+} can be enhanced and separated as high as 94% in the system by using a simple filtration method. The detection limit of colorimetric detection of Hg^{2+} was found to be 1.6×10^{-8} mol/L.

A thermo-responsive hyperbranched polyethylenimine with isobutyramide groups (HPEI-IBAm) capped with AuNPs was developed for Hg^{2+} detection and separation [17]. The nanocomposite could enrich about 94.5% of Hg^{2+} among other heavy metal ions. Later, the Hg^{2+} could be removed by centrifugation method. Also, the color change of the solution from red to beige was used for the naked-eye detection of Hg^{2+}. The LOD was estimated as 8.8 nM and 1.1 nM by color and UV-Vis absorbance change, respectively. The thermo-responsive nanocomposites (HPEI-IBAm) are more sensitive than simple AuNPs.

For colorimetric detection and removal of trace amounts of Hg^{2+}, Chitosan–gold nanocomposite-based paper strips were developed by L. Hu *et al.* [18]. The test papers were developed by submerging normal filter paper into the composite dispersion. The color of the nanocomposite both in solution and paper strips changed to yellow from dark red only with Hg^{2+}. The LOD was calculated as 3.2×10^{-9} $molL^{-1}$ and 5.0×10^{-8} $molL^{-1}$ for the solution and test paper, respectively. This chitosan–gold nanocomposite sensor was non-toxic and used to detect trace amounts of Hg^{2+} in fruit or vegetable juice samples.

In another colorimetric approach, nanocomposite MoS_2-Au as peroxidase mimetic was prepared by a solvothermal method using microwave synthesis [19]. Hg^{2+} encourages composites peroxidase-like activity for the uninterrupted detection of Hg^{2+}. A LOD of 5 nM was observed with a linear response ranging from 20 nM to 20 μM. This colorimetric method was also applied to determine Hg^{2+} in different water samples. The selectivity was found to be outstanding among other interfering cations.

Chromium Detection

Chromium is considered a carcinogenic cation and an important pollutant to the environment. It exists in water as $HCrO_4^-$, $HCrO_4^{2-}$ or $Cr_2O_7^{2-}$. Chromium detection through optical sensors is an advantage over other analytical methods due to easy preparation and low cost. Truskewycz *et al.* [20] developed an effective Cr(VI) sensor, a novel hydrogel composite ZnO-QCD composed of CDs/PVP/ZnO. The composite displayed a decrease in fluorescence only with Cr^{6+} among other cations, and LOD was estimated as 1.2 μM. The ZnO-QCD composite material was very cheap and useful for high-throughput monitoring and environmental diagnostic applications upon proper modification into a 96-well plate.

Silver clusters/hydrogel based nanocomposite [Ag-clusters/poly[(N-isopropyl acrylamide)-co-acrylic acid (IPA-co-AA)] photoluminescent probe was reported to detect Cr(VI) with red emission in aqueous solutions. Photoreduction process was used to obtain thermo-responsive nanocomposites. The probe was non-toxic, environmentally friendly, clean, and quick approach. The photoluminescent metal nanoclusters were generated by using stimuli-responsive hydrogel templates to detect temperature and Cr(VI) for the very first time. To prepare Ag-clusters, the submicron particles of the thermo-responsive hydrogel (IPA-co-AA) were estimated as good templates; the Ag clusters were formed from Ag atoms which were formed from Ag ions, which are distributed and immobilized in IPA-co-AA hydrogen network structures by stable means. The formation of hydrogel composites of fluorescent Ag clusters/ IPA-co-AA was characterized through XPS, XRD, FT-IR and TEM techniques. They exhibited a red color emission under 365nm UV light, and they were observed to be in milky white color at ordinary temperature. This particular emission of the composites was observed from Ag clusters. The emission in PL intensity was quenched with the increase in temperature, which suggests aggregation-caused quenching (ACQ), and these changes were reversible. The reversible and repeatable hydrogel nanoparticles of IPA-co-AA copolymer support the reversibility of the probe. Cr(VI) capability as a strong electron acceptor and Ag-clusters electron-donating ability cause a decrease in the PL performance. The LOD was estimated as 1 ppb for Cr(VI). The hydrogel nanocomposites of Ag clusters/ IPA-co-AA could be a selective sensor for quantitative detection of Cr(VI). Many potential applications of these nanocomposites in the fields of temperature sensors, laboratory analysis and environmental monitoring, *etc.*, can be promised.

Silver Detection

A "turn-on" dual emission ratiometric fluorescent probe has been reported to detect Ag^+ by combining gold nanoclusters, silicon spheres and carbon dots. Carbon dots-silicon spheres (CSs) were invented *in-situ* from carbon dots by the hydrothermal method on SiO_2 spheres. The gold nanoclusters, stabilized with glutathione, were coupled to the surface of CSs *via* covalent bonds to form CSs-AuNCs (CSA). The sensitivity and selectivity of ratiometric probe CSA were much better for Ag^+ than any other metals, and the response could be evaluated with the naked eye. The fluorescence change of CSA probe was observed to change into pink from purple and from blue under UV light with the increase in the concentration of Ag^+. An Ag^+ sensor was fabricated as an accurate, fast, selective and highly sensitive sensor with the LOD as low as 1.6nM. The CSA composites, with a single excitation wavelength of 380nm, displayed double emission at 448nm and 610nm. The CSAs were stable for several days, even in UV light, which demonstrates better bleaching resistance. The comparison of the zeta potentials of CSA composites before and after the addition of Ag^+ determines the stability of CSA composites. The fluorescence intensity at various pH values has been explored, obtaining the maximum intensity at pH 7. CSA fluorescent probe can be applied for real samples analysis due to excellent reliability and accuracy.

NANOCOMPOSITES FOR ANION SENSORS

Fluoride Ion Detection

In the past decades, a wide range of anion sensing chemical probes have been developed, and they show different selectivity with a wide variety of toxic anions [21 - 23]. Fluoride ions can be especially fascinating due to their vital presence in medicine, biology and environmental sciences. Fluoride toxicity is also known to increase bone density [24, 25]; scientists are also exploring the role of fluoride ion in osteoporosis and fluorosis. Although the traditional analysis of fluoride ion determination remains important, the optical detection of fluoride ions in water has great advantage for the analytical chemist.

A novel polymeric nanocomposite based on a hybrid of inorganic and organic material *via* a surface modification strategy has been designed and synthesized by R. Sedghi *et al*. The nanocomposites of TiO_2/poly(acrylamide-co-methylene bis acrylamide-co-2-(3-(4-nitrophenyl)thioureido)ethyl methacrylate) comprise a polymeric shell on the TiO_2-NPs obtained using acrylamide (AM) as a monomer, 2-(3-(4-nitrophenyl)thioureido)ethyl methacrylate (NPhM) as a co-monomer probe, and methylenbisacrylamide (MBA) as a cross-linker [26]. MBA provides a 3D structure for the maximum interaction between the fluoride ions and the polymeric shell. The nanocomposite has smooth surface morphology and its geometry is spherical. This nanocomposite was utilized for the selective and sensitive colorimetric recognition of fluoride ions in organic solutions at ultra-trace levels. The average size of TiO_2 nanoparticles and polymeric shell on the surface was determined as ~20 nm and ~120 nm, respectively. The visible color change from pale yellow to orange upon the addition of F^- ion was detected by naked eye as well as UV-Vis measurement (Fig. **2**).

The high surface area of the polymer-nanocomposite material and the repeating structure of the polymer may be the attributes for the trace level detection of F^- ion. The polymer nanocomposite selectively bound to F^- with a detection limit of 3 μM; also, no interference from CH_3COO^-, $H_2PO_4^-$, $H_2SO_4^-$, I^-, Cl^- and Br^- was observed. In the presence of more than 1 equiv. of F^- ions, the formation of 1:2 hydrogen bonds between F^- ions and the NH group of nanocomposites was noticed. At 2 equiv. of F^-, deprotonation of thiourea group of the chemosensor occurred.

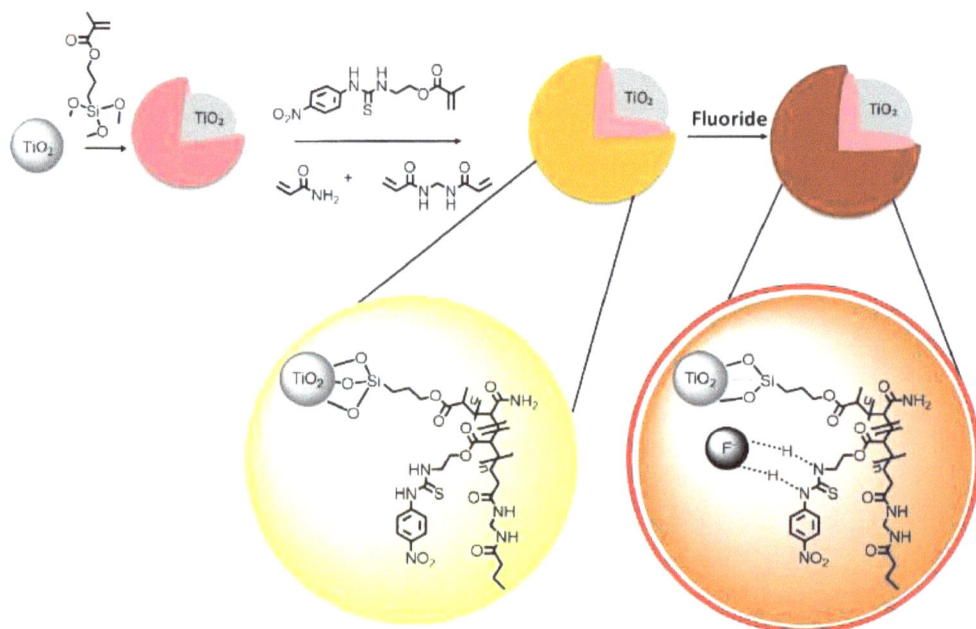

Fig. (2). TiO$_2$-NPs nanocomposite-based colorimetric sensor for fluoride detection [26].

M. Honjo *et al.* described the preparation of a novel nanocomposite for the anion sensing in aqueous media. These nanocomposites (Tb-BTC@Lipo) comprise liposomes and fluorescent coordination polymer (Tb-BTC) in a confined environment [27]. For the luminescent coordination polymer nanoparticles, they selected [Tb(BTC)(H$_2$O)$_6$] (Tb-BTC; BTC=benzene-1,3,5-tricarboxylate) having a 1-D chain structure, as it has a strong fluorescence and could be prepared in an aqueous medium without heating. The high water dispersibility of the nanocomposite is due to the formation of nanocrystals of Tb-BTC, which are grown anisotropically. This anisotropic growth of the nanocrystals could be attributed to the incorporation of liposomes into the AmB channels (Fig. **3**).

The nanocomposite (Tb-BTC@Lipo) exhibits a high sensitivity to trace the amount of fluoride in water as compared to only Tb-BTC. The higher surface area of the anisotropic Tb-BTC nanocrystals confined within the liposomes is responsible for high sensitivity. The high quenching efficiency with F$^-$ ion among other anions reveals the binding affinity of Tb^{3+} to F$^-$. 1D chains have spaces between them, which could be accessed by the smallest ion radius, F$^-$. The accessibility of F$^-$ ions to the surface and the inside is improved due to the anisotropic nanoplate of Tb-BTC, which accelerates the exchange of BTC^{3-} to F$^-$ ions. Hence, the detection and capturing of optimized elements of biological small molecules in living organisms could be made possible with these nanocomposites.

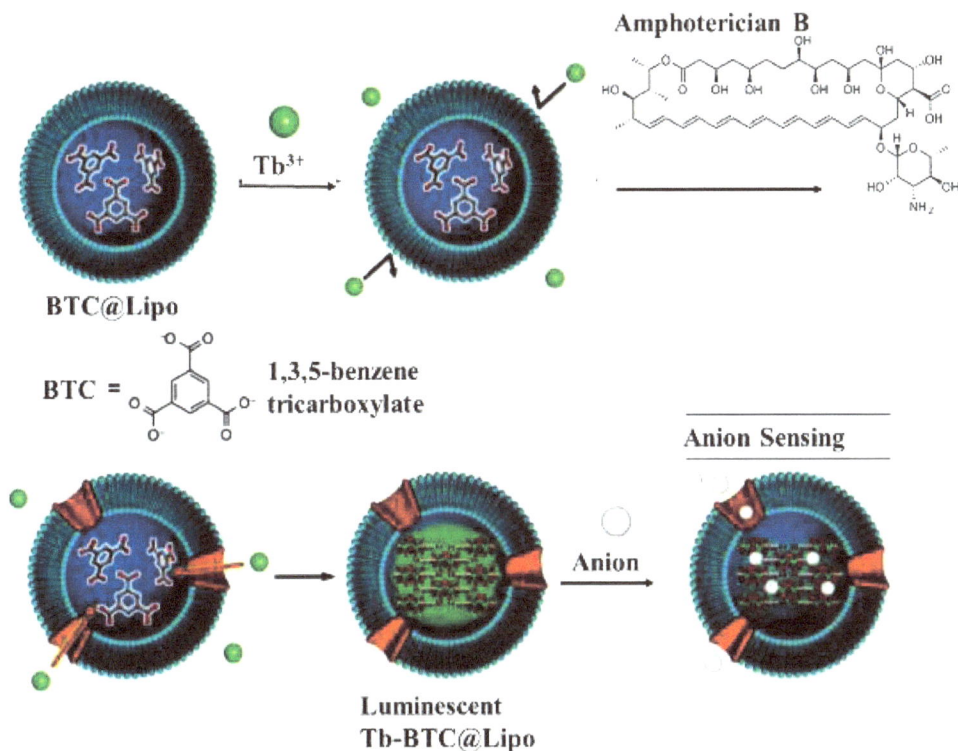

Fig. (3). Anion sensing behavior and schematic depiction of fluorescent Tb-BTC@Lipo (reproduced with permission from Dalton Trans., 2017,46,7141-7144 RSC Publications).

Sulfide Detection

There is an increased interest in the higly sensitive and selective detection of toxic thiols using nanoscale materials [28]. Among thiols, Hydrogen sulfide (H_2S) detection is more important due to environmental toxicity and its essential role in various biological processes [29]. Optical sensors are advantageous among different sensors due to the on-field detection and easy sample preparation. In this regard, nanocomposite-based H_2S sensing materials have not been much explored. H_2S can exist in solution in the form of S^{2-} and SH^-.

H. Sammi and co-workers synthesized GQDs:Eu-ZIF-8 nanocomposites, employed as the susceptible and selective platform for detecting S^{2-} pollutants in ethanol medium [30]. One-step simple solvothermal approach was employed to synthesize highly luminescent GQDs:Eu from the pre-synthesized graphene oxide (GO) layers and purified by gel permeation chromatography. Then, GQDs:Eu-ZIF-8 nanocomposites were synthesized upon incremental addition of ZIF-8 (zeolitic imidazole frameworks) NCs to the GQDs:Eu solution. These nanocomposites are formed by the surface adsorption between the ZIF-8 and GQDs:Eu. The fluorescence behaviour behavior of the nanocomposites may be altered due to the host-guest interaction between the GQDs and ZIF-8, and this altered fluorescence behaviour pattern is used for the sensing of pollutant species. The GQDs:Eu-ZIF-8 nanocomposites are highly sensitive towards the detection of S^{2-} with a very high turn-on fluorescence over other ions with LOD of 0.12 ppm. Hence, these nanocomposites are viable to detect S^{2-} ions in solution. Volume independent turn-on fluorescence behavior of the sensor is observed with an increase in the amount of S^{2-} at the concentration of 1 ppm.

Zou and co-workers prepared gellan gum-silver nanoparticle-based bio-nano composites and used them as a colorimetric sensor for H_2S [31]. In the synthesis of AgNPs, gellan gum acts as both a capping as well as a reducing agent. The yellow color was displayed by the prepared GG-AgNPs solution showing absorbance at 420 nm. The treatment of HCl solution with sodium sulfide produced H_2S gas, which on reaction with GG-AgNPs detected H_2S colorimetrically. At pH 7, the absorption peak at 420 nm decreased with the disappearance of yellow color. The detection limit was found to be 0.81 μM. A sensory hydrogel of GG-AgNPs was prepared to monitor the spoilage of meat. It shows a color change from yellow to colorless from chicken breast and silver carp fish in a packaging system, thus, it is used as an *in situ* and non-destructive method for the detection of H_2S generated from packing materials. GG-AgNPs show high selectivity towards H_2S among the several volatile compounds generated from spoiled silver carp and chicken breast. The potential application of hydrogel was explored in smart food packaging due to its low toxicity and low cost.

Cyanide Detection

Cyanide (CN^-) ion is known as a highly toxic anion and lethal substance to humans and animals, owing to its ability to bind with hemoglobin [32]. With reference to the WHO, the minimum permissible limit of cyanide in drinking water is 70 μg/L. Hence, it is essential to estimate the trace amount of cyanides in various samples, such as food products, water samples, blood, serum and urine samples [33]. Although many nanoparticles-based sensors are known for the optical detection of cyanide ion in water, the use of nano-composite materials is less known [34].

Cuc-Ags PVA nanocomposite films were developed and used for colorimetric CN^- detection. The performance of the nanocomposite films was determined by dipping the film into the cyanide spiked water solution [34]. A color transition from yellowish-brown to colorless was observed with the naked eye (Fig. **4**). The colorimetric detection was in line with the solution phase studies. The LOD of CN⁻ was found to be 0.15 μM with good selectivity over other anions. The real-life application of CN⁻ detection in apricot seed extract was also established by using CucAgs NPs.

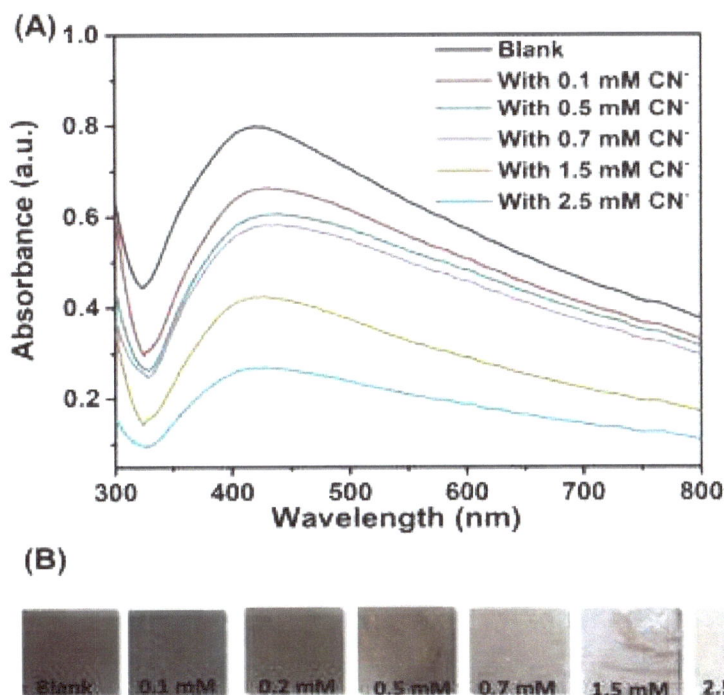

Fig. (4). (A) Change in the absorption spectrum of Cu_cAg_s PVA NC films with varying amounts of cyanide ion; **(B)** Change in color of NC films with different concentrations of cyanide ion (Reproduced with permission from Mater. Chem. Phys., 260 (2021) 124132. Elsevier).

In another report, methyl orange nanocomposites functionalized by chitosan (CFMO) with copper (CFMOCuNPs) and iron (CFMOFeNPs) were reported as sensors for CN and AsO_3^{-4} in aqueous solutions [35]. The prepared chitosan-based metal composites were proceeded to nano-size by chemical reduction methods. The chitosan functionalized metal nanocomposites were characterized by scanning electron microscopy (SEM), IR spectroscopy, EDX spectra and high-resolution powder XRD. The metal-induced cationic N,N-dimethylbenzamine displayed the colorimetric responses with anions CN and AsO_3^{-4} among the test anions *via* electrostatic attraction. These nanocomposite probes exhibited high selectivity and sensitivity towards CN⁻ and AsO_3^{-4} with LOD of 1 and 2 ppm in an aqueous medium, respectively, with a color change of golden yellow and pink.

CONCLUDING REMARKS

In conclusion, the recent developments in nanocomposite-based fluorescent and colorimetric sensors to detect toxic ions, such as mercury, chromium, arsenate, fluoride, cyanide and sulfide, have been summarized. The structural compositions of nanocomposites, detection mechanism, spectral properties, sensitivity and selectivity, are described in detail. The nanocomposite sensors have more advantages over conventional probes due to low-cost detection platforms, rapid detection, and also the possibility to make portable devices for commercialization.

Nanocomposite sensors exhibit better biocompatibility and biodegradability, and less cytotoxicity, compared to the precursor nanomaterials. For example, chitosan is known for its less toxic substance; it tends to reduce the toxicity of synthetic nanomaterial when used in combination. Nanocomposite colorimetric sensor could be used for "real-time monitoring" for toxic anions and cations in environmental and biological samples. It is evident from the examples discussed in the chapter that nanocomposites have attracted less attention over small/macro molecule-based sensors. This may be due to the lack of a proper combination of nanomaterials and matrix that make a stable component for the sensing application. In the near future, a synergistic combination of various nanomaterials and matrix will lead to more promising examples involving the effective detection of anions and cations.

CONSENT FOR PUBLICATION

Not applicable.

CONFLICT OF INTEREST

The author declares no conflict of interest, financial or otherwise.

ACKNOWLEDGEMENTS

All the authors acknowledge the Department of Chemistry, NIT Kurukshetra, for the research facilities. DAJ gratefully acknowledges the funding agency SERB-DST (EMR/2017/004085) and the Department of Biotechnology (BT/PR23533/NNT/28/1301/2017) for the research fund.

REFERENCES

[1]   Idumah, C.I.; Ezeani, E.O.; Nwuzor, I.C. A review: advancements in conductive polymers nanocomposites, Polym.-Plast. Technol. Mater., (2020) Ahead of Print. **2020**.

[2]   Vera, M.; Mella, C.; Urbano, B.F. Smart polymer nanocomposites, recent advances and perspectives. *J. Chil. Chem. Soc.,* **2020**, *65*(4), 4973-4981.
       [http://dx.doi.org/10.4067/S0717-97072020000404973]

[3]   Govan, J. Recent advances in magnetic nanoparticles and nanocomposites for the remediation of water resources. *Magnetochemistry,* **2020**, *6*(4), 49.
[http://dx.doi.org/10.3390/magnetochemistry6040049]

[4]   Huynh, T-P. Chemical and biological sensing with nanocomposites prepared from nanostructured copper sulfides. *Nano Futures,* **2020**, *4*(3), 032001.
[http://dx.doi.org/10.1088/2399-1984/ab9a28]

[5]   Baker, A.; Khan, M.S.; Iqbal, M.Z.; Khan, M.S. Tumor-targeted Drug Delivery by Nanocomposites. *Curr. Drug Metab.,* **2020**, *21*(8), 599-613.
[http://dx.doi.org/10.2174/1389200221666200520092333] [PMID: 32433002]

[6]   Dhas, N.; Kudarha, R.; Garkal, A.; Ghate, V.; Sharma, S.; Panzade, P.; Khot, S.; Chaudhari, P.; Singh, A.; Paryani, M.; Lewis, S.; Garg, N.; Singh, N.; Bangar, P.; Mehta, T. Molybdenum-based hetero-nanocomposites for cancer therapy, diagnosis and biosensing application: Current advancement and future breakthroughs. *J. Control. Release,* **2021**, *330*, 257-283.
[http://dx.doi.org/10.1016/j.jconrel.2020.12.015] [PMID: 33345832]

[7]   Ma, M.; Li, H.; Xiong, Y.; Dong, F. Rational design, synthesis, and application of silica/graphene-based nanocomposite: A review. *Mater. Des.,* **2021**, *198*, 109367.
[http://dx.doi.org/10.1016/j.matdes.2020.109367]

[8]   Qi, M.; Li, S.; Du, Y.; Ye, P.; Han, H.; He, Q.; Wang, M. Research progress of the graphene-based nano-composites. *Gongneng Cailiao,* **2017**, *48*, 3042-3049.

[9]   Sharma, N.; Rana, V.S. A review on polysaccharide based nanocomposite hydrogel systems fabrication using diverse reinforcing materials. *J. Polym. Compos.,* **2020**, *8*, 6-17.
[http://dx.doi.org/10.37591/jopc.v8i1.3732]

[10]   Wang, J.; Shen, H.; Xia, Y.; Komarneni, S. Light-activated room-temperature gas sensors based on metal oxide nanostructures: A review on recent advances, Ceram. *Int., (2020) Ahead of Print.,*

[11]   Sedghi, R.; Kazemi, S.; Heidari, B. Novel selective and sensitive dual colorimetric sensor for mercury and lead ions derived from dithizone-polymeric nanocomposite hybrid. *Sens. Actuators B Chem.,* **2017**, *245*, 860-867.
[http://dx.doi.org/10.1016/j.snb.2017.01.203]

[12]   Sedghi, R.; Heidari, B.; Behbahani, M. Synthesis, characterization and application of poly(acrylamide-co-methylenbisacrylamide) nanocomposite as a colorimetric chemosensor for visual detection of trace levels of Hg and Pb ions. *J. Hazard. Mater.,* **2015**, *285*, 109-116.
[http://dx.doi.org/10.1016/j.jhazmat.2014.11.049] [PMID: 25497023]

[13]   Yan, Z.; Hu, L.; Nie, L.; You, J. One-pot preparation of graphene–Ag nano composite for selective and environmentally-friendly recognition of trace mercury(ii). *RSC Advances,* **2016**, *6*(111), 109857-109861.
[http://dx.doi.org/10.1039/C6RA16810H]

[14]   Guo, L.; Song, Y.; Cai, K.; Wang, L. "On-off" ratiometric fluorescent detection of Hg^{2+} based on N-doped carbon dots-rhodamine B@TAPT-DHTA-COF. *Spectrochim. Acta A Mol. Biomol. Spectrosc.,* **2020**, *227*, 117703.
[http://dx.doi.org/10.1016/j.saa.2019.117703] [PMID: 31685421]

[15]   Zhang, D.; Jiang, W.; Zhao, Y.; Dong, Y.; Feng, X.; Chen, L. Carbon dots rooted PVDF membrane for fluorescence detection of heavy metal ions. *Appl. Surf. Sci.,* **2019**, *494*, 635-643.
[http://dx.doi.org/10.1016/j.apsusc.2019.07.141]

[16]   Yan, Z.; Xue, H.; Berning, K.; Lam, Y-W.; Lee, C-S. Identification of multifunctional graphene-gold nanocomposite for environment-friendly enriching, separating, and detecting Hg^{2+} simultaneously. *ACS Appl. Mater. Interfaces,* **2014**, *6*(24), 22761-22768.
[http://dx.doi.org/10.1021/am506875t] [PMID: 25458522]

[17]   Liu, Y.; Xu, L.; Liu, J.; Liu, X. Simultaneous enrichment, separation and detection of mercury(ii) ions

using cloud point extraction and colorimetric sensor based on thermoresponsive hyperbranched polymer–gold nanocomposite. *Anal. Methods*, **2015**, *7*(24), 10151-10161.
[http://dx.doi.org/10.1039/C5AY02406D]

[18]   Hu, L.; Zhu, B.; Zhang, L.; Yuan, H.; Zhao, Q.; Yan, Z. Chitosan-gold nanocomposite and its functionalized paper strips for reversible visual sensing and removal of trace Hg^{2+} in practice. *Analyst (Lond.)*, **2019**, *144*(2), 474-480.
[http://dx.doi.org/10.1039/C8AN01707G] [PMID: 30426976]

[19]   Ma, C.; Ma, Y.; Sun, Y.; Lu, Y.; Tian, E.; Lan, J.; Li, J.; Ye, W.; Zhang, H. Colorimetric determination of Hg^{2+} in environmental water based on the Hg^{2+}-stimulated peroxidase mimetic activity of MoS_2-Au composites. *J. Colloid Interface Sci.*, **2019**, *537*, 554-561.
[http://dx.doi.org/10.1016/j.jcis.2018.11.069] [PMID: 30471610]

[20]   Truskewycz, A.; Beker, S.A.; Ball, A.S.; Murdoch, B.; Cole, I. Incorporation of quantum carbon dots into a PVP/ZnO hydrogel for use as an effective hexavalent chromium sensing platform. *Anal. Chim. Acta*, **2020**, *1099*, 126-135.
[http://dx.doi.org/10.1016/j.aca.2019.11.053] [PMID: 31986269]

[21]   Ghosh, A.; Jose, D.A.; Kaushik, R. Anthraquinones as versatile colorimetric reagent for anions. *Sens. Actuators B Chem.*, **2016**, *229*, 545-560.
[http://dx.doi.org/10.1016/j.snb.2016.01.140]

[22]   Jose, D.A.; Kumar, D.K.; Ganguly, B.; Das, A. Urea and thiourea based efficient colorimetric sensors for oxyanions. *Tetrahedron Lett.*, **2005**, *46*(32), 5343-5346.
[http://dx.doi.org/10.1016/j.tetlet.2005.06.011]

[23]   Jose, D.A.; Kumar, D.K.; Ganguly, B.; Das, A. Efficient and simple colorimetric fluoride ion sensor based on receptors having urea and thiourea binding sites. *Org. Lett.*, **2004**, *6*(20), 3445-3448.
[http://dx.doi.org/10.1021/ol048829w] [PMID: 15387519]

[24]   Jose, D.A.; Kumar, D.K.; Kar, P.; Verma, S.; Ghosh, A.; Ganguly, B.; Ghosh, H.N.; Das, A. Role of positional isomers on receptor-anion binding and evidence for resonance energy transfer. *Tetrahedron*, **2007**, *63*(48), 12007-12014.
[http://dx.doi.org/10.1016/j.tet.2007.09.008]

[25]   Jose, D.A.; Kar, P.; Koley, D.; Ganguly, B.; Thiel, W.; Ghosh, H.N.; Das, A. Phenol- and catechol-based ruthenium(II) polypyridyl complexes as colorimetric sensors for fluoride ions. *Inorg. Chem.*, **2007**, *46*(14), 5576-5584.
[http://dx.doi.org/10.1021/ic070165+] [PMID: 17569524]

[26]   Sedghi, R.; Javadi, H.; Heidari, B.; Rostami, A.; Varma, R.S. Efficient Optical and UV-Vis Chemosensor Based on Chromo Probes-Polymeric Nanocomposite Hybrid for Selective Recognition of Fluoride Ions. *ACS Omega*, **2019**, *4*(14), 16001-16008.
[http://dx.doi.org/10.1021/acsomega.9b02098] [PMID: 31592470]

[27]   Honjo, M.; Koshiyama, T.; Fukunaga, Y.; Tsuji, Y.; Tanaka, M.; Ohba, M. Sensing of fluoride ions in aqueous media using a luminescent coordination polymer and liposome composite. *Dalton Trans.*, **2017**, *46*(22), 7141-7144.
[http://dx.doi.org/10.1039/C7DT01071K] [PMID: 28534572]

[28]   Jose, D.A.; Sakla, R.; Sharma, N.; Gadiyaram, S.; Kaushik, R.; Ghosh, A. Sensing and Bioimaging of the Gaseous Signaling Molecule Hydrogen Sulfide by Near-Infrared Fluorescent Probes. *ACS Sens.*, **2020**, *5*(11), 3365-3391.
[http://dx.doi.org/10.1021/acssensors.0c02005] [PMID: 33166465]

[29]   Kaushik, R.; Sakla, R.; Ghosh, A.; Selvan, G.T.; Selvakumar, P.M.; Jose, D.A. Selective Detection of H_2S by Copper Complex Embedded in Vesicles through Metal Indicator Displacement Approach. *ACS Sens.*, **2018**, *3*(6), 1142-1148.
[http://dx.doi.org/10.1021/acssensors.8b00174] [PMID: 29856208]

[30]   Sammi, H.; Kukkar, D.; Singh, J.; Kukkar, P.; Kaur, R.; Kaur, H.; Rawat, M.; Singh, G.; Kim, K-H.

Serendipity in solution–GQDs zeolitic imidazole frameworks nanocomposites for highly sensitive detection of sulfide ions. *Sens. Actuators B Chem.,* **2018**, *255*, 3047-3056.
[http://dx.doi.org/10.1016/j.snb.2017.09.129]

[31] Zhai, X.; Li, Z.; Shi, J.; Huang, X.; Sun, Z.; Zhang, D.; Zou, X.; Sun, Y.; Zhang, J.; Holmes, M.; Gong, Y.; Povey, M.; Wang, S. A colorimetric hydrogen sulfide sensor based on gellan gum-silver nanoparticles bionanocomposite for monitoring of meat spoilage in intelligent packaging. *Food Chem.,* **2019**, *290*, 135-143.
[http://dx.doi.org/10.1016/j.foodchem.2019.03.138] [PMID: 31000029]

[32] Kaushik, R.; Sakla, R.; Ghosh, A.; Dama, S.; Mittal, A.; Jose, D.A. Copper Complex-Embedded Vesicular Receptor for Selective Detection of Cyanide Ion and Colorimetric Monitoring of Enzymatic Reaction. *ACS Appl. Mater. Interfaces,* **2019**, *11*(50), 47587-47595.
[http://dx.doi.org/10.1021/acsami.9b17316] [PMID: 31741372]

[33] Kaushik, R.; Ghosh, A.; Singh, A.; Gupta, P.; Mittal, A.; Jose, D.A. Selective Detection of Cyanide in Water and Biological Samples by an Off-the-Shelf Compound. *ACS Sens.,* **2016**, *1*(10), 1265-1271.
[http://dx.doi.org/10.1021/acssensors.6b00519]

[34] Kumar, R.; Kaushik, R.; Kumar, R.; Jose, D.A.; Sharma, P.K.; Sharma, A. Facile synthesis of CucAgs based nanoparticles and nanocomposites as highly selective and sensitive colorimetric cyanide sensor. *Mater. Chem. Phys.,* **2021**, *260*, 124132.
[http://dx.doi.org/10.1016/j.matchemphys.2020.124132]

[35] Ejeromedoghene, O.; Adewuyi, S.; Amolegbe, S.A.; Akinremi, C.A.; Moronkola, B.A.; Salaudeen, T. Electrovalent chitosan functionalized methyl-orange/metal nanocomposites as chemosensors for toxic aqueous anions. *Nano-Structures & Nano-Objects,* **2018**, *16*, 174-179.
[http://dx.doi.org/10.1016/j.nanoso.2018.06.004]

CHAPTER 14

Nanocomposites for Humidity Sensor: An Overview

Pratibha Singh[1], Chandra Shekhar Kushwaha[1] and **Saroj Kr Shukla[1,*]**

[1] *Department of Polymer Science, Bhaskaracharya College of Applied Sciences, University of Delhi, Delhi-110075, India*

Abstract: The present chapter describes the synthesis and applications of different classes of nanocomposites in humidity sensing applications, along with their innovative surface and responsive properties. The uses of nanocomposite-based humidity sensors in different fields like environmental monitoring, packaging, and the medical field have been described with suitable examples and illustrations. Furthermore, the humidity sensing mechanism of nanocomposite based humidity sensors are explained with sensing parameters and with future requirements

Keywords: Humidity sensors, Hybrid materials, Sensing mechanism, and applications.

INTRODUCTION

The synergistic properties in hybrid materials with nano confinements have yielded several advantageous features for effective chemical and biochemical sensing with precise parameters. Some of the notable advantageous features of this hybrid, in general also called nanocomposite, are surface area, selective catalysis, porosity, responsive and induced physical properties. These properties are also frequently used in sensing humidity in gas, liquid, and solid substances as optical, electrical, and mechanical transducers [1, 2].

Although humidity sensing is the oldest analytical tool for monitoring atmospheric prediction since the 12th century BC by Shuang Ruller, in China, after measuring the change in mass change due to adsorption humidity on charcoal due to variation in humidity contents. However, in the current era, the importance of humidity sensors has drastically increased due to the importance of water molecules in several sustaining reactions of living and nonliving bodies [3, 4]. Thus, the importance of humidity sensors encouraged scientists to use a wide

---

*Corresponding Author S. K. Shukla**: Department of Polymer Science, Bhaskaracharya College of Applied Sciences, University of Delhi, Delhi-110075, India, E-mail: sarojshukla2003@yahoo.co.in

Manorama Singh, Vijai K Rai and Ankita Rai (Eds.)

range of materials like ceramic, polymers, and carbonaceous materials with certain limitations.

However, for advancing the humidity sensing properties, the different composites materials like metal oxide to metal oxides, metal to metal oxides, polymer to biopolymer, metal oxides to polymers, carbon nanostructure to the polymer are used for precised monitoring of humidity. For example, Shukla *et al.* have increased the humidity sensing range of pristine polyaniline after grafting it with cellulose by 33% [5]. Some other significant synergistic effect of material on humidity sensing is given in Table **1**.

Table 1. Evolution of significant synergistic properties in nanocomposites.

S. N.	Composition	Synergistic Properties	Remark	Ref.
1	WO3 and MWCNT	Synergistic evolution of the *p-p* junction	Enhanced sensitivity in wider range humidity	[6]
2	TiO_2, $(K,Na)NbO_3$	Evolved hetero structure	Improved sensitivity by two to four order	[7]
3	ZnO and Polyvinyledene fluoride	Improved surface oxygen vacancy defects	Improved response and recovery times of 30s and 51s recovery, respectively	[8]
4	SnO_2 and TiO_2	Heavy ion beam induced dense electronic excitation	Long term reproducibility and recovery	[9]
5	Molybdenum disulfide and polyvinyl pyrrolidone	Inkjet printable to fabricate humidity sensors.	High sensitivity, ultrafast response/recovery behavior, and good reproducibility	[10]

The finding reveals the importance of making nano-sized composites in humidity sensing in order to fulfill the requirement in different filed. Furthermore, the existing literature indicates the effective increment in humidity sensing technology due to advances in nanocomposites with certain limitations like wide range sensitivity from gas to solid. This synergized advancement in humidity sensing using nanocomposites is presented in this chapter, along with findings and challenges.

OVERVIEW ON NANOCOMPOSITE BASED HUMIDITY SENSORS

In general, the nanocomposites are multicomponent hybrid materials with at least one component in nanosize, *i.e.,* 1 to 100 nm. Further, depending on the materials, the nanocomposite is classified into the following three categories: a) metal matrix, b) ceramic matrix, and c) polymer matrix. In resultant materials, all matrices incorporate their unique features, such as metal stretchability, ceramic stability, and polymer processability. As a result, all features contribute to humidity sensing; however, ceramic and polymer-based nanocomposites are most commonly used in humidity sensing applications. Another important aspect is to control dispersion and agglomeration of dispersed phase during composite formations along with size confinements [11, 12]. Therefore, several methods are used for making a composite, but broadly they are grouped as *ex-situ* or *in-situ* methods for composite formation. In the *ex-situ* method, both components are prepared separately, and after that, they are composed with a defined matrix using different energy like chemical energy in solution method, thermal in melt mixing, and electrical energy in electrical depositions. However, in the *in-situ* method, the composite is prepared through the process of the development of a matrix in the presence of components. Both methods are still in use for the preparation of different composites with their inherited properties. The basic comparison of both methods is given in Table **2**, along with examples.

Table 2. Comparison between *ex-situ* and *in-situ* methods [13].

Ex-situ	*In-situ*
Time consuming, bulk interaction, suitable for large scale production with disadvantages of requisite nanoparticles, dispersibility and control of long term aggregations	The simple and effective route to control long-term aggregation and good spatial distribution along with the presence of unreacted educts.

The involved fundamental principle in humidity sensing is the monitoring of different induced electrical, optical and magnetic properties against different humidity levels.

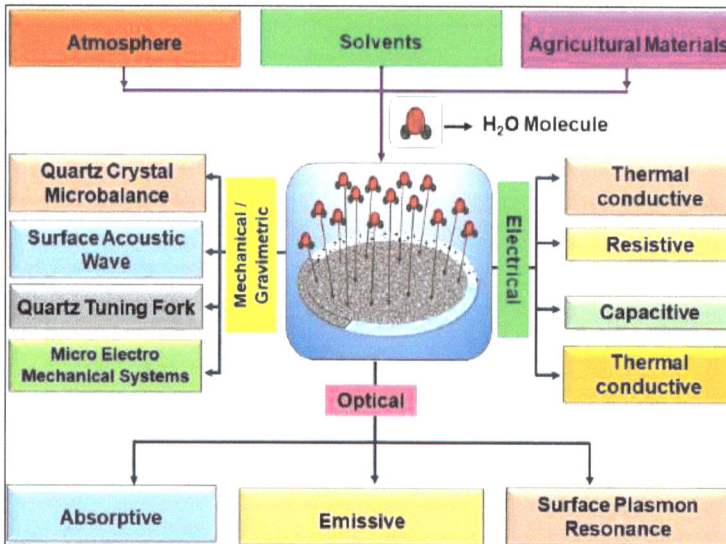

Fig. (1). Different class of humidity sensing principles [14].

In general, the different kinds of humidity in the surrounding is indicated in the following terms:

Absolute Humidity

It represents the number of water molecules present in a particular volume of gas, which is expressed as mg/ml or g/m^3 (Equation 1).

$$\text{Absolute humidity} = \frac{\text{Mass of water vapour}}{\text{Volume}} \tag{1}$$

Relative Humidity

It indicates the ratio for the relative mass of the water vapour present in a fixed volume of gas to the maximum water-retaining capacity of the gas at a specific temperature and pressure, as represented through Equations 2 & 3 and also calculated by partial pressure.

$$\text{Relative humidity} = \frac{\text{Water vapour pressure}}{\text{Saturated vapour pressure}} \tag{2}$$

$$\text{Relative humidity (\%)} = \frac{\text{Mass vapour present}}{\text{Mass when saturated}} \times 100 \tag{3}$$

Dew Point

It is a characteristic temperature to condense gas into a liquid and is also an indicator to not hold further water vapour as well full saturation. Thus, the cooling beyond the dew points condenses the water vapour into liquid form.

All three parameters are important in the context of environmental monitoring due to changes in the properties of the materials are generated due to the interaction of water molecules on the surface of sensing materials. For example, the adsorption of humidity on a ceramic surface develops hydroxyl and hydronium ions; the presence of these ions generates induced current as well as reduces the resistance of the composite. This monitoring of induced resistance and current are the basic principle of amperometry and chemi-resistive sensors [15]. An illustrative example of resistive humidity sensors is depicted in Fig. (**2**) with the generation of ionized ions on PANI composite with the suitability for humidity sensing up to 95 relative humidity.

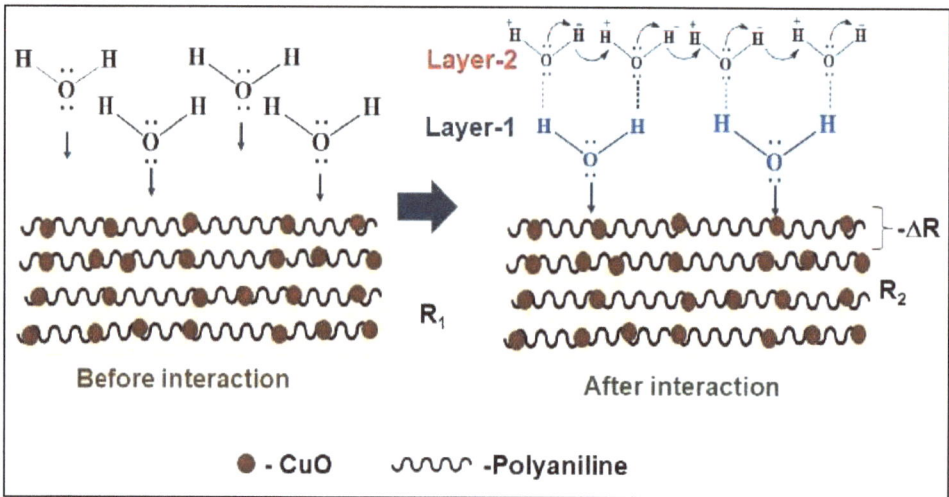

Fig. (2). Basic principles of electrochemical humidity sensing [15].

Similarly, the adsorption of water molecules changes the refractive index of a sensing substrate; thus, the developed hybrid refractive index changes the light permeation intensity through the optical fiber. Therefore, a change in output power has been explored in the sensing of humidity present surrounding, and a simple setup has been developed using different ceramic coating [16]. The experimental setup has been illustrated in Fig. (**3**).

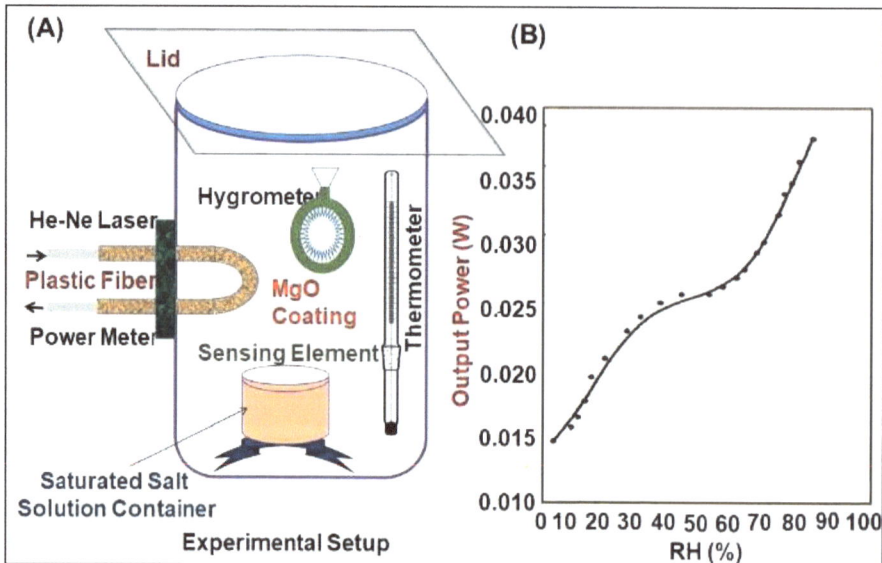

Fig. (3). Optical fiber-based humidity sensors [16].

The gravimetric responses are explored in humidity sensing in terms of the frequency of crystal adsorbed by water molecules. This change in frequency, according to Equation 4, is explored for precise sensing of water molecules.

$$\Delta m = c.\Delta f \tag{4}$$

The abbreviated terms are 'm' for mass of a water molecule, 'c' for constant, and 'f' for crystal frequency.

NANOCOMPOSITES FOR HUMIDITY SENSING

The nanoconfined hybrid materials with tunable hydrophilicity, porosity, surface roughness and interactive sites are promising for use in efficient and wide range humidity sensing. In this regard, the individual constituents of composites have their own impact on humidity sensing applications. In an example, the presence of ZnO added the hydrophilic nature along with ionizability in the polymer matrix of polypyrrole, which increases humidity adsorption properties along with effective channelization of sensing responses. The illustration of the sensing mechanism and role of zinc oxide is depicted in Fig. (**4**).

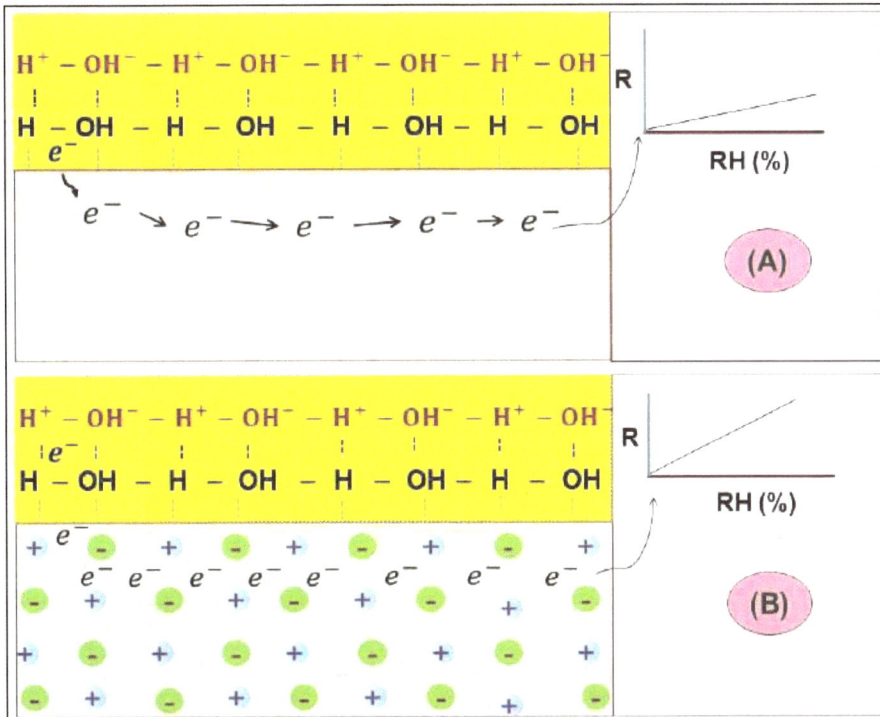

Fig. (4). Impact of ZnO in polypyrrole for electrochemical humidity sensing [14].

Thus, most of the nanocomposites are used in humidity sensing with their own merits and demerits after exploiting different transducers. However, the composites fall into three categories for long-term research and humidity sensing applications.

Ceramic Nanocomposite

Ceramics are stable, high-temperature hybrid materials that contain both metallic and non-metallic constituents and can have amorphous, crystalline, or polycrystalline nature. The addition of nano confinements in ceramic has added several advantageous features like high surface area and unusual oxidation states associated with magnetic, optical, and electrical properties. After applying various transducers, viz. electrical and optical, these qualities have validated the ceramic as a possible candidate for humidity sensing, with several businesses having commercialised the ceramic-based sensor [17]. The sensing mechanism is based on associative and non-dissociative adsorption interaction of water molecules on the ceramic surface, which changes the electrical resistance, capacitance, and impedance for sensing applications. Further, the oxidizing, reducing nature of ceramic dictates the proton generation ability after surface interaction between water and ceramic surface, while the porosity governs the sensing range due to relative induced capillary condensation as per Equation 5.

$$rk = \frac{2\gamma M}{\rho RT ln\left(\dfrac{Ps}{P}\right)} \tag{5}$$

rk is Kelvin radius, P is water vapour pressure, Ps is water vapour pressure at saturation, γ is surface tension, ρ is density, and M is the molecular weight of H_2O. The value of *rk* decides the condensation of water molecules taking place at particular temperatures and the vapor pressures of water molecules. Here, the equilibrium between the adsorption and condensation of humidity controls the behaviour of sensing.

Processing and form of ceramic samples in the pellets, disks, films, bars and single crystal are important aspects for ceramic in humidity sensing. In this regard, different processing techniques like sol-gel, screen printing, solution casting, plasma spray, and spray pyrolysis are used for ceramic-based sensing [18]. The size and thickness of ceramic nanocomposites are other dimensions of ceramic in humidity sensing for effective applications. Some of the significant ceramic nanocomposite-based sensors are listed in Table **3**, along with methods and parameters.

Table 3. Ceramic nanocomposite based humidity sensors.

S. No	Composition	Transducer	Synergistic Properties	Ref.
1	Mixed metal nanocomposites of Cu, Sr, Cd	Impedance	Efficient response, recovery and stability	[19]
2	In_2O_3 and KIT-6	Resistance	Swift response and recovery time for 17 s and 19 s, along with magnificent stability in the 11–98% RH range.	[20]
3	$BaTiO_3$	Ionic conductivity	Improve sensing properties due to optimised hydrophilicity	[21]
4	ZnO/WS_2	Capacitive	More improvements in sensing properties, *i.e.*, response, sensitivity, and hysteresis, than pristine ZnO sensors.	[22]
5	$ZnCr_2O_4–K_2CrO_4$	Resistance	Linear relative humidity range 25-90%. The impedance at 100 Hz, from 10^5 Ω at 10% RH to 10^3 Ω at 93% RH.	[23]
6	Fe_2TiO_5	Resistance	Linear relative humidity range = 40–90%. Response time=16 s, low hysteresis = 8.39% at 25 °C and 2.64% at 50 °C.	[24]
7	$TiO_2/(K,Na)NbO_3$	Impedance	The linear humidity range is 12-94% RH. Response time 25 s and recovery time 38 s. Hysteresis less than 5%.	[7]
8	TiO_2 thin film	Optical	Linear humidity range is 10 - 95%, RH sensitivity is 0.78, response time is 36s and recovery time is 73s.	[25]
9	MgO thin film	Optical	The linear sensor sensitivity in 5–95.	[26]
10	$CsPb_2Br_5/BaTiO_3$	Capacitive	Linear sensing range = 5 – 95% RH, Humidity sensitivity = 21426 pF RH%$^{-1}$ with superior linearity (0.991). Response/recovery time (5 s) and very low hysteresis of 1.7%.	[27]
11	$(Na_{0.25}Nb_{0.75})_xTi_{1-x}O_2$, where x = 0, 0.01, 0.02, and 0.05	Capacitive	Linear sensing range = 11 – 96% RH, Sensitivity of sensor = 13,705 pF/%RH. Humidity sensitivity is tuned by a small dc bias of 0–4 V.	[28]

Carbon Nanocomposite

Currently, carbon nanocomposites have gained significant importance among scientists in the fabrication of different types of humidity sensors due to size and shape-dependent surface and bulk properties. Despite the fact that carbon-based nanomaterials are hydrophobic by nature, chemical functionalization using covalent and non-covalent interactions has been widely explored for the use of CNCs in humidity sensing applications after the incorporation of metals, metal oxides, and biopolymers. Dispersion is another issue with carbon nanomaterials in composite forms, thus numerous innovations are being investigated, such as the use of surfactants, mixing, and heating, in order to develop suitable CNCs for efficient humidity sensing applications [29, 30]. *In situ* polymerization is another strategy to prepare the CNCs with specific structural design and properties with tunable interactivity and responsiveness after grafting with polymeric or non-polymeric chain. Thus, developed CNCs have been explored as a potential candidate in humidity sensing due to surface interacting properties for advanced sensing applications. In an example, Yu *et al.* has prepared the graphite carbon nitride and zinc oxide composite with selective defect of oxygen with higher adsorption energy towards water molecules. Breath monitoring with high humidity responses, low hysteresis, and good linearity has been investigated further [31]. Some other important CNCs based sensors are tabulated in Table **3**, along with their properties and significances.

Table 4. Carbon-based composite for humidity sensors.

S. No	Compositions	Transducer	Synergistic Properties	Ref.
1	Graphene/SWCNT	Conductance	Linear humidity range of 5–80%, Response/recovery time is 198/110 ms. The maximum sensitivity is ~1820% at the RH value of 80%.	[32]
2	Fe3O4/Au/CNT	Resistance	The linear humidity sensing range is 10-70%RH. The response and recovery time is 200s and 250s. The sensitivity is 11.4%.	[33]
3	g-C_3N_4	Resistance	The linear humidity sensing range is 11-97% RH. The response and recovery time is 0.3 s and 11 s. The sensitivity is 6457548.8 Ω/%RH.	[34]
4	Graphene oxide (GO)	Quartz crystal microbalance (QCM)	Leaner sensing range = 23 - 92% RH, response/recovery time =10 s/11 s,	[35]
5	MWCNTs-Chitosan	Quartz crystal microbalance (QCM)	The linear humidity sensing range is 11–95% RH. The sensitivity is 46.7 Hz/% RH; It has humidity hysteresis of ~ 1.1% RH, quick response, and recovery time 75 s and 34 s, respectively.	[36]
6	RGO)-Polyethylene Oxide (PEO)	Quartz crystal microbalance (QCM)	The linear humidity sensing range is 0–84% RH. The sensitivity is 20 Hz/%RH. Response and recovery times are 11 s and 7 s at 84% @ 90% RH, respectively, and negligible hysteresis ~1.21%RH.	[37]

(Table 4) cont.....

S. No	Compositions	Transducer	Synergistic Properties	Ref.
7	Carboxymethyl cellulose (CMC)/ (CNTs)	Optical	The linear humidity sensing range is 35–85% RH. Humidity sensitivity is 230.95 pm/%RH and temperature sensitivity is 26.35 pm/°C. Response and recovery time are 2.67 s and 3.28 s, respectively.	[38]
8	rGO/Fe$_2$O$_3$	Impedance	The linear humidity sensing range is 11–97% RH. The response times are 56, 34, and 63 s at of 43, 75, and 97% RH, respectively	[39]
9	PVA/CNTs	Optical	The linear humidity sensing range is 30–90% RH. Sensitivity is 0.0369 dB/%RH Response and recovery time are 2.53 s and 2.67 s.	[40]
10	cellulose nanofiber/carbon nanotube (CNF/CNT)	Current-voltage (I-V) curves	The linear humidity sensing range is 11–95% RH. Response and recovery times are 321 and 435 s, respectively. Sensor response is 65.0% ($\Delta I/I_0$) and hysteresis is 8.4% RH.	[41]
11	o-MWCNTs coated paper	Current-voltage (I-V) curves	The linear humidity sensing range is 11–95% RH. Response and recovery times are 470 and 500 s, respectively. Sensor response is 33.0% ($\Delta I/I_0$) and hysteresis is 7.6% RH	[42]

Polymer Nanocomposite

Processability, functionality, and long-term stability of polymers are notable features of polymers to be used in humidity sensing along with their inherited disadvantages of conducting nature as well as hydrophobic behavior [43]. However, these behaviors are improved with the use of conductive biopolymers along with the incorporation of metals and metal oxides. The insertion of heterogeneity in the polymer chain is another significant improvement in the properties of the polymer to be used in humidity sensing applications. Shukla *et al.* have prepared electro-chemical humidity sensors using CuO encapsulated polyaniline prepared by *in-situ* polymerization method. The composite exhibited improved hydrophilicity, electrical conductivity and adsorption properties for effective humidity sensing. The proposed sensor exhibited the highly improved sensing properties, *i.e.*, sensing range of 10–95% RH, sensitivity 4.5 ohm/RH, response time 40 s, and recovery time 55 s, and the sensing behavior are shown in Fig. (**5**) [15].

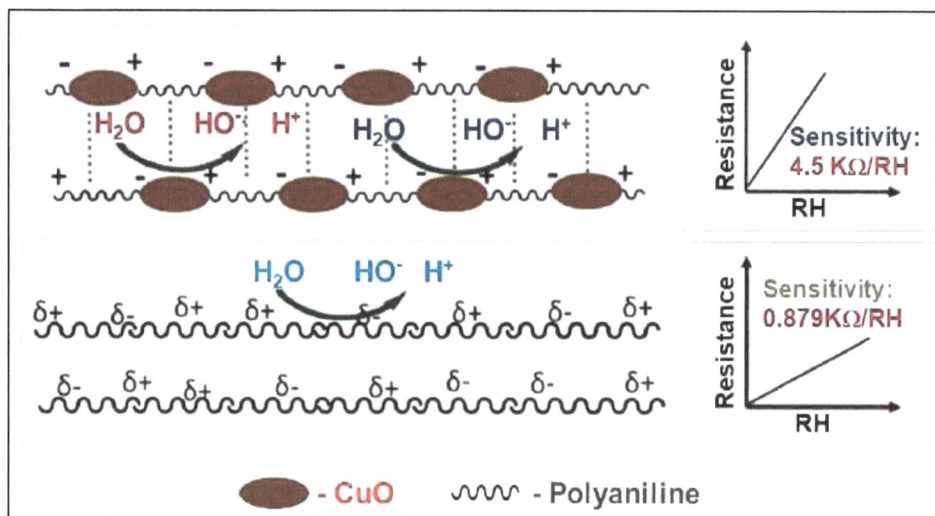

Fig. (5). Humidity sensing behavior of CuO/PANI nanocomposites [15].

The heterogeneity in hybrid structure also allows coordinating nature of sensing substrate with water molecules as ligand molecules. Thus, coordinating interaction between a water molecule and sensing substrate allows quenching in luminesce intensity along with light permeability. This interaction allows optical humidity sensing for quality control of organic solvents [44]. Some other important explored polymer composites are listed in Table **5**.

Table 5. Polymer composites for humidity sensors.

S. No	Composition	Transducer	Linear Sensing Range (%RH)	Synergistic Properties	Ref.
1	Cellulose/KOH ionic film	Impedance	11.3–97.3	The response and recovery time is 6.0 and 10.8 s, respectively. Very small hysteresis ~ 0.57%,	[45]
2	MCM-41/PPy	Capacitance	11–95	The response and recovery time is 915 s and 100 s. The sensitivity is 119 pF/RH.	[46]
3	poly (ether-ether ketone)/polyvinyl butyral (SPEEK/PVB)	Resistance	11–98	Small hysteresis ~2.68%, fast response time <1 s and rapid recovery time 5 s. mouth-breathing rate inhalation is about 2.8 s.	[47]
4	MCM-41/PEDOT	Capacitance	11–95	Response-recovery times are 165 s and 115 s.	[48]

(Table 5) cont.....

S. No	Composition	Transducer	Linear Sensing Range (%RH)	Synergistic Properties	Ref.
5	Polyvinylidene fluoride (PVDF-TiO_2)	Capacitance	30–80	Response and recovery times are 45 s and 11 s, respectively. The average of hysteresis is 1.05%.	[49]
6	CuO/PANI	Resistive	10–95	Response and recovery times are 40 s and 55 s, respectively. The sensitivity of the sensor is 0.879 $k\Omega/RH$.	[15]
7	PANI/NiO	Resistive	5–90	Response and recovery times are 60 s and 90 s, respectively. The sensitivity of the sensor is 7.929 $k\Omega/RH$ and the stability is 90 days.	[50]
8	Co-dispersed-PANI	Optical	20–100	Response and recovery times are 8 s and 60 s, respectively.	[51]
9	SnO2/ PANI	Resistive	5–95	Response and recovery times are 26 s and 30 s, respectively. The sensitivity of the sensor is 0.22%RH.	[52]
10	Polypyrrole-grafted-Cellulose	Resistive	5–95	Response and recovery times are 25 s and 20 s, respectively. The sensitivity of the sensor is 0.20%RH, and stability of 40 days.	[5]

CONCLUDING REMARKS

The impact of nanocomposite in humidity sensing has been described along with synergistic properties like hydrophilicity, adsorption and responsive nature. The structural optimization in hybrid nanomaterials has significantly improved the sensing properties of humidity sensors in terms of different sensing parameters, along with suitability in different fields like packaging, breath monitoring, and agriculture field.

CONSENT FOR PUBLICATION

Not applicable.

CONFLICT OF INTEREST

The authors declare no conflict of interest, financial or otherwise.

ACKNOWLEDGEMENTS

Pratibha Singh and CSK thank UGC, New Delhi, and CSIR-New Delhi [8/642(0002)/2016-EMR-I] for providing fellowship. The authors are also thankful to Dr. Balaram Pani, Principal, Bhaskaracharya College of Applied Sciences, University of Delhi, for maintaining socio academic culture in the college.

REFERENCES

[1]   Vieira, S.M.C.; Beecher, P.; Haneef, I.; Udrea, F.; Milne, W.I.; Namboothiry, M.A.G.; Carroll, D.L.; Park, J.; Maeng, S. Use of nanocomposites to increase electrical "gain" in chemical sensors. *Appl. Phys. Lett.,* **2007**, *91*(20), 203111.
[http://dx.doi.org/10.1063/1.2811716]

[2]   Komarneni, S. Nanocomposites. *J. Mater. Chem.,* **1992**, *2*(12), 1219-1230.
[http://dx.doi.org/10.1039/jm9920201219]

[3]   Chen, Z.; Lu, C. Humidity sensors: A review of materials and mechanisms. *Sens. Lett.,* **2005**, *3*(4), 274-295.
[http://dx.doi.org/10.1166/sl.2005.045]

[4]   Farahani, H.; Wagiran, R.; Hamidon, M.N. Humidity sensors principle, mechanism, and fabrication technologies: a comprehensive review. *Sensors (Basel),* **2014**, *14*(5), 7881-7939.
[http://dx.doi.org/10.3390/s140507881] [PMID: 24784036]

[5]   Shukla, S.K. Synthesis and characterization of polypyrrole grafted cellulose for humidity sensing. *Int. J. Biol. Macromol.,* **2013**, *62*, 531-536.
[http://dx.doi.org/10.1016/j.ijbiomac.2013.10.014] [PMID: 24141068]

[6]   Duong, V.T.; Nguyen, C.T.; Luong, H.B.; Nguyen, D.C.; Nguyen, H.L. Ultralow-detection limit ammonia gas sensors at room temperature based on MWCNT/WO$_3$ nanocomposite and effect of humidity. *Solid State Sci.,* **2021**, *113*, 106534.
[http://dx.doi.org/10.1016/j.solidstatesciences.2021.106534]

[7]   Si, R.; Xie, X.; Li, T.; Zheng, J.; Cheng, C.; Huang, S.; Wang, C. TiO$_2$/(K,Na)NbO$_3$ Nanocomposite for Boosting Humidity-Sensing Performances. *ACS Sens.,* **2020**, *5*(5), 1345-1353.
[http://dx.doi.org/10.1021/acssensors.9b02586] [PMID: 32268729]

[8]   Arularasu, M.V.; Harb, M.; Vignesh, R.; Rajendran, T.V.; Sundaram, R. PVDF/ZnO hybrid nanocomposite applied as a resistive humidity sensor. *Surf. Interfaces,* **2020**, *21*, 100780.
[http://dx.doi.org/10.1016/j.surfin.2020.100780]

[9]   Kumar, V.; Chauhan, V.; Ram, J.; Gupta, R.; Kumar, S.; Chaudhary, P.; Yadav, B.C.; Ojha, S.; Sulania, I.; Kumar, R. Study of humidity sensing properties and ion beam induced modifications in SnO$_2$-TiO$_2$ nanocomposite thin films. *Surf. Coat. Tech.,* **2020**, *392*, 125768.
[http://dx.doi.org/10.1016/j.surfcoat.2020.125768]

[10]   Jin, X.F.; Liu, C.R.L.; Chen, L.; Zhang, Y.; Zhang, X.J.; Chen, Y.M.; Chen, J.J. Inkjet-printed MoS2/PVP hybrid nanocomposite for enhanced humidity sensing. *Sens. Actuators A Phys.,* **2020**, *316*, 112388.
[http://dx.doi.org/10.1016/j.sna.2020.112388]

[11]   Pandey, N.; Shukla, S.K.; Singh, N.B. Water purification by polymer nanocomposites: an overview. *Nanocomposites,* **2017**, *3*(2), 47-66.
[http://dx.doi.org/10.1080/20550324.2017.1329983]

[12]   Gopinath, K.P.; Rajagopal, M.; Krishnan, A.; Sreerama, S.K. A Review on Recent Trends in Nanomaterials and Nanocomposites for Environmental Applications. *Curr. Anal. Chem.,* **2020**, *17*(2),

202-243.
[http://dx.doi.org/10.2174/1573411016666200102112728]

[13]   Guo, Q.; Ghadiri, R.; Weigel, T.; Aumann, A.; Gurevich, E.; Esen, C.; Medenbach, O.; Cheng, W.; Chichkov, B.; Ostendorf, A. Comparison of *in Situ* and *ex Situ* Methods for Synthesis of Two-Photon Polymerization Polymer Nanocomposites. *Polymers (Basel),* **2014**, *6*(7), 2037-2050.
[http://dx.doi.org/10.3390/polym6072037]

[14]   Shukla, S.K.; Kushwaha, C.S.; Shukla, A.; Dubey, G.C. Integrated approach for efficient humidity sensing over zinc oxide and polypyrole composite. *Mater. Sci. Eng. C,* **2018**, *90*, 325-332.
[http://dx.doi.org/10.1016/j.msec.2018.04.054] [PMID: 29853098]

[15]   Singh, P.; Shukla, S.K. Structurally optimized cupric oxide/polyaniline nanocomposites for efficient humidity sensing. *Surf. Interfaces,* **2020**, *18*, 100410.
[http://dx.doi.org/10.1016/j.surfin.2019.100410]

[16]   Shukla, S.K.; Tiwari, A.; Parashar, G.K.; Mishra, A.P.; Dubey, G.C. Exploring fiber optic approach to sense humid environment over nano-crystalline zinc oxide film. *Talanta,* **2009**, *80*(2), 565-571.
[http://dx.doi.org/10.1016/j.talanta.2009.07.026] [PMID: 19836521]

[17]   Blank, T.A.; Eksperiandova, L.P.; Belikov, K.N. Recent trends of ceramic humidity sensors development: A Review. *Sens. Actuators B Chem.,* **2016**, *228*, 416-442.
[http://dx.doi.org/10.1016/j.snb.2016.01.015]

[18]   Banerjee, R.; Manna, I. *Ceramic nanocomposites*; Woodhead Publishing: Cambridge, UK, **2013**.
[http://dx.doi.org/10.1533/9780857093493]

[19]   Shaheen, K.; Shah, Z.; Khan, B.; Adnan, B.; Omer, M.; Alamzeb, M.; Suo, H. Electrical, Photocatalytic, and Humidity Sensing Applications of Mixed Metal Oxide Nanocomposites. *ACS Omega,* **2020**, *5*(13), 7271-7279.
[http://dx.doi.org/10.1021/acsomega.9b04074] [PMID: 32280868]

[20]   Jakhar, S.; Duhan, S.; Nain, S. Novel one step hydrothermal synthesis of cubic Ia3d large pore 3D mesoporous In_2O_3/KIT-6 hybrid nanocomposite with humidity sensing applications. *J. Porous Mater.,* **2020**, *27*(5), 1253-1263.
[http://dx.doi.org/10.1007/s10934-020-00897-x]

[21]   Kumar, A.; Wang, C.; Meng, F.Y.; Liang, J.G.; Xie, B.F.; Zhou, Z.L.; Zhao, M.; Kim, N.Y. Aerosol deposited $BaTiO_3$ film based interdigital capacitor and squared spiral capacitor for humidity sensing application. *Ceram. Int.,* **2021**, *47*(1), 510-520.
[http://dx.doi.org/10.1016/j.ceramint.2020.08.158]

[22]   Dwiputra, M.A.; Fadhila, F.; Imawan, C.; Fauzia, V. The enhanced performance of capacitive-type humidity sensors based on ZnO nanorods/WS_2 nanosheets heterostructure. *Sens. Actuators B Chem.,* **2020**, *310*, 127810.
[http://dx.doi.org/10.1016/j.snb.2020.127810]

[23]   Bayhan, M.; Kavasoğlu, N. A study on the humidity sensing properties of $ZnCr_2O_4$-K_2CrO_4 ionic conductive ceramic sensor. *Sens. Actuators B Chem.,* **2006**, *117*(1), 261-265.
[http://dx.doi.org/10.1016/j.snb.2005.11.053]

[24]   Nikolic, M.V.; Vasiljevic, Z.Z.; Lukovic, M.D.; Pavlovic, V.P.; Vujancevic, J.; Radovanovic, M.; Krstic, J.M.; Vlahovic, B.; Pavlovic, V.B. Humidity sensing properties of nanocrystalline pseudobrookite (Fe_2TiO_5) based thick films. *Sens. Actuators B Chem.,* **2018**, *277*, 654-664.
[http://dx.doi.org/10.1016/j.snb.2018.09.063]

[25]   Shukla, S.K.; Bharadvaja, A.; Parashar, G.K.; Mishra, A.P.; Dubey, G.C.; Tiwari, A. Fabrication of ultra-sensitive optical fiber based humidity sensor using TiO_2 thin film. *Adv. Mater. Lett.,* **2012**, *3*, 365-370.
[http://dx.doi.org/10.5185/amlett.2012.5350]

[26]   Shukla, S.K.; Parashar, G.K.; Mishra, A.P.; Misra, P.; Yadav, B.C.; Shukla, R.K.; Bali, L.M.; Dubey,

G.C. Nano-like magnesium oxide films and its significance in optical fiber humidity sensor. *Sens. Actuators B Chem.,* **2004**, *98*(1), 5-11.
[http://dx.doi.org/10.1016/j.snb.2003.05.001]

[27] Cho, M.; Kim, S.; Kim, I.; Kim, E.; Wang, Z.; Kim, N.; Kim, S.; Oh, J. Perovskite-Induced Ultrasensitive and Highly Stable Humidity Sensor Systems Prepared by Aerosol Deposition at Room Temperature. *Adv. Funct. Mater.,* **2020**, *30*(3), 1907449.
[http://dx.doi.org/10.1002/adfm.201907449]

[28] Li, T.Y.; Si, R.J.; Sun, J.; Wang, S.T.; Wang, J.; Ahmed, R.; Zhu, G.B.; Wang, C.C. Giant and controllable humidity sensitivity achieved in (Na+Nb)co-doped rutile TiO_2. *Sens. Actuators B Chem.,* **2019**, *293*, 151-158.
[http://dx.doi.org/10.1016/j.snb.2019.05.019]

[29] Gopiraman, M.; Kim, I.S. Carbon Nanocomposites: Preparation and Its Application in Catalytic Organic Transformations, in: Nanocomposites - Recent Evolutions. In: *IntechOpen*; , **2019**.

[30] Tulliani, J.M.; Inserra, B.; Ziegler, D. Carbon-based materials for humidity sensing: A short Review. *Micromachines (Basel),* **2019**, *10*(4), 232.
[http://dx.doi.org/10.3390/mi10040232] [PMID: 30935138]

[31] Yu, S.; Chen, C.; Zhang, H.; Zhang, J.; Liu, J. Design of high sensitivity graphite carbon nitride/zinc oxide humidity sensor for breath detection. *Sens. Actuators B Chem.,* **2021**, *332*, 129536.
[http://dx.doi.org/10.1016/j.snb.2021.129536]

[32] Cai, B.; Yin, H.; Huo, T.; Ma, J.; Di, Z.; Li, M.; Hu, N.; Yang, Z.; Zhang, Y.; Su, Y. Semiconducting single-walled carbon nanotube/graphene van der Waals junctions for highly sensitive all-carbon hybrid humidity sensors. *J. Mater. Chem. C Mater. Opt. Electron. Devices,* **2020**, *8*(10), 3386-3394.
[http://dx.doi.org/10.1039/C9TC06586E]

[33] Lee, J.; Mulmi, S.; Thangadurai, V.; Park, S.S. Magnetically aligned iron oxide/gold nanoparticle-decorated carbon nanotube hybrid structure as a humidity sensor. *ACS Appl. Mater. Interfaces,* **2015**, *7*(28), 15506-15513.
[http://dx.doi.org/10.1021/acsami.5b03862] [PMID: 26112318]

[34] Meng, W.; Wu, S.; Wang, X.; Zhang, D. High-sensitivity resistive humidity sensor based on graphitic carbon nitride nanosheets and its application. *Sens. Actuators B Chem.,* **2020**, *315*, 128058.
[http://dx.doi.org/10.1016/j.snb.2020.128058]

[35] Ho, C.Y.; Wu, Y.S. Diamine decorated graphene oxide film on quartz crystal microbalance for humidity-sensing analysis. *Appl. Surf. Sci.,* **2020**, *510*, 145257.
[http://dx.doi.org/10.1016/j.apsusc.2020.145257]

[36] Qi, P.; Xu, Z.; Zhang, T.; Fei, T.; Wang, R. Chitosan wrapped multiwalled carbon nanotubes as quartz crystal microbalance sensing material for humidity detection. *J. Colloid Interface Sci.,* **2020**, *560*, 284-292.
[http://dx.doi.org/10.1016/j.jcis.2019.10.080] [PMID: 31670101]

[37] Wang, S.; Xie, G.; Su, Y.; Su, L.; Zhang, Q.; Du, H.; Tai, H.; Jiang, Y. Reduced graphene oxide-polyethylene oxide composite films for humidity sensing *via* quartz crystal microbalance. *Sens. Actuators B Chem.,* **2018**, *255*, 2203-2210.
[http://dx.doi.org/10.1016/j.snb.2017.09.028]

[38] Li, J.; Zhang, J.; Sun, H.; Yang, Y.; Ye, Y.; Cui, J.; He, W.; Yong, X.; Xie, Y. An optical fiber sensor based on carboxymethyl cellulose/carbon nanotubes composite film for simultaneous measurement of relative humidity and temperature. *Opt. Commun.,* **2020**, *467*, 125740.
[http://dx.doi.org/10.1016/j.optcom.2020.125740]

[39] Morsy, M.; Mokhtar, M.M.; Ismail, S.H.; Mohamed, G.G.; Ibrahim, M. Humidity Sensing Behaviour of Lyophilized rGO/Fe2O3 Nanocomposite. *J. Inorg. Organomet. Polym. Mater.,* **2020**, *30*(10), 4180-4190.
[http://dx.doi.org/10.1007/s10904-020-01570-1]

[40]   Li, J.; Liu, X.; Sun, H.; Wang, L.; Zhang, J.; Deng, L.; Ma, T. An optical fiber sensor coated with electrospinning polyvinyl alcohol/carbon Nanotubes composite film. *Sensors (Basel),* **2020**, *20*(23), 6996.
[http://dx.doi.org/10.3390/s20236996] [PMID: 33297437]

[41]   Zhu, P.; Ou, H.; Kuang, Y.; Hao, L.; Diao, J.; Chen, G. Cellulose Nanofiber/Carbon Nanotube Dual Network-Enabled Humidity Sensor with High Sensitivity and Durability. *ACS Appl. Mater. Interfaces,* **2020**, *12*(29), 33229-33238.
[http://dx.doi.org/10.1021/acsami.0c07995] [PMID: 32608963]

[42]   Zhao, H.; Zhang, T.; Qi, R.; Dai, J.; Liu, S.; Fei, T. Drawn on Paper: A Reproducible Humidity Sensitive Device by Handwriting. *ACS Appl. Mater. Interfaces,* **2017**, *9*(33), 28002-28009.
[http://dx.doi.org/10.1021/acsami.7b05181] [PMID: 28767212]

[43]   Dubey, N.; Kushwaha, C.S.; Shukla, S.K. A review on electrically conducting polymer bionanocomposites for biomedical and other applications. *Int. J. Polym. Mater.,* **2020**, *69*(11), 709-727.
[http://dx.doi.org/10.1080/00914037.2019.1605513]

[44]   Wu, Y.; Ji, J.; Zhou, Y.; Chen, Z.; Liu, S.; Zhao, J. Ratiometric and colorimetric sensors for highly sensitive detection of water in organic solvents based on hydroxyl-containing polyimide-fluoride complexes. *Anal. Chim. Acta,* **2020**, *1108*, 37-45.
[http://dx.doi.org/10.1016/j.aca.2020.02.043] [PMID: 32222242]

[45]   Wang, Y.; Zhang, L.; Zhou, J.; Lu, A. Flexible and Transparent Cellulose-Based Ionic Film as a Humidity Sensor. *ACS Appl. Mater. Interfaces,* **2020**, *12*(6), 7631-7638.
[http://dx.doi.org/10.1021/acsami.9b22754] [PMID: 31961643]

[46]   Qi, R.; Lin, X.; Dai, J.; Zhao, H.; Liu, S.; Fei, T.; Zhang, T. Humidity sensors based on MCM-41/polypyrrole hybrid film *via in-situ* polymerization. *Sens. Actuators B Chem.,* **2018**, *277*, 584-590.
[http://dx.doi.org/10.1016/j.snb.2018.09.062]

[47]   Li, X.; Zhuang, Z.; Qi, D.; Zhao, C. High sensitive and fast response humidity sensor based on polymer composite nanofibers for breath monitoring and non-contact sensing. *Sens. Actuators B Chem.,* **2021**, *330*, 129239.
[http://dx.doi.org/10.1016/j.snb.2020.129239]

[48]   Qi, R.; Zhang, T.; Guan, X.; Dai, J.; Liu, S.; Zhao, H.; Fei, T. Capacitive humidity sensors based on mesoporous silica and poly(3,4-ethylenedioxythiophene) composites. *J. Colloid Interface Sci.,* **2020**, *565*, 592-600.
[http://dx.doi.org/10.1016/j.jcis.2020.01.062] [PMID: 31991287]

[49]   Mallick, S.; Ahmad, Z.; Touati, F.; Shakoor, R.A. Improvement of humidity sensing properties of PVDF-TiO_2 nanocomposite films using acetone etching. *Sens. Actuators B Chem.,* **2019**, *288*, 408-413.
[http://dx.doi.org/10.1016/j.snb.2019.03.034]

[50]   Singh, P.; Kushwaha, C.S.; Shukla, S.K.; Dubey, G.C. Synthesis and Humidity Sensing Properties of NiO Intercalated Polyaniline Nanocomposite. *Polym. Plast. Technol. Mater.,* **2019**, *58*(2), 139-147.
[http://dx.doi.org/10.1080/03602559.2018.1466170]

[51]   Vijayan, A.; Fuke, M.; Hawaldar, R.; Kulkarni, M.; Amalnerkar, D.; Aiyer, R.C. Optical fibre based humidity sensor using Co-polyaniline clad. *Sens. Actuators B Chem.,* **2008**, *129*(1), 106-112.
[http://dx.doi.org/10.1016/j.snb.2007.07.113]

[52]   Shukla, S.K.; Shukla, S.K.; Govender, P.P.; Agorku, E.S. A resistive type humidity sensor based on crystalline tin oxide nanoparticles encapsulated in polyaniline matrix. *Mikrochim. Acta,* **2016**, *183*(2), 573-580.
[http://dx.doi.org/10.1007/s00604-015-1678-2]

SUBJECT INDEX

A

Acid 35, 36, 38, 39, 40, 47, 49, 51, 64, 71, 72, 86, 87, 99, 103, 121, 122, 123, 124, 125, 145, 149, 150, 151, 154, 174, 179, 192, 196, 205, 218, 219, 220, 228
  acrylic 87, 121
  ascorbic 35, 36, 47, 71, 123, 125, 179, 196, 219, 228
  caffeic 38, 49, 218
  calcination Caffeic 40
  carboxylic 103
  chloroauric 121
  dimercaptosuccinic 174
  etidronic 47
  ferroceneboronic 192
  folic 124
  indolylboronic 123
  lactic 220
  methacrylic 86, 99, 122, 124
  phthalic 205
  picric 64, 149, 150, 151
  thio salicylic 154
  uric 35, 36, 39, 47, 51, 71, 72, 87, 124, 196, 219
Activity 17, 20, 37, 110, 163, 190, 207, 229
  anthropogenic 163
  antimicrobial 17, 20
  photocatalytic 20
Adsorption process 19, 102
Amperometric 216, 218, 219, 220
  response 220
  sensor 216, 218, 219
Amperometry technique 214, 215
Analyte oxytetracycline 116
Analyte(s) 23, 46, 132, 206, 225
  electrochemical 46
  sensing 206, 225
  solution systems 23
  transferrin 132
Analytical 31, 208
  sensing performance 208

techniques 31
Anemia 167, 169
Anti-knocking agents 166
Antimicrobial agents 131
Artificial recognition materials 113
Aspiring futuristic tools 110
Atomic force microscopy (AFM) 2, 193, 207

B

Barette-Joyner-Halenda (BJH) 204
Bimetallic 30, 33, 35, 39, 40, 197
  carbon composites 30
  carbon nanocomposites 33, 35, 39, 40
  nanocomposites 197
Biosensors 10, 30, 37, 48, 51, 52, 70, 71, 72, 85, 110, 111, 112, 190, 195, 198
  electrochemical 70, 71, 72
  enzymatic electrochemical microfluidic 51
  non-enzymatic glucose 195
  sweat glucose 52

C

Calcination methods 38
Cancer, pancreatic 167
Carbon 22, 84, 117, 123, 165, 171, 199, 214
  based functional moieties 84
  paste electrode (CPE) 22, 117, 123, 165, 171, 199, 214
Catalysis, enzymatic 96
Chemical 68, 235
  oxidative polymerization 68
  reduction methods 235
Chitosan 52, 71, 124, 127, 130, 229, 235, 236
  biopolymer 130
  glucose oxidase 52
Cholesterol oxidase 71
Chromatography 31, 112, 131, 213, 233
  gel permeation 233
  high-performance liquid 131
Chronoamperometry methods 217

V

W

X

Z

www.ingramcontent.com/pod-product-compliance
Lightning Source LLC
Chambersburg PA
CBHW050818220326
41598CB00006B/253